MODERN CARBOHYDRATE CHEMISTRY

FOOD SCIENCE AND TECHNOLOGY

A Series of Monographs, Textbooks, and Reference Books

1. Flavor Research: Principles and Techniques, *R. Teranishi, I. Hornstein, P. Issenberg, and E. L. Wick (out of print)*
2. Principles of Enzymology for the Food Sciences, *John R. Whitaker*
3. Low-Temperature Preservation of Foods and Living Matter, *Owen R. Fennema, William D. Powrie, and Elmer H. Marth*
4. Principles of Food Science

   Part I: Food Chemistry, *edited by Owen R. Fennema*
   Part II: Physical Methods of Food Preservation, *Marcus Karel, Owen R. Fennema, and Daryl B. Lund*
5. Food Emulsions, *edited by Stig Friberg*
6. Nutritional and Safety Aspects of Food Processing, *edited by Steven R. Tannenbaum*
7. Flavor Research: Recent Advances, *edited by R. Teranishi, Robert A. Flath, and Hiroshi Sugisawa*
8. Computer-Aided Techniques in Food Technology, *edited by Israel Saguy*
9. Handbook of Tropical Foods, *edited by Harvey T. Chan*
10. Antimicrobials in Foods, *edited by Alfred Larry Branen and P. Michael Davidson*
11. Food Constituents and Food Residues: Their Chromatographic Determination, *edited by James F. Lawrence*

12. Aspartame: Physiology and Biochemistry, *edited by Lewis D. Stegink and L. J. Filer, Jr.*
13. Handbook of Vitamins: Nutritional, Biochemical, and Clinical Aspects, *edited by Lawrence J. Machlin*
14. Starch Conversion Technology, *edited by G. M. A. van Beynum and J. A. Roels*
15. Food Chemistry: Second Edition, Revised and Expanded, *edited by Owen R. Fennema*
16. Sensory Evaluation of Food: Statistical Methods and Procedures, *Michael O'Mahony*
17. Alternative Sweeteners, *edited by Lyn O'Brien Nabors and Robert C. Gelardi*
18. Citrus Fruits and Their Products: Analysis and Technology, *S. V. Ting and Russell L. Rouseff*
19. Engineering Properties of Foods, *edited by M. A. Rao and S. S. H. Rizvi*
20. Umami: A Basic Taste, *edited by Yojiro Kawamura and Morley R. Kare*
21. Food Biotechnology, *edited by Dietrich Knorr*
22. Food Texture: Instrumental and Sensory Measurement, *edited by Howard R. Moskowitz*
23. Seafoods and Fish Oils in Human Health and Disease, *John E. Kinsella*
24. Postharvest Physiology of Vegetables, *edited by J. Weichmann*
25. Handbook of Dietary Fiber: An Applied Approach, *Mark Dreher*
26. Food Toxicology, Parts A and B, *Jose M. Concon*
27. Modern Carbohydrate Chemistry, *Roger W. Binkley*

Other Volumes in Preparation

MODERN CARBOHYDRATE CHEMISTRY

Roger W. Binkley

Department of Chemistry
Cleveland State University
Cleveland, Ohio

MARCEL DEKKER, INC. New York • Basel

Library of Congress Cataloging in Publication Data

Binkley, Roger W.
Modern carbohydrate chemistry / Roger W. Binkley.
p. cm. -- (Food science and technology ; 27)
Includes index.
ISBN 0-8247-7789-1
1. Carbohydrates. I. Title. II. Series: Food science and technology (Marcel Dekker, Inc.) ; 27.
QD321.B59 1988
547.7'8--dc19 87-26552
CIP

MARCEL DEKKER, INC.
270 Madison Avenue, New York, New York 10016

Current printing (last digit):
10 9 8 7 6 5 4 3 2 1

PRINTED IN THE UNITED STATES OF AMERICA

This book is dedicated to Wendell W. Binkley who has devoted his professional life to research, teaching, and service in the field of carbohydrate chemistry.

Wendell Binkley teaching a student to use a polarimeter. April, 1986, Cleveland, Ohio.

Preface

This book has been designed to provide a current understanding of the chemistry of carbohydrates both for persons new to the field and for those with some experience with these compounds. In preparing this book, it has been assumed that the interested reader will already have studied organic chemistry. For those whose familiarity with carbohydrates is limited, the first five chapters supply a general background and introduce basic terminology and fundamental concepts. For chemists with this background, the remaining nine chapters, which constitute the major portion of the text, will be of greater interest because they contain a concise description of the current development of synthetic and mechanistic carbohydrate chemistry.

I hope this book will be useful to a variety of chemists. Especially important are undergraduate and graduate students who desire to know more about carbohydrate chemistry than is contained in the one or two chapters on the subject found in most organic chemistry textbooks. Among others who may find this work valuable are organic chemists who, perhaps to support their research programs, need basic information about the chemistry of carbohydrates. They should find this book useful to consult for current understanding on a particular topic and as a source of references to further reading. Finally, it would be fitting if carbohydrate chemists, who are primarily responsible for the ideas and information described here, would be aided when, from time to time, they need to review a particular reaction mechanism or to see what reagents and approaches have been used to solve a synthetic problem.

As with other scientific fields, carbohydrate chemistry has grown rapidly in recent years. This growth provides, in large part, the rationale for a new book on carbohydrate chemistry. Much of what is in common use by carbohydrate chemists today was unknown in 1970, the date of publication of the most recent text describing the basic chemistry of carbohydrates. At this time, there is a need to bring things up to date.

It is difficult when writing about an area as broad as carbohydrate chemistry to decide how best to include the necessary, basic information without presenting an overwhelming amount of material. This dilemma forces the author to make certain compromises and requires that the chemistry of some compounds and types of compounds not be discussed; such discussion can be omitted without significantly sacrificing the basic understanding of carbohydrate chemistry. Several groups of compounds which fall into this category are nucleosides, nucleotides, polymeric carbohydrates, and their derivatives.

Finally, I would like to acknowledge and thank those who have made the preparation of this book possible. The Standard Oil Company (Ohio) has been a generous supporter of my research in carbohydrate chemistry for many years. This support and encouragement has been instrumental in allowing me to gain the necessary background to undertake this writing. Also, The Cleveland State University has provided similar support and has allowed me a professional leave during the 1985-86 academic year to concentrate my efforts on this writing. Most recently, my good friends and colleagues, Thomas W. Flechtner and Frederick K. Oplinger, read this manuscript and offered many helpful suggestions.

Roger W. Binkley

Contents

1 Introduction

I. THE ROLE OF CARBOHYDRATES

II. CARBOHYDRATE CHEMISTRY: PAST, PRESENT, FUTURE

I. THE ROLE OF CARBOHYDRATES

Carbohydrates are easily identified as one of the most important classes of chemical compounds. They hold this elevated position because they are a collection of substances upon which all plant and animal life depend. This dependence is basic and multifaceted. Only a few of the roles played by carbohydrates need to be considered in order for their value to be appreciated. First, at perhaps the most fundamental level, carbohydrates are the molecules in which energy from the sun, our primary energy source, is stored by photosynthesis (Equation 1). In addition, all animals use

$$CO_2 + H_2O \xrightarrow{\text{PHOTOSYNTHESIS}} \text{CARBOHYDRATES} + O_2 \qquad (1)$$

carbohydrates for a variety of functions. For example, D-glucose [1], the most abundant simple carbohydrate, is the vehicle for

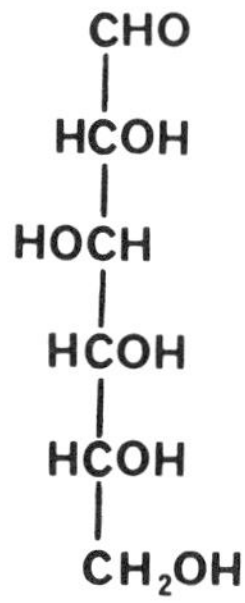

1

energy transport in mammals. Mammals also convert D-glucose [1] into a polymeric form, glycogen (Figure 1), which is stored primarily in muscle tissue and in the liver to be metabolized when it is needed. The nucleic acids are carbohydrate-containing polymers which control the biosynthesis of proteins and are responsible for the transfer of genetic information. In plants, the dependence on carbohydrates is different but perhaps even greater than in animals. The carbohydrate polymers cellulose (Figure 2) and hemicellulose are the primary structural materials used for support by plants; in fact, carbohydrates are second in abundance only to water among plant constituents. The several examples of carbohydrates and their biological functions noted here are only a few of the many which exist. Other compounds certainly deserve to be mentioned; however, those selected are sufficient to provide an indication of the significance of carbohydrates to life processes.

Although carbohydrates clearly function at the most fundamental level in living organisms, they also play an increasingly important role in modern industrial society. Carbohydrates form the basis for the manufacture of sugar, starch, paper, wood pulp,

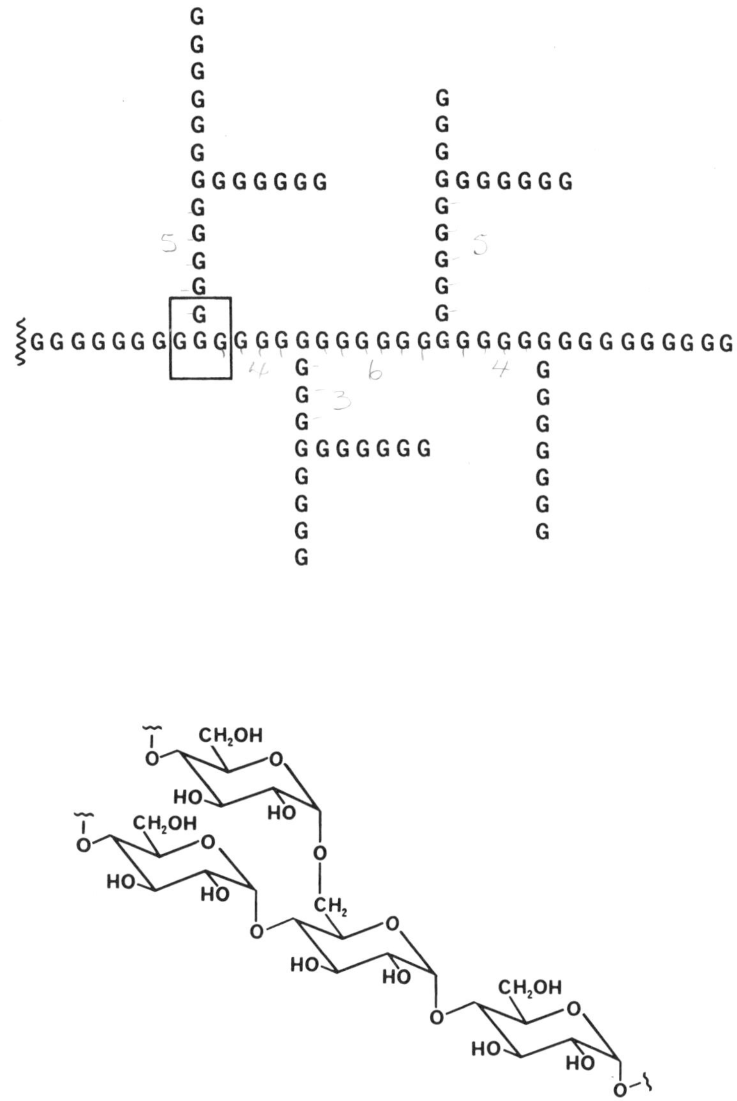

FIG. 1. (top) Representation of the structure of a portion of the polysaccharide glycogen. Each "G" is a D-glucose residue. (bottom) Structural formula for that portion of glycogen enclosed in the box.

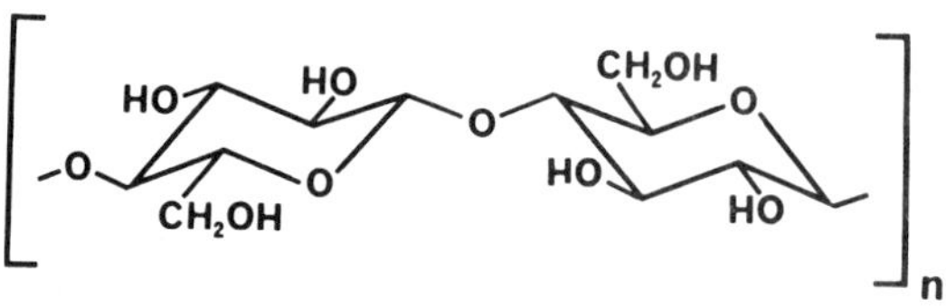

FIG. 2. Representation of the structure of the polysaccharide cellulose. n > 1200.

and the products derived from these materials. Carbohydrates also are required for the production of many food products, pharmaceuticals, and special chemicals. The variety and extent to which carbohydrate-containing materials are being incorporated into industrial processes is continually increasing.

II. CARBOHYDRATE CHEMISTRY: PAST, PRESENT, AND FUTURE

With the range of biological and societal functions which depend upon carbohydrates, it is not surprising that they were among the first organic compounds to be explored. Much of the early understanding of stereochemistry, for example, resulted from investigation of carbohydrates. Early studies of carbohydrates also contributed greatly to the body of knowledge of carbonyl compounds and alcohols. It is initially surprising, therefore, that for much of the middle and latter part of the twentieth century, when organic chemistry was growing and maturing rapidly, carbohydrates did not have a central role in the development. One reason for this situation was that the basic synthetic reactions which were being discovered and the reaction mechanisms which were being elucidated were more easily studied using less highly functionalized molecules. During this period of development for organic chemistry, a parallel growth was occurring in the chemistry of carbohydrates. Much of this growth resulted from transfer of new synthetic reactions and mechanistic understanding from other areas of chemistry to carbohydrate chemistry. Significant fundamental contributions also were made by the relatively small group of chemists who had specialized in carbohydrates.

During the 1970s a renewed interest in carbohydrates began to develop among a large group of chemists who were not specialists in the field. This interest was due primarily to two changes that were occurring. First, as the field of chemistry became more sophisticated, more complex problems, especially biochemical ones, were investigated. In particular, biochemical processes were being examined more frequently on a molecular level. The ability to understand biochemistry at this level naturally called for an increased understanding in the basic chemistry of the molecules involved. Research on carbohydrates, as well as on other classes of compounds, benefited from this situation.

A second reason for renewed interest in carbohydrate chemistry came from the synthetic challenge offered by complicated organic molecules. In recent years synthetic chemistry had reached a level of sophistication in which it was reasonable to undertake the syntheses of quite complex structures. The target molecules, which usually were biologically active or related to active compounds, typically possessed specific stereochemistry at a number of chiral centers. The syntheses of these molecules required reactions which were highly stereoselective or starting materials with known stereochemistry at a number of chiral centers or both. Carbohydrates, with their well-defined stereochemistry at numerous chiral centers, were logical choices for starting materials and frequently were used. As a result, interest and research in carbohydrate chemistry expanded.

Because of the expanding need to use carbohydrates and understand better their chemistry, the study of these compounds appears to be in the early stages of a period of significant growth. The drive to unravel and control complex biochemical processes is a consuming one whose success is dependent, in part, on the continued development of carbohydrate chemistry. The expansion and improvement of industrial uses of carbohydrates is likewise coupled to understanding their chemistry. The period which we are now entering promises to be one from which we will emerge with a much greater knowledge of the chemistry and biochemistry of carbohydrates. As this occurs, a better appreciation of the value of these complicated but fascinating compounds is inevitable.

2 Definitions

I. CARBOHYDRATES
   A. Monosaccharides
   B. Oligosaccharides
   C. Polysaccharides
II. CARBOHYDRATE DERIVATIVES

I. CARBOHYDRATES

Carbohydrates are a diverse group of substances for which it is difficult to find a comprehensive definition. One way of describing these compounds, which is satisfactory in nearly all situations, is that carbohydrates are the polyhydroxy aldehydes, ketones, alcohols, acids, and their derivatives, and the polymers derived from these compounds.

Carbohydrates are subdivided in several ways. One method of subdivision is based upon the nature of the functional group or groups, other than the hydroxyl group, present in the molecule. Some examples of this type of classification are given in Table 1.

Table 1 Classes of Carbohydrate Derivatives

Name	Functional Group	Example
glycose	aldehyde or ketone	any monosaccharide (e.g., D-glucose [1] or D-fructose [4])
aldose	aldehyde	D-glucose [1]
ketose	ketone	D-fructose [4]
glycitol or alditol	alcohol	D-glucitol [10]
glyconic or aldonic acid	carboxylic acid	D-gluconic acid [11]
glycaric or aldaric acid	dicarboxylic acid	D-glycaric acid [12]
uronic acid	carboxylic acid, aldehyde	D-glucuronic acid [13]

A second and extremely important type of subdivision, one based approximately on molecular size, creates three groups of carbohydrates—the monosaccharides, the oligosaccharides, and the polysaccharides.

A. Monosaccharides

The monosaccharides are the polyhydroxy aldehydes and ketones. These compounds also are known as aldoses (the aldehydes) and ketoses (the ketones). Common aldoses include D-glucose [1], D-galactose [2], and D-ribose [3] while D-fructose [4] is the most common ketose. The smallest compounds which clearly qualify as monosaccharides are the two enantiomers of glyceraldehyde [5] and [6] and 1,3-dihydroxyacetone [7]. In principle, monosaccharides could have quite long carbon atom chains but, in practice, those with more than seven or eight carbon atoms are rarely encountered.

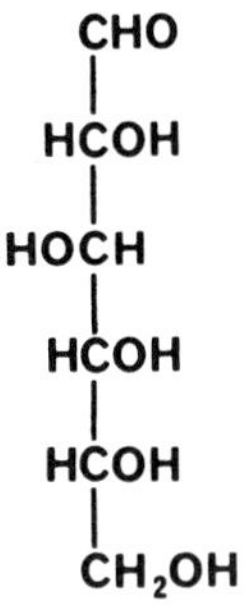

1

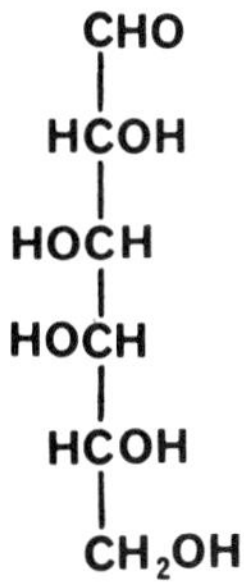

2

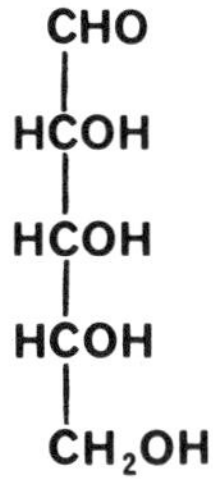

3

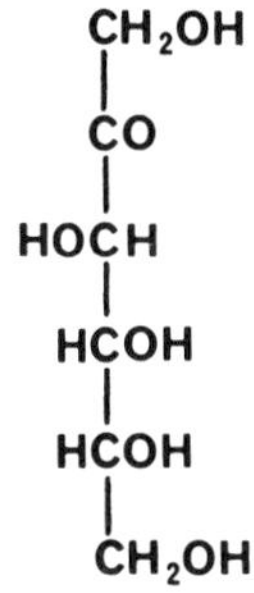

4

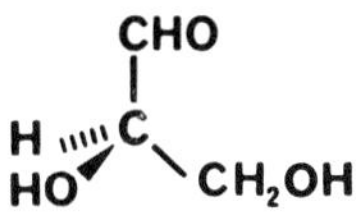

5

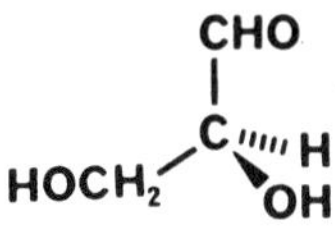

6

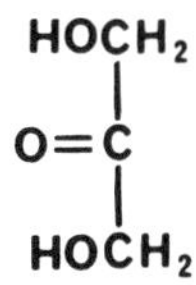

7

It should be noted at this point that it is often convenient to represent structures of carbohydrates in different ways. Several common methods are found in this and the preceding chapter. These and other ways for picturing molecular structure are considered in detail in Chapter 3 (Structural Representation) and Chapter 6 (Conformational Analysis). Also, a complete explanation of the meaning of the symbol "D", present in many of the names encountered thus far, is found in Chapter 4 (Stereochemical Designation).

Monosaccharides are further categorized according to the number of carbon atoms in the central chain. For example, a monosaccharide with a central chain of six carbon atoms is a hexose. If this compound is also an aldehyde (e.g., D-glucose [1] and D-galactose [2]), then it is an aldohexose. In a like manner, aldotriose, aldotetrose, and aldopentose refer to aldoses with three, four, and five carbon atoms, respectively, in the central chain. Thus, D-ribose [3] is an aldopentose.

A set of names, analogous to those just encountered for aldoses, also exists for ketoses. Ketoses with three, four, five, and six carbon atoms in the central chain are ketotrioses, ketotetroses, ketopentoses, and ketohexoses, respectively. An alternate terminology frequently used for indicating the presence of the ketone group is with the "ulose" ending. This approach gives rise to the names triulose, tetrulose, pentulose, and hexulose for the ketones with three, four, five, and six carbon atoms, respectively. D-Fructose [4], therefore, can be described as a ketohexose or a hexulose.

When monosaccharide units are joined together by acetal linkages, oligosaccharides and polysaccharides result. Under proper conditions, hydrolysis of these more complex carbohydrates regenerates the constituent monosaccharides but monosaccharide hydrolysis does not occur. Experimentally, therefore, monosaccharides are carbohydrates which cannot be converted hydrolytically into smaller molecules.

B. Oligosaccharides

Oligosaccharides are compounds formed when small numbers (2–10) of monosaccharides are coupled together. The coupling is of a specific type; that is, an oxygen atom forms a bridge between the units coupled together. This oxygen atom must be part of an acetal or ketal group. Examples of these types of linkages are shown in compounds 8 and 9. The number of monosaccharide units present in an oligosaccharide usually is indicated by a prefix such as di, tri, tetra, or penta (for two, three, four, and five units, respectively). Thus, sucrose [8] is a disaccharide while raffinose [9] is a trisaccharide. The term "sugar" often is used to refer to

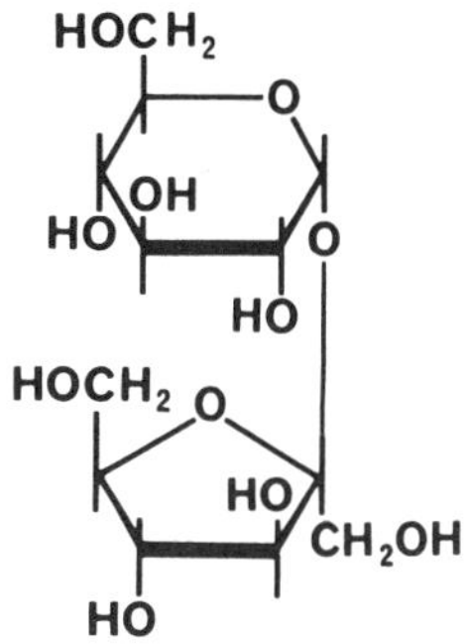

8

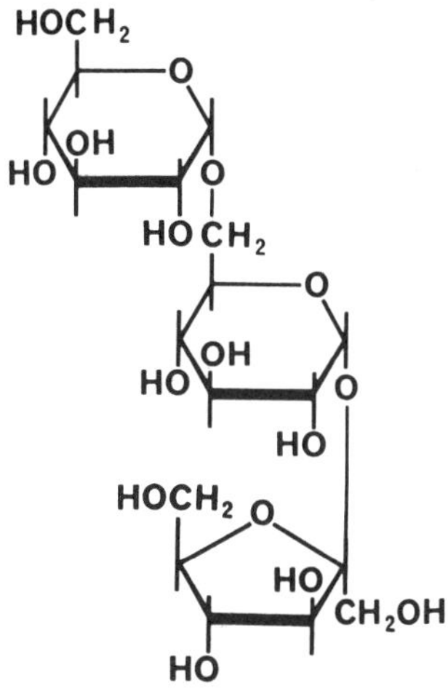

9

any member of the group of mono- and oligosaccharides; hence, compounds 1–9 are all sugars.

C. Polysaccharides

Attaching a larger number (>10) of monosaccharide residues together by acetal linkages produces a polysaccharide. The polysaccharides cellulose, which contains more than 2500 glucose units, and glycogen, which often contains an ever greater number of glucose residues, are two of the biologically important carbohydrates encountered in Chapter 1. One of the reasons for identifying special groups of carbohydrates (monosaccharides, oligosaccharides, polysaccharides) is that these subclasses, particularly the monosaccharides, provide a framework which is used to organize and classify carbohydrate derivatives.

II. CARBOHYDRATE DERIVATIVES

Monosaccharides have an organizational position among carbohydrates similar to that occupied by alkanes among less heavily functionalized organic molecules; that is, various groups of carbohydrates are named from their relationship to the monosaccharides. For instance, compounds resulting from carbonyl to alcohol reduction in aldoses are called alditols; thus, reduction of D-glucose [1] produces D-glucitol [10] (Equation 1). This

(1)

1 10

pattern is followed in generating names for other monosaccharide derivatives. A compilation of these names and examples of the structures they represent is given in Table 1.

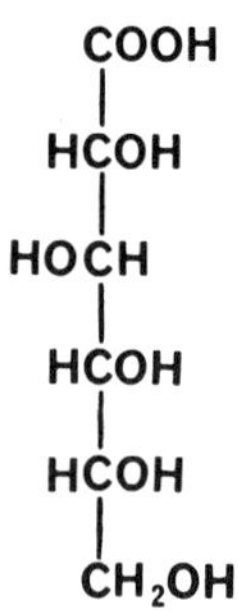

11

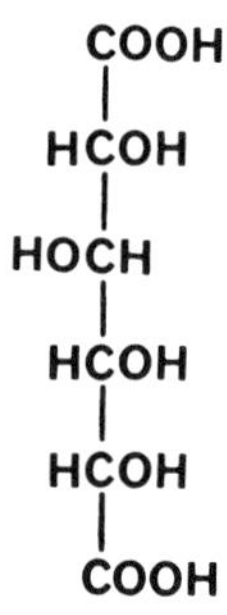

12

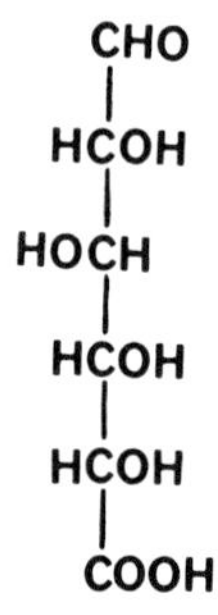

13

3 Structural Representation

I. ACYCLIC SYSTEMS

Effectively representing configuration in chiral molecules, which are three-dimensional structures, with two-dimensional drawings is a problem for many organic compounds, especially carbohydrates. In an effort to solve this problem, a number of types of drawings have been developed; each has its own advantages and disadvantages. Where acyclic carbohydrates are concerned, either wedge-slash or Fischer projection formulas normally are drawn. Examples

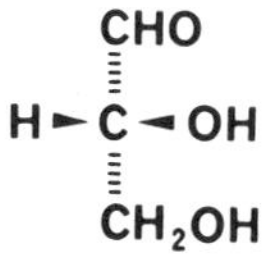

1a

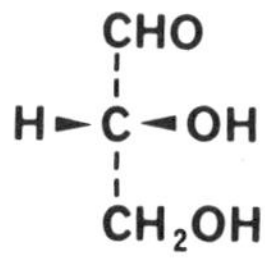

1b

FIG. 1. Wedge-slash 1a and wedge-dash 1b formulas for D-glyceraldehyde 1.

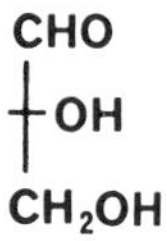

1c

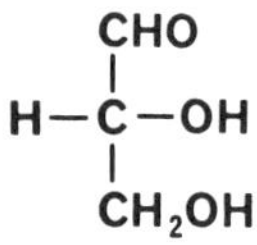

1d

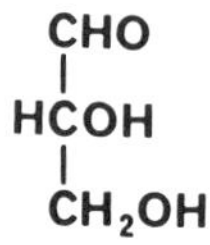

1e

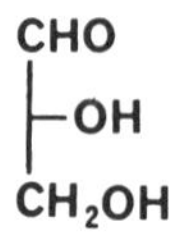

1f

FIG. 2. Fischer projection formulas for D-glyceraldehyde 1.

of these two types of structural representation, along with variations which are sometimes used, are shown for D-glyceraldehyde [1] in Figures 1 and 2.

A. Wedge-Slash Formulas

The wedge-slash formula 1a (Figure 1) is an attractive form of structural representation because it not only conveys configurational information but it often allows qualitative judgment about intramolecular group interactions. This type of formula also has the advantage that it is widely used in many areas of chemistry and, consequently, most chemists recognize the wedge to mean a bond directed toward the viewer from the plane of the page and the series of slashes to indicate a bond originating in this plane and projecting away from the viewer. A common modification of the wedge-slash formula is realized by replacing the slashes with dashed lines [1b] (Figure 1). This replacement normally does not lead to ambiguity but since dashed lines sometimes indicate partial bonding, structural formulas of the wedge-dash type will not be used here.

B. Fischer Projections

Carbohydrate chemists often favor Fischer projections over wedge-slash formulas because Fischer projections are easily written and describe in a compact manner the configuration at a number of chiral centers. The simplicity of the Fischer projection can be appreciated by comparing the two representations for D-glucitol [2a and 2b] shown in Figure 3. A number of modifications of the basic Fischer projection have been introduced. Some of these are shown in Figure 2 for D-glyceraldehyde [1c-1f]. Each of the formulas in Figure 2 is intended to have the same structural significance; that is, each indicates that the hydroxyl group and hydrogen atom attached to the chiral center are projecting from the plane of

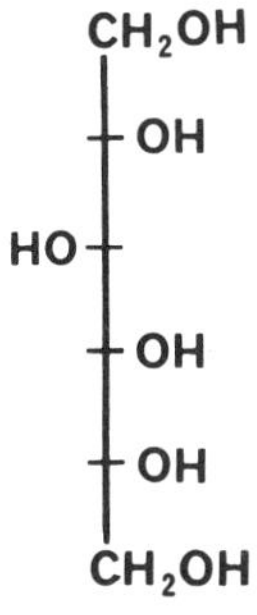

2a

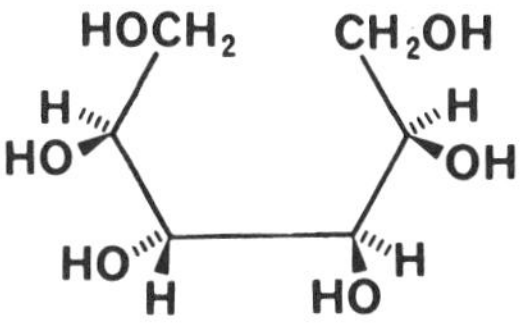

2b

FIG. 3. Fischer projection 2a and wedge-slash 2b formulas for D-glucitol 2.

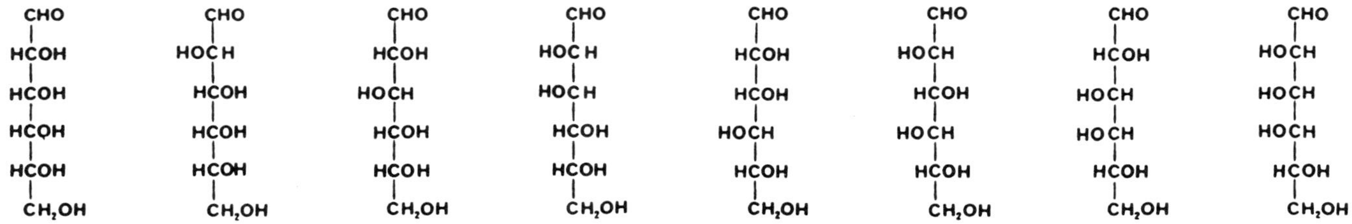

D-ALLOSE D-ALTROSE D-GLUCOSE D-MANNOSE D-GULOSE D-IDOSE D-GALACTOSE D-TALOSE

D-RIBOSE D-ARABINOSE D-XYLOSE D-LYXOSE

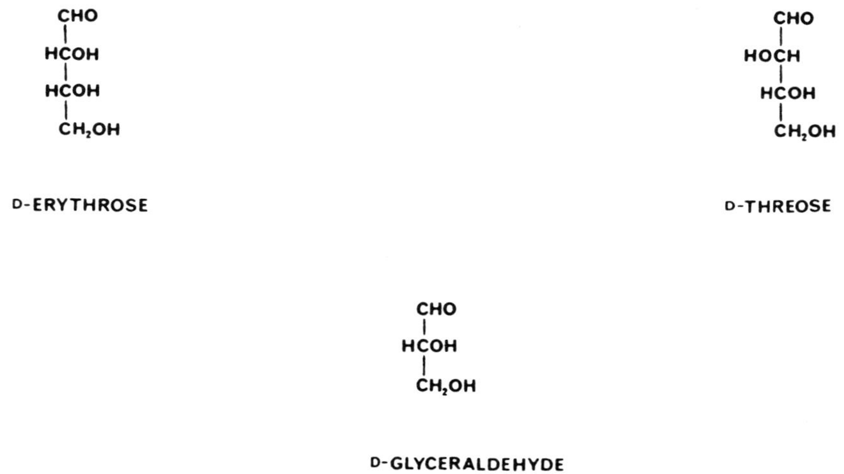

FIG. 4. The D-aldoses with three to six carbon atoms.

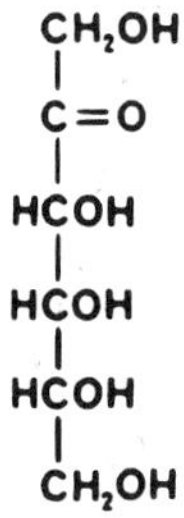

D-PSICOSE

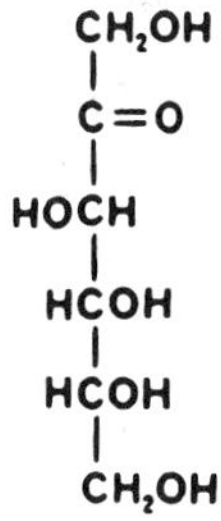

D-FRUCTOSE

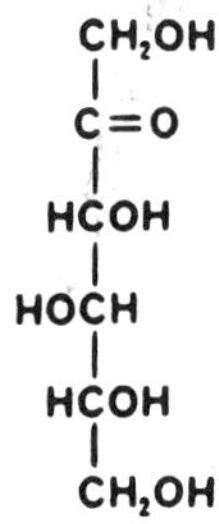

D-SORBOSE

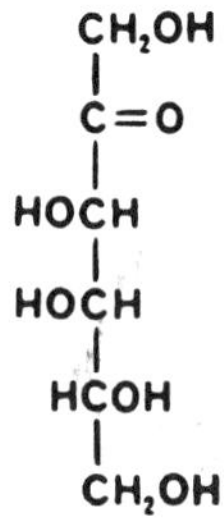

D-TAGATOSE

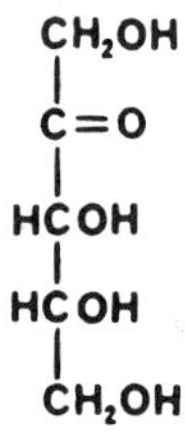

D-ERYTHRO-PENTULOSE

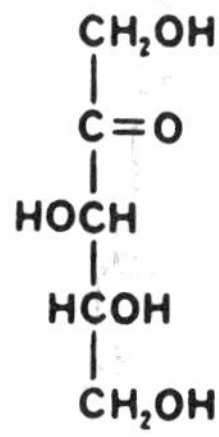

D-THREO-PENTULOSE

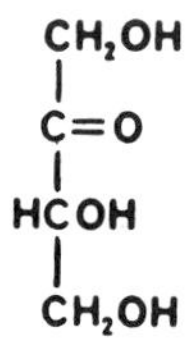

D-GLYCERO-TETRULOSE

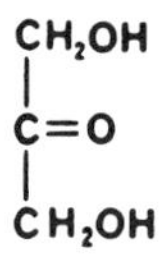

1,3-DIHYDROXY-2-PROPANONE
(1,3-DIHYDROXYACETONE)

FIG. 5. The D-family of 2-ketoses with three to six carbon atoms.

the page toward the viewer while the groups attached to the vertical bonds are pointing away from the viewer. The forms of the Fischer projection in which a chiral center is represented by the intersection of two lines [1c] or the meeting point of two lines [1f] are particularly attractive because they clearly indicate the stereochemical intent of the formula. Other forms [1d and 1e], unless clearly labeled, can be confused with structures for which no stereochemical meaning is intended.

The simplicity of the Fischer projection is an asset in quickly recognizing and easily remembering common configurational patterns. This simplicity also allows a number of formulas to be written quickly and with ease; thus, Fischer projections conveniently are shown for all fifteen D-aldoses with three to six carbon atoms in Figure 4 and for the eight members of the D-family of 2-ketoses with three to six carbon atoms in Figure 5.

II. CYCLIC SYSTEMS

Monosaccharides typically exist as cyclic structures which are intramolecular hemiacetals and hemiketals. An early method for representing these ring systems involved modification of Fischer projections in the manner shown in Figure 6. These modified Fischer projections with unusually long, nonlinear bonds now largely have been replaced by alternative representations. One alternative, shown in Figure 7, is an extension of the wedge-slash approach to cyclic structures. A second, pictured in Figure 8, is a formula, known as a Haworth projection, which was developed specifically to represent ring systems frequently encountered in carbohydrates. Both Haworth projection and wedge-slash formulas are intended to convey only configurational information. These formulas typically are used when the conformation of the ring system is uncertain. When conformational information is available

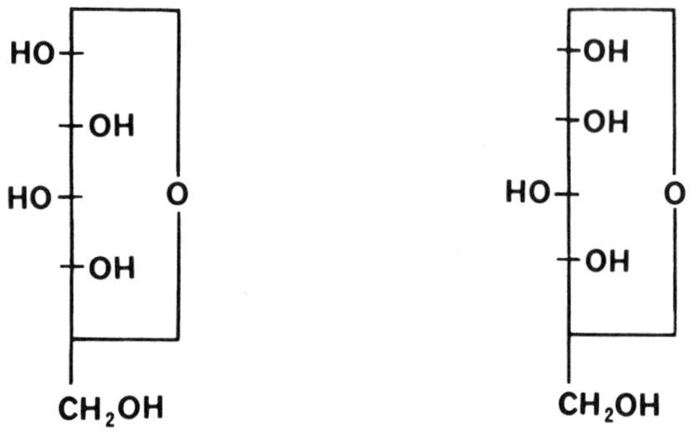

FIG. 6. Fischer projection formulas for D-glucose [3] modified to show cyclic structures.

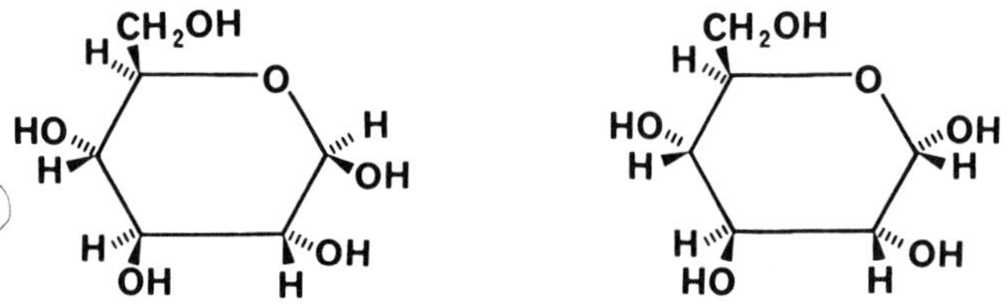

FIG. 7. Wedge-slash formulas for pyranoid and furanoid rings.

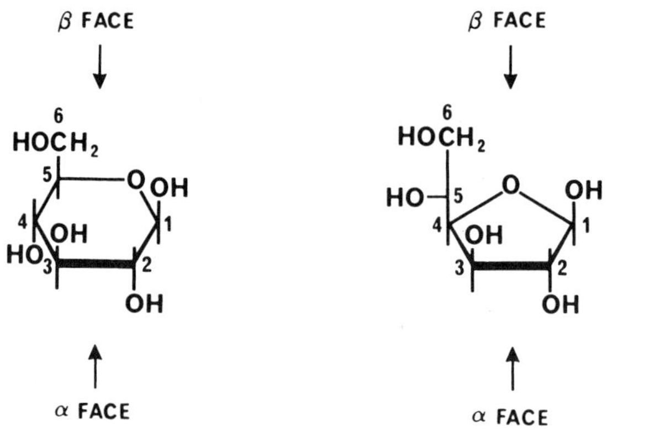

FIG. 8. Haworth projection formulas for pyranoid and furanoid rings.

or can be reasonably assumed, formulas patterned after the representations (boat, chair, envelope, twist boat, etc.) used for carbocyclic systems usually are adopted. Haworth projection and wedge-slash formulas are discussed in this chapter. Conformation formulas will be considered in Chapter 6 (Conformational Analysis).

A. Pyranoid and Furanoid Ring Systems

Both furanoid (five-membered) and pyranoid (six-membered) rings are conveniently represented either by wedge-slash or Haworth projection formulas.

1. Wedge-Slash Formulas

The extension of wedge-slash formulas to cyclic systems (Figure 7) is relatively simple and, thus, requires little explanation. The viewer is considered to be looking down on the ring. Groups projecting up toward the viewer are connected to the ring by wedges and those projecting away from the viewer are attached to the ring by slashes. The standard ring orientation is with the hemiacetal or hemiketal carbon atom at the extreme right-hand side of the structure.

2. Haworth Projections

The pyranoid ring in its Haworth projection is pictured as a planar structure which is seen by a viewer who is above and in front of the ring (Figure 8). The standard ring orientation has the hemiacetal carbon atom (C-1) on the extreme right of the formula. The numbering of the ring proceeds in a clockwise fashion. The furanoid ring also is viewed from above and in front and has the hemiacetal carbon atom on the extreme right (Figure 8). If there is a group containing one or more asymmetric centers attached to a pyranoid or furanoid ring (note the furanoid ring in Figure 8), a Fischer projection is generally used to represent the configuration at the exocyclic asymmetric carbon atoms. Cyclic hemiketals are treated in a similar fashion to cyclic hemiacetals. The ring orientation has the hemiketal carbon atom at the extreme right of the formula.

B. α and β Anomers

Since the formation of a cyclic structure from an acyclic aldose or ketose generates a new chiral center (the hemiacetal or hemiketal carbon atom), such ring formation produces two new diastereomeric compounds. The diastereomers produced when D-glucose [3] forms a pyranoid ring are shown in Equation 1. These

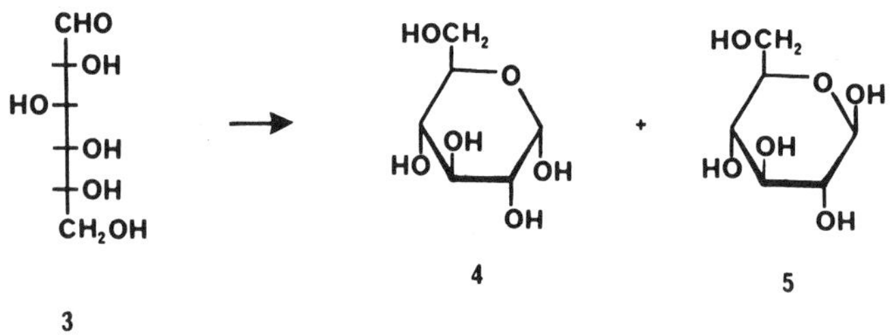

(1)

diastereomers [4 and 5] are referred to an anomers. A carbon atom which was part of a carbonyl group before cyclization becomes an anomeric carbon atom. Consequently, C-1 in compound 4 is such an atom, as is C-1 in compound 5; in fact, in cyclic aldoses C-1 is always the anomeric carbon atom. In cyclic ketoses the location of the anomeric carbon atom depends upon the position of the carbonyl group prior to cyclization; for example, in the 2-ketoses, C-2 is the anomeric carbon atom. Anomers are distinguished from each other by the Greek letters α and β. For a cyclic D-aldose oriented in the standard way, the α anomer has the hydroxyl group on C-1 projecting below the plane of the ring (e.g., compound 4) while the β anomer has the hydroxyl group directed above the plane of the ring (compound 5).

When considering the possible pathways of approach of a reagent to a cyclic structure, it is often desirable to use terms which distinguish between the two sides or "faces" of the ring. In situations where such a distinction is needed, the terms "α face" and "β face" are used. The α face is the side of the ring where an α substituent at C-1 would be located and the β face that side which would have a β anomeric subtituent. The α and β faces of the pyranoid and furanoid rings in Figure 8 are labeled.

4 Stereochemical Designation: The D, L System

Since a typical carbohydrate contains a number of chiral centers, an effective method for designating configuration in this type of compound is of considerable importance. The R and S designators, used extensively in other areas of organic chemistry, probably would have been adopted for carbohydrates if carbohydrates were structurally simpler molecules; however, due to their stereochemical complexity, an older system, one specifically designed for these molecules, has been retained.

This older system uses the two enantiomers of glyceraldehyde as its reference compounds and differentiates between them with the symbols D and L; thus, all monosaccharides are viewed as being derived from either D-glyceraldehyde [1] or L-glyceraldehyde [2]. (To indicate that these symbols refer to configurational

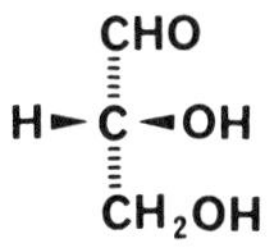

1

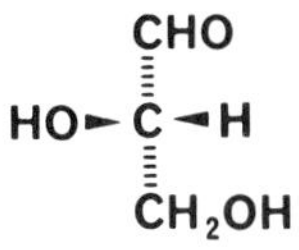

2

designation, D and L are written with two lines underneath or as small capital letters.) For example, if a CHOH group is inserted between the chiral center and the CHO group in D-glyceraldehyde [1], the aldotetroses 3 and 4 are formed. These compounds, D-erythrose [3] and D-threose [4], are D-sugars since they are

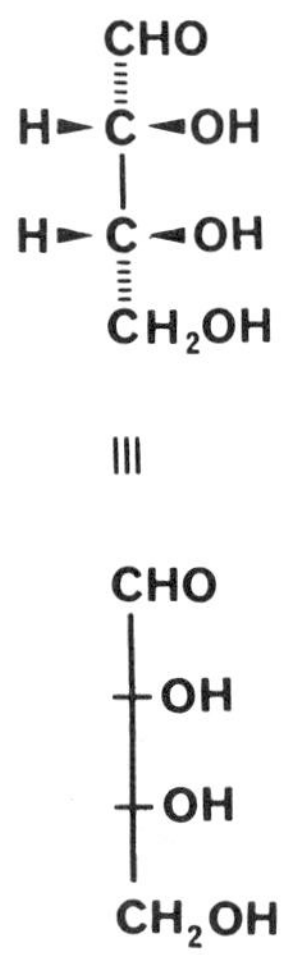

3

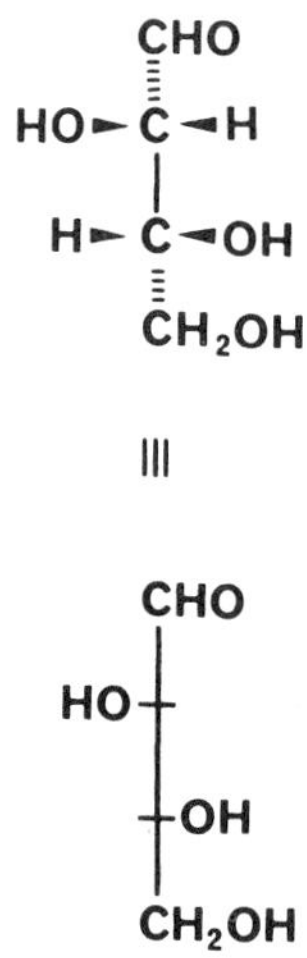

4

derived from D-glyceraldehyde [1]. Repeating this CHOH addition process on compounds 3 and 4 produces four aldopentoses: D-ribose [5], D-arabinose [6], D-xylose [7], and D-lyxose [8].

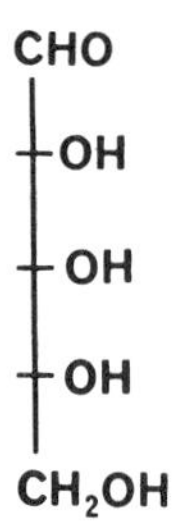

5

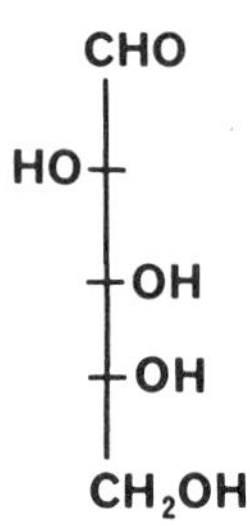

6

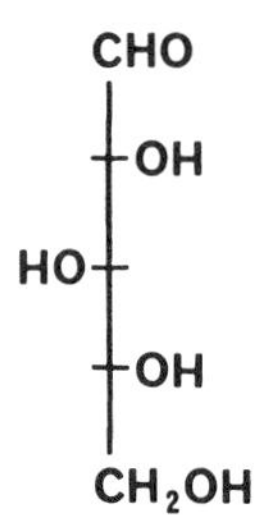

7

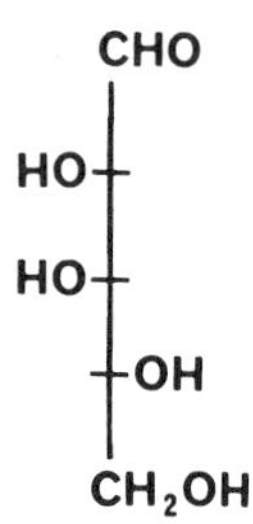

8

In this fashion, the D-family of aldoses shown in Figure 4 of Chapter 3 is constructed. The enantiomeric L-aldoses can be formed in a similar manner from L-glyceraldehyde [2]. (It is worth noting that the D and L designators indicate nothing about the rotation of plane polarized light. They are only stereochemical designators.)

In general, a compound is a member of the D-family if the asymmetric carbon atom most remote from the carbonyl (reference) group has the same configuration as D-glyceraldehyde [1]. When a compound is written in its Fischer projection, with the reference group at the top (or for ketoses as close to the top as possible), assignment to the D or L series is determined by the configuration of the asymmetric carbon atom farthest from the top. Compounds in the D series have the hydroxyl group on the right and those in the L series have it on the left. Some examples of D and L series compounds with various reference groups are provided by D-allose [9], L-allose [10], D-galactonic acid [11], and D-psicose [12]. When two possible reference groups are

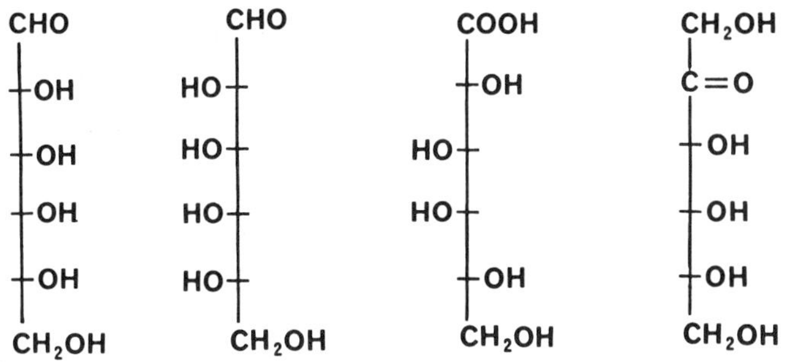

9 10 11 12

present in the same molecule, the order of preference is CHO (first), COOH (second), CO (third); thus, CHO is the reference group in D-glucuronic acid [13] rather than COOH.

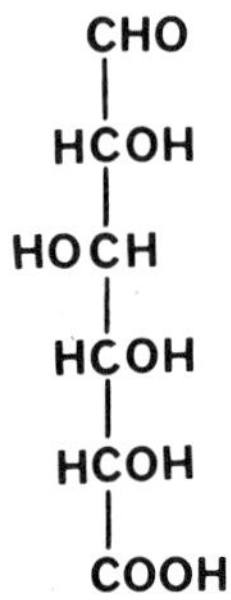

13

Since membership in either the D or L family is determined by the stereochemistry at the chiral center most distant from the reference group, alterations in molecular structure which occur away from this chiral center, in general, do not change the D or L designation of the compound. For example, if the C-4 hydroxyl group in D-erythrose [*3*] is replaced by a chlorine atom (Figure 1), the resulting compound [*14*] still is a member of the D family; in fact, one reason for retaining the D, L system for carbohydrates is this consistency in relationship among family members. This type of consistency is less certain when the *R*, *S* convention is used, because *R*, *S* designation is determined by group priority and group priority can change without changing configuration. For example, consider the effect on *R*, *S* designation of the chlorine atom for hydroxyl group substitution just described for D-erythrose [*3*]. The *R*, *S* system requires a change in the stereochemical designation at C-3 as a result of this substitution because a new group priority is established (Figure 1).

In molecules with more than one chiral center, the D or L designator becomes a key element in describing stereochemistry for the entire molecule. Consider compounds *9* and *10*. The name allose gives the relationship among all the chiral centers in each

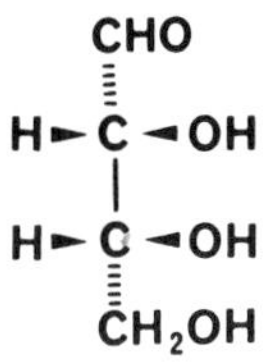

3

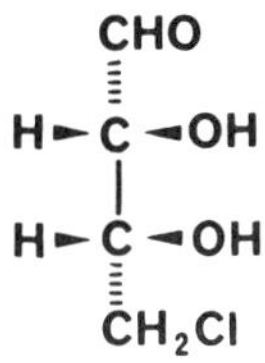

14

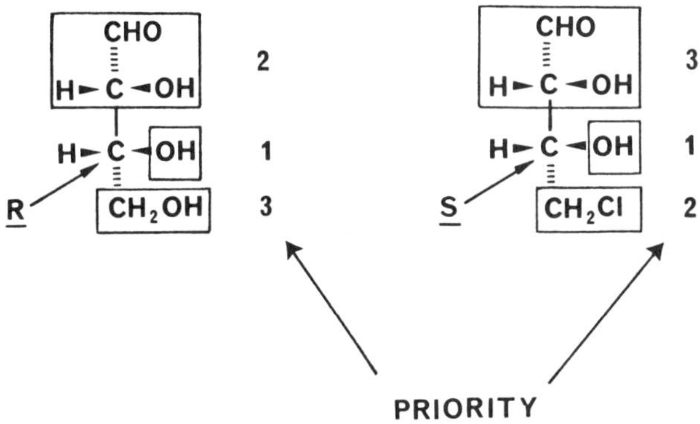

FIG. 1. Effect on D, L and R, S designations of replacement of the C-4—hydroxyl group in D-erythrose with a chlorine atom.

molecule. As seen in the Fischer projections for 9 and 10, allose specifies the configuration at each chiral center to be such that all hydroxyl groups are on the same side of the projection formula. The terms D and L specify, in effect, which side of the projection the hydroxyl groups are on by fixing the stereochemistry at a particular chiral center (that farthest from the reference group). In a similar manner, a single name in association with D or L completely defines the structure, including stereochemistry, of each of the aldoses with six or fewer carbon atoms (Figure 4 in Chapter 3).

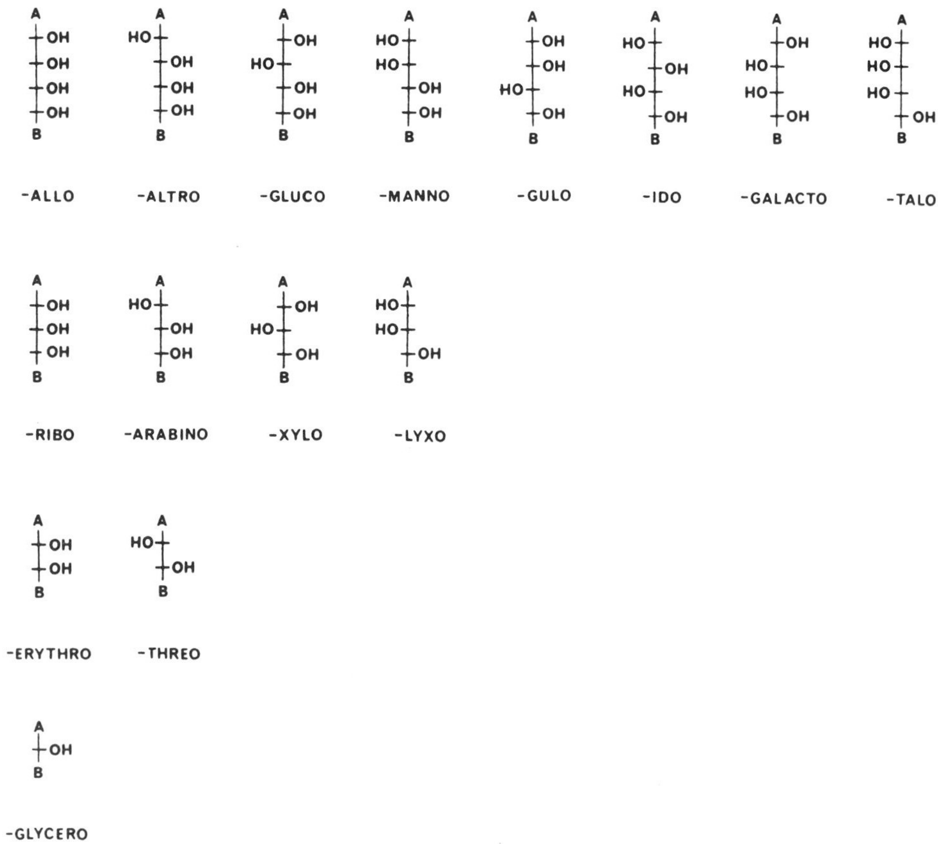

FIG. 2. Prefixes for designating configuration at consecutive but not necessarily contiguous asymmetric carbon atoms. The principal functional group is "A".

The scope of the D, L system for describing stereochemical relationships extends well beyond simple aldoses and ketoses. It is applicable whenever a backbone chain of asymmetric centers (not necessarily contiguous) can be recognized in a molecule. The prefixes glycero, erythro, threo, arabino, lyxo, ribo, xylo, allo, altro, galacto, gluco, gulo, ido, manno, and talo are used to indicate the number of chiral centers and the stereochemical relationship among them. The meaning of each prefix is shown in Figure 2. The extended use of the D, L system can be appreciated by considering several examples. For instance, in compound 15 there are three chiral centers showing a ribo pattern. The chiral center most remote from the reference group has the D configuration; therefore, the term D-ribo completely describes the stereochemistry in this molecule. In a like manner, compounds 16 and 17 have L-xylo and D-glycero configurations, respectively.

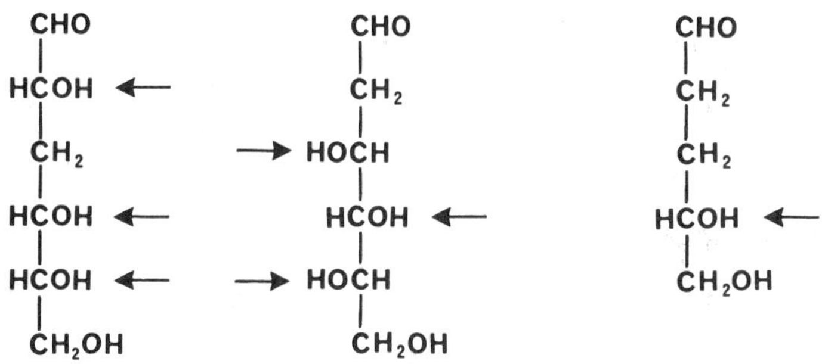

15 16 17

Combining the proper prefix with either D or L next to the root name of the molecule specifies the stereochemistry at as many as four chiral centers.

The D, L system can be used when atoms other than oxygen are attached to the carbon atoms in the central chain. Compounds 18–20, which have D-xylo [18], D-ribo [19], and D-arabino [20] configurations, provide examples in which chlorine, sulfur, and carbon atoms, respectively, are attached to the monosaccharide backbone.

For compounds with more than four asymmetric carbon atoms, a combination of prefixes is used. The stereochemistry in

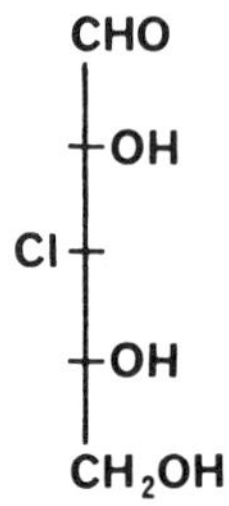

18

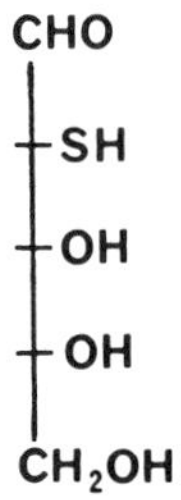

19

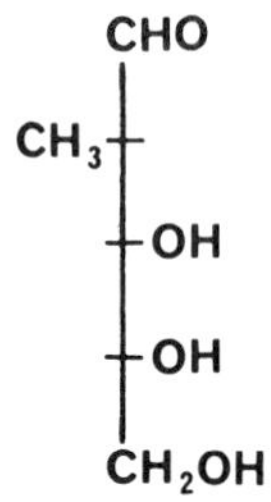

20

compound 21, for example, is described with the prefixes L-threo and D-gluco. In deciding which configurational prefixes to use,

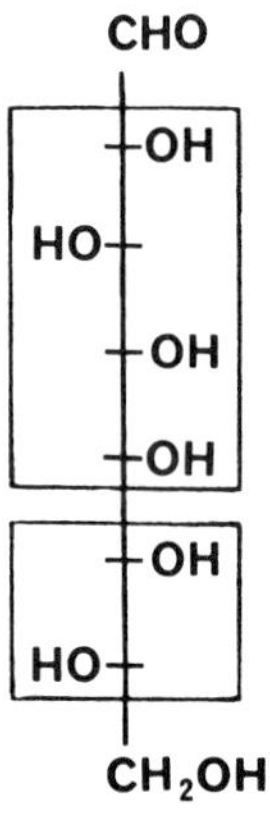

21

the molecule is divided so that there is a four carbon atom prefix for the centers closest to the reference group.

5 Naming of Carbohydrates

I. CONSTRUCTING NAMES

- A. Introduction
- B. α- and β-D-Glucopyranose [1, 2] and α- and β-D-Glucofuranose [3, 4]
- C. 1,2,3,4,6-Penta-*O*-acetyl-α-D-glucopyranose [5]
- D. 4-*O*-Acetyl-1,2-di-*O*-benzoyl-6-*O*-p-tolylsulfonyl-α-D-galactopyranose [6]
- E. 4,6-*O*-Benzylidene-β-D-glucopyranose [7] and 1,2:5,6-Di-*O*-isopropylidene-α-D-glucofuranose [8]
- F. Methyl α-D-Mannopyranoside [9]
- G. Methyl 6-Deoxy-α-D-mannopyranoside [10], 6-Amino-6-deoxy-α-D-glucopyranose [11], and 6-Deoxy-6-iodo-β-D-galactopyranose [12]
- H. Methyl 3-Deoxy-α-D-*ribo*-hexopyranoside [13]
- I. 1,6-Anhydro-β-D-glucopyranose [14]
- J. Methyl 4,6-Di-*O*-acetyl-2,3-dideoxy-α-D-*erythro*-hex-2-enopyranoside [15] and Methyl 4,6-Di-*O*-acetyl-2,3-dideoxy-α-D-*erythro*-hexopyranoside [16]

I. CONSTRUCTING NAMES

A. Introduction

The names associated with carbohydrates are often rather long but the process for constructing these names usually is a simple one. Although carbohydrate nomenclature can become complex in special situations, these complicated naming problems will not be considered here; rather, the goal of this chapter is to assist the reader in understanding and becoming able to construct names for the types of molecules typically encountered. For an extensive set of rules governing the naming of carbohydrates, the Rules of Carbohydrate Nomenclature (1) can be consulted.

A simple method for beginning to develop familiarity with carbohydrate nomenclature is to examine a few names. Actually, most of the basic names for monosaccharides already have been encountered (see Figures 3 and 4 in Chapter 3) and, fortunately, these names provide much of the fundamental terminology which is used. Since most carbohydrates exist as cyclic structures, the emphasis in the present chapter (and from this point forward) will be on cyclic compounds.

B. α- and β-D-Glucopyranose [1, 2] and α- and β-D-Glucofuranose [3, 4]

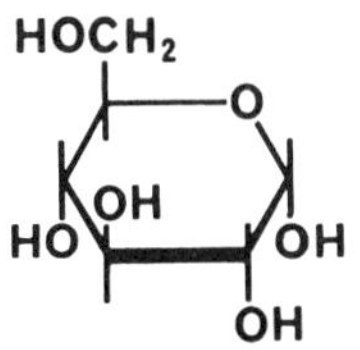

1

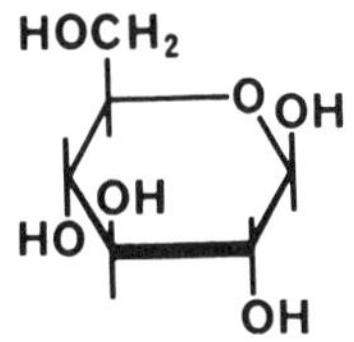

2

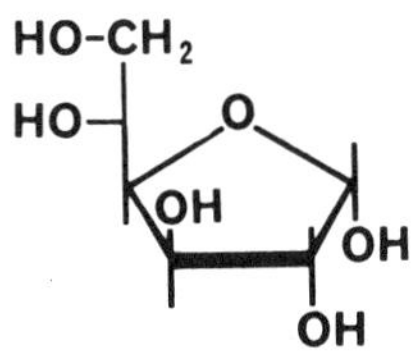

3

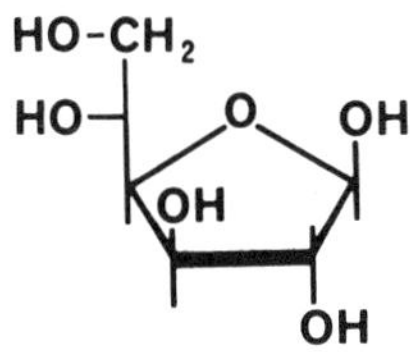

4

The first compounds to be considered are the common cyclic structures formed from D-glucose. In α-D-glucopyranose [1] the "pyranose" portion of the name refers to the pyranoid ring in the molecule. The D designator specifies the configuration at C-5 (the chiral center farthest from the reference group) while stereochemistry at C-1 is given by the Greek letter α and at C-2, C-3, and C-4 by the "gluco" prefix. If the configuration at C-1 in α-D-glucopyranose [1] is inverted, β-D-glucopyranose [2] is produced. When D-glucose forms a five-membered ring, "furanose" is the term which identifies this new ring system; thus, compounds 3 and 4 are α-D-glucofuranose [3] and β-D-glucofuranose [4].

C. 1,2,3,4,6-Penta-*O*-acetyl-α-D-glucopyranose [5]

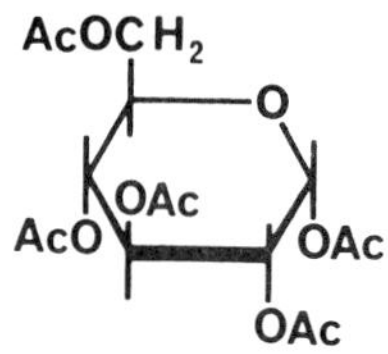

5

When the hydroxyl groups in compound 1 are acetylated, the resulting D-glucose derivative is 1,2,3,4,6-penta-*O*-acetyl-α-D-glucopyranose [5]. The "*O*-acetyl" portion of the name establishes both that CH_3CO groups are being described and that each of these groups is attached to an oxygen atom. The total number of acetyl groups is specified by the term "penta", and the numbers 1,2,3,4,6 identify which oxygen atoms bear acetyl groups. Oxygen atoms have the same numbers as the carbon atoms to which they are bonded. (The acetyl group is one of many substituent groups which can be attached to an oxygen atom in a carbohydrate. A list of common groups of this type is found in Table 1.)

D. 4-*O*-Acetyl-1,2-di-*O*-benzoyl-6-*O*-*p*-tolylsulfonyl-α-D-galactopyranose [6]

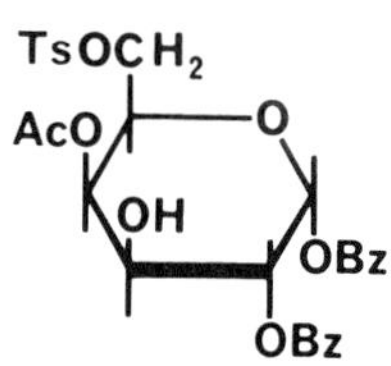

6

There is a preferred order for listing substituent groups. Consider 4-*O*-acetyl-1,2-di-*O*-benzoyl-6-*O*-*p*-tolylsulfonyl-α-D-galactopyranose [6], which has benzoyl groups attached to O-1 and O-2,

Table 1 Common Substituent Groups Attached to Oxygen Atoms

Structure	Group Name (Abbreviation)
CH_3CO	acetyl (Ac)
C_6H_5CO	benzoyl (Bz)
$C_6H_5CH_2$	benzyl (Bzl)
C_6H_5CH	benzylidene (attached to two oxygens)
CH_3CH_2	ethyl (Et)
CH_3CH	ethylidene (attached to two oxygens)
$(CH_3)_2C$	isopropylidene (attached to two oxygens)
CH_3SO_2	mesyl or methylsulfonyl (Ms)
CH_3	methyl (Me)
CH_2	methylene (attached to two oxygens)
NO_2	nitro (NO_2)
$CH_3C_6H_5SO_2$	tosyl or p-tolylsulfonyl (Ts)
CH_3SO_2	triflyl or trifluoromethylsulfonyl (Tf)
C_5H_9O	tetrahydropyranyl (THP)

an acetyl group to O-4, and a *p*-tolylsulfonyl group to O-6. In constructing the name for compound 6, groups are listed in alphabetical order. Prefixes such as di, tri, tetra, penta, and hexa are ignored in establishing the alphabetical order.

E. 4,6-*O*-Benzylidene-β-D-glucopyranose [7] and 1,2:5,6-Di-*O*-isopropylidene-α-D-glucofuranose [8]

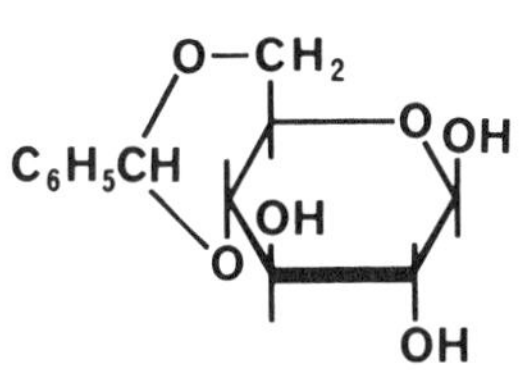

7

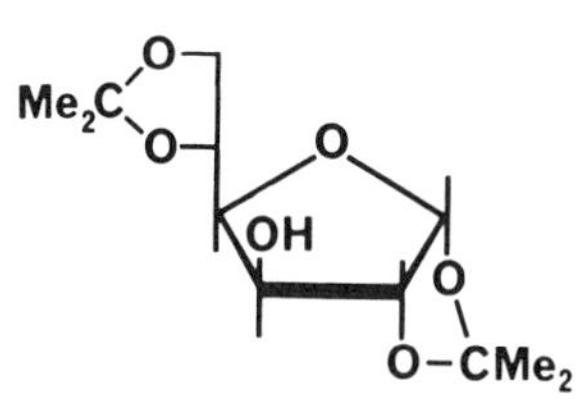

8

A group such as a cyclic acetal or ketal can be attached to more than one oxygen atom in a monosaccharide unit. This situation exists for the benzylidene group in 4,6-*O*-benzylidene-β-D-glucopyranose [7]. Attachment to two oxygen atoms is signified both by the group name and by identifying the oxygen atoms (O-4 and O-6) to which the group is bonded. In 1,2:5,6-di-*O*-isopropylidene-α-D-glucofuranose [8], the two points of connection (1,2 and 5,6) for each isopropylidene group are specified.

F. Methyl α-D-Mannopyranoside [9]

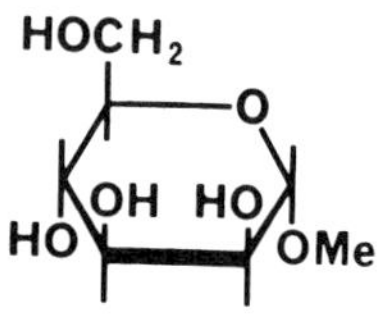

9

When the hydrogen atom in the anomeric hydroxyl group is replaced with an alkyl group, a glycoside such as methyl α-D-mannopyranoside [9] is formed. In naming glycosides, the alkyl group (called the aglycon) is named first and is a separate word. The presence of a glycosidic linkage is also indicated by the "side" ending on the name.

G. Methyl 6-Deoxy-α-D-mannopyranoside [10], 6-Amino-6-deoxy-α-D-glucopyranose [11], and 6-Deoxy-6-iodo-β-D-galactopyranose [12]

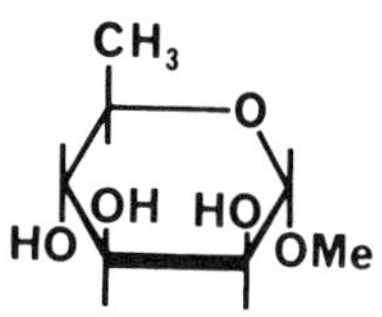

10

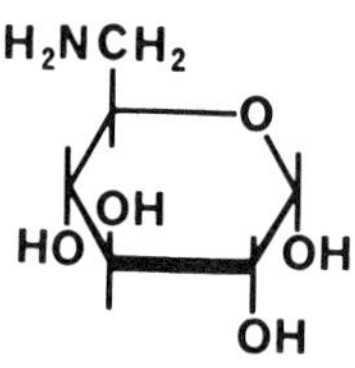

11

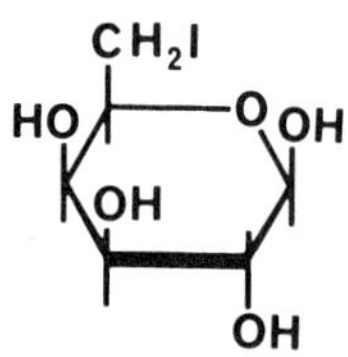

12

If a hydroxyl group is replaced by a substituent that does not bond to the carbon atom chain by a C-O bond, the term "deoxy" appears in the compound name. The deoxy term by itself is used if the replacing substituent is a hydrogen atom, as is the case with methyl 6-deoxy-α-D-mannopyranoside [10], but if other replacing groups are present, additional identification is required. For example, in 6-amino-6-deoxy-α-D-glucopyranose [11] and 6-deoxy-6-iodo-β-D-galactopyranose [12], the substituent names are included. A list of common substituent groups that can be attached directly to the carbon atom chain (and when thus attached, require the "deoxy" term in the name) is found in Table 2.

Table 2 Common Substituent Groups Attached Directly to the Carbon Atom Chain

Structure	Group Name
NH_2	amino
Br	bromo
Cl	chloro
F	fluoro
CH_3CH_2	ethyl
I	iodo
CH_3	methyl
NO_2	nitro[a]

[a]Attachment to the carbon atom chain is at the nitrogen atom.

H. Methyl 3-Deoxy-α-D-*ribo*-hexopyranoside [13]

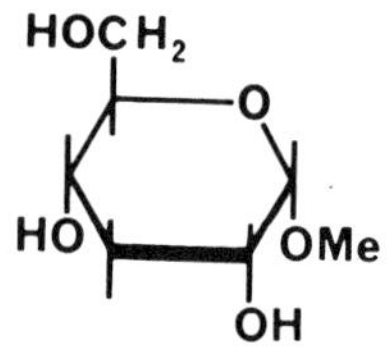

13

If hydroxyl group replacement causes a loss of chirality at one or more centers, the new stereochemical pattern is identified by one of the prefixes (italicized) shown in Figure 2 of Chapter 4. The name methyl 3-deoxy-α-D-*ribo*-hexopyranoside [13] conveys the following information: "hexopyranoside" signifies that the compound under consideration is a six-carbon atom containing glycoside with a pyranoid ring; α-D specifies stereochemistry at C-C-1 and C-5; *ribo* gives the configuration at the remaining chiral centers (C-2 and C-4); and 3-deoxy indicates that no oxygen atom is bonded to C-3. An achiral center is passed over in choosing the proper configurational prefix.

I. 1,6-Anhydro-β-D-glucopyranose [14]

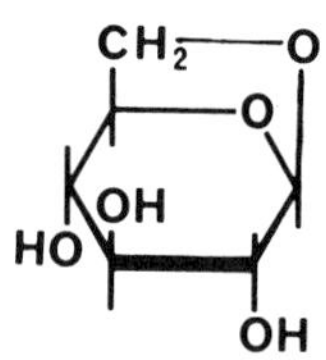

14

A compound which appears to have been formed from the intra-molecular elimination of the elements of water from two hydroxyl groups in a monosaccharide or monosaccharide derivative contains "anhydro" in its name. 1,6-Anhydro-β-D-glucopyranose [14] is such a compound. The numbers 1,6 identify the carbon atoms bonded together by the anhydro linkage. (The words "which appears to have been" in the first sentence in this paragraph are necessary because anhydro compounds can be formed by reactions other than loss of water.)

J. Methyl 4,6-Di-*O*-acetyl-2,3-dideoxy-α-D-*erythro*-hex-2-enopyranoside [15] and Methyl 4,6-Di-*O*-acetyl-2,3-dideoxy-α-D-*erythro*-hexopyranoside [16]

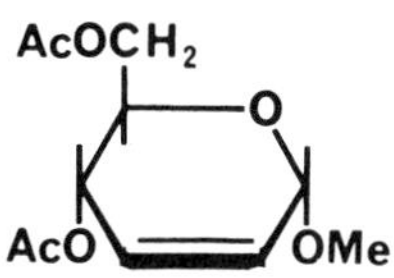

15

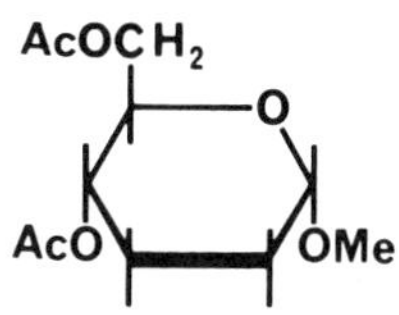

16

In a molecule containing a double bond, the name is generated by first naming the saturated parent compound; thus, in order to name compound 15, the name for 16 first is determined. Compound 16 is methyl 4,6-di-*O*-acetyl-2,3-dideoxy-α-D-*erythro*-hexopyranoside. Compound 15 is named by inserting "2-en" after the "hex" portion of the name. Compound 15 is methyl 4,6-di-*O*-acetyl-2,3-dideoxy-α-D-*erythro*-hex-2-enopyranoside.

K. *O*-α-D-Galactopyranosyl-(1→4)-α-D-galactopyranose [17] and *O*-α-D-Glucopyranosyl-(1→4)-*O*-α-D-galactopyranosyl-(1→4)-β-D-galactopyranose [18]

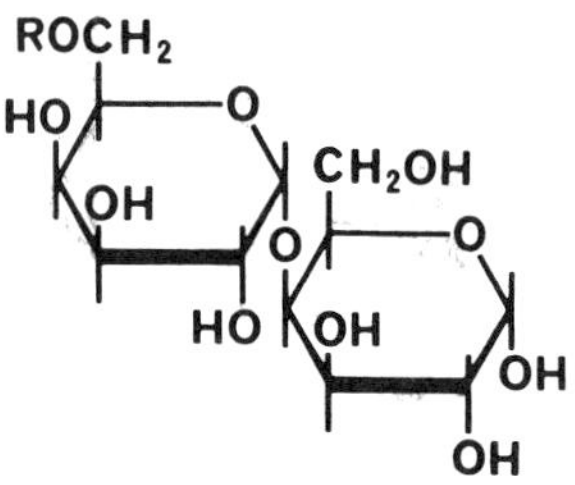

17 R = H

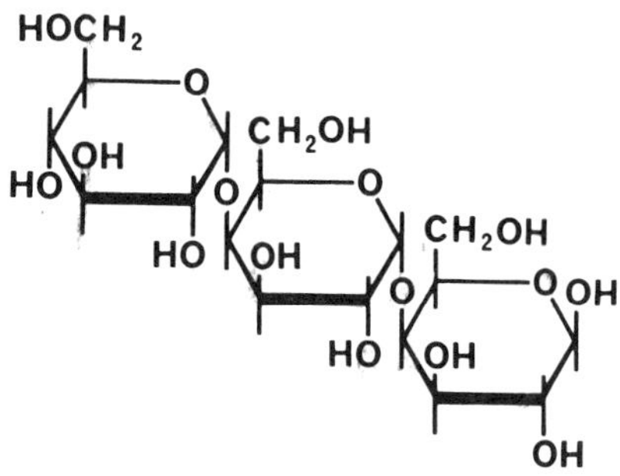

18

When a cyclic monosaccharide residue is attached at its anomeric carbon atom to another carbohydrate, the monosaccharide residue is named as a substituent. The terminal "e" of the monosaccharide name is replaced by "yl", as in the disaccharide 17, which is named 4-*O*-α-D-galactopyranosyl-α-D-galactopyranose. An equally acceptable name for this disaccharide is *O*-α-D-galactopyranosyl-(1→4)-α-D-galactopyranose [17]. The atom numbers at the points of connection between the two rings are linked by an arrow (1→4). Use of arrows allows the construction of tri- and higher oligosaccharide names. The trisaccharide 18 is *O*-α-D-glucopyranosyl-(1→4)-*O*-α-D-galactopyranosyl-(1→4)-β-D-galactopyranose.

L. *O*-(6-*O*-Methyl-α-D-galactopyranosyl)-(1→4)-α-D-galactopyranose [19]

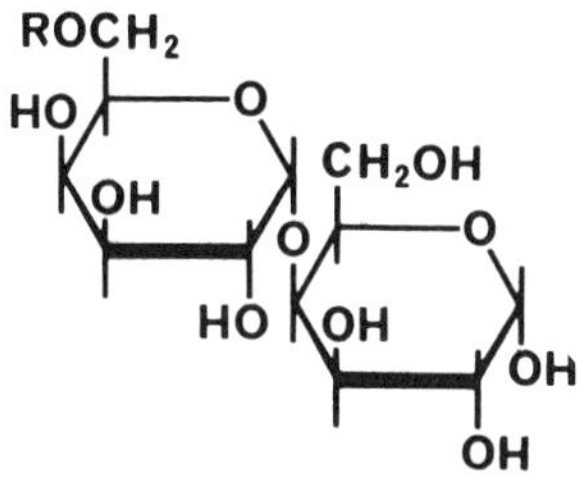

19 R = CH_3

It is sometimes necessary to insert parentheses for clarity if one of the monosaccharide residues in an oligosaccharide contains a substituent; thus, the use of parentheses to enclose "6-*O*-methyl-α-D-galactopyranosyl" in *O*-(6-*O*-methyl-α-D-galactopyranosyl)-(1→4)-α-D-galactopyranose [19] makes clear the location of the methyl group.

M. 2,3,4,6-Tetra-*O*-acetyl-α-D-glucopyranosyl Bromide [20]

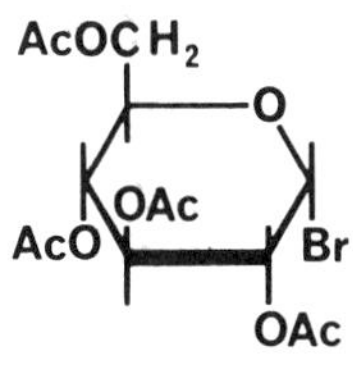

20

The glycosyl terminology is also used for monosaccharide derivatives in which the anomeric hydroxyl group has been replaced by a group which does not have an oxygen atom bonded to the anomeric carbon atom. The glycosyl halide 2,3,4,6-tetra-*O*-acetyl-α-D-glucopyranosyl bromide [20] is an example of such a molecule.

N. 4,6-*O*-(*S*)-Benzylidene-α-D-glucopyranose [21]

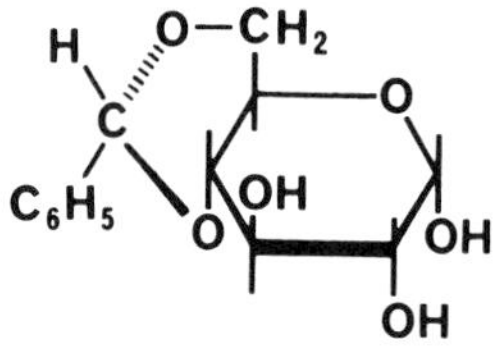

7R

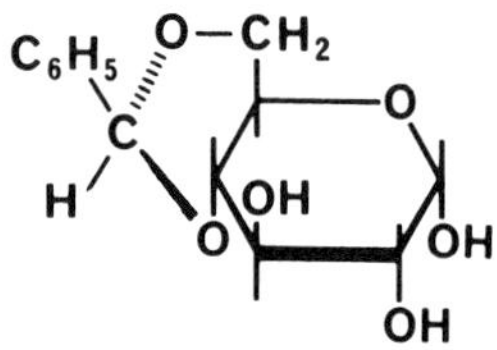

21

As a final observation, it should be noted that situations arise in which the D,L system cannot be used to designate stereochemistry. Such a situation, in fact, exists for compound 7. The name given 7, 4,6-*O*-benzylidene-α-D-glucopyranose, does not distinguish it from its diastereomer, compound 21. Attempting to differentiate between compounds 7 and 21 using the D,L system would lead to ambiguity since the chiral benzylic carbon atom has two oxygens bonded to it. This chiral center has no clear relationship to D or L-glyceraldehyde. The solution to this problem is to give *R* and *S* designations to the benzylic carbon atoms; compound 7 is 4,6-*O*-(*R*)-benzylidene-α-D-glucopyranose [7*R*] and compound 21 is 4,6-*O*-(*S*)-benzylidene-α-D-glucopyranose.

II. SUBSTITUENT GROUPS

Two types of substituent groups in carbohydrates have been identified: those which replace the hydrogen atoms of a hydroxyl group (Table 1) and those which bond directly to the carbon atom chain (Table 2). A few groups, such as methyl and ethyl, belong in both categories. It is often desirable when referring to a group attached to an oxygen atom to use a term that represents the substituent group and the oxygen atom together. This is done by adding "oxy" to the end of the name; thus, methylsulfonyl (mesyl, CH_3SO_2) becomes methylsulfonyloxy (mesyloxy, CH_3SO_2O). For a few very common groups, such as acetyl, methyl, and ethyl, the addition of the "oxy" is accompanied by the deletion of the "yl" to give acetoxy, methoxy, and ethoxy, respectively.

III. CONCLUSION

Although a much more extensive discussion of carbohydrate nomenclature certainly is possible, the analysis of the names of the representative compounds 1-21 provides sufficient background to be able to understand the names used for most compounds and to construct a wide variety of names.

REFERENCE

1. Rules of Carbohydrate Nomenclature, *J. Org. Chem.*, *28*,281 (1963) and *J. Chem. Soc.*, 5307 (1962).

6 Conformational Analysis

I. SATURATED PYRANOID RING SYSTEMS

A. Basic Concepts and Terminology

One might expect a similarity to exist between the conformations favored by substituted pyranoid rings and those adopted by substituted cyclohexanes. This expectation is largely fulfilled. Although the presence of an oxygen atom alters ring geometry and electron distribution, this perturbation is not sufficient to destroy the basic similarities between the pyranoid ring and its carbocyclic analog. A pyranoid ring exists in solution at room temperature in conformational equilibrium between two chairlike forms of unequal energy. The two chair conformations for methyl 2,3,4-tri-*O*-acetyl-β-D-xylopyranoside [1] are shown in Figure 1. Other possible conformations, such as the boat and skew (twist-boat) forms are of higher energy. (There are, in fact, six boat and six skew forms possible for a pyranoid ring (1).) Two of these boat conformations are shown in Figure 2 and two skew forms are pictured in Figure 3. Most molecules rarely assume these higher energy conformations: nevertheless, there is sufficient energy available in typical chemical reactions to permit non-chairlike

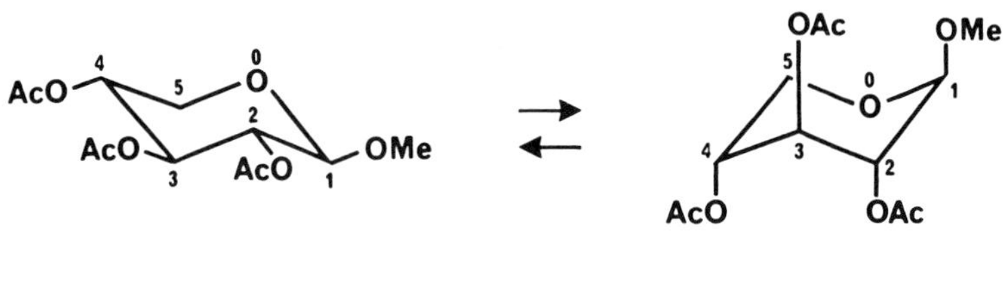

$^4C_1$ $^1C_4$

FIG. 1. The two chair conformations for methyl 2,3,4-tri-*O*-acetyl-β-D-xylopyranoside [1].

$^{1,4}B$ $B_{1,4}$

FIG. 2. Two boat conformations for methyl 2,3,4-tri-*O*-acetyl-β-D-xylopyranoside [1].

conformations to be adopted momentarily. Boat and skew forms, therefore, can be important conformations in chemical reactions.

A terminology which is simple and effective in describing the conformations adopted by pyranoid rings has evolved in recent years (2). To understand the symbols used, it is instructive to focus attention first on two chair conformations such as those shown in Figure 1. These conformations are each represented by a capital "C". In each structure the ring oxygen atom and carbon atoms 2, 3, and 5 are approximately in a plane. The conformation

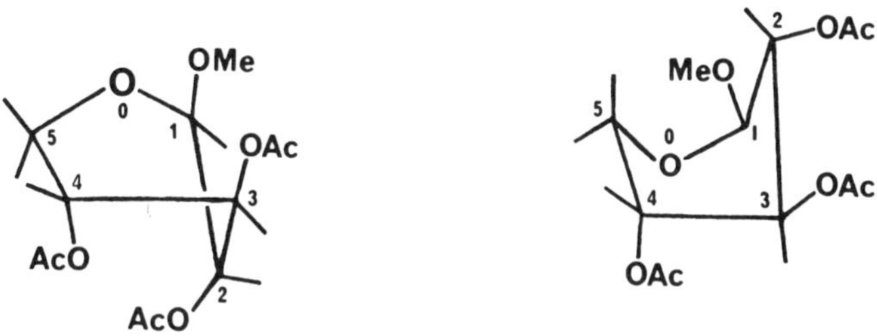

$^{0}S_{2}$ $^{2}S_{0}$

FIG. 3. Two skew conformations for methyl 2,3,4-tri-*O*-acetyl-β-D-xylopyranoside [1].

on the left in Figure 1 (all acetoxy groups equatorial) has C-4 above this plane and C-1 below the plane; thus, it is designated ${}^4C_1$. The other conformation (all acetoxy groups axial) has C-1 above and C-4 below the ring plane; consequently, it bears the designation ${}^1C_4$. It is necessary also to know whether the compound under consideration belongs to the D or L family. The need for this information becomes apparent when comparing the chair conformations of a member of the D family, shown in Figure 1, with the corresponding conformations of its enantiomer (a member of the L family), pictured in Figure 4. The conformation with all acetoxy groups equatorial is ${}^4C_1$ in the D series and ${}^1C_4$ in the L series; thus, the complete designation for the "all equatorial" conformation in Figure 1 is ${}^4C_1$ (D) and that in Figure 4 is ${}^1C_4$ (L). Usually the family to which a compound under consideration belongs is clear and the D or L designation does not need to accompany the conformational symbol.

A similar procedure is used for describing boat and skew conformations, represented by "B" and "S", respectively. The boat conformation shown on the left in Figure 2 has C-1 and C-4 both above the plane defined by the ring oxygen atom, C-2, C-3, and C-5; thus, it is a ${}^{1,4}B$ (D) conformation, while that on the right (Figure 2, with C-1 and C-4 below the plane, is a $B_{1,4}$ (D)

FIG. 4. The two chair conformations for methyl 2,3,4-tri-*O*-acetyl-β-D-xylopyranoside [1].

conformation. In Figure 3, the skew conformation on the left has C-1, C-3, C-4, and C-5 in a plane with the ring oxygen atom (number 0 [zero]) above the plane and C-2 below. Such an arrangement is a $^{0}S_{2}$ (D) conformation. The $^{2}S_{0}$ (D) conformation also is shown in Figure 3.

Two other approaches to designating conformations of cyclic carbohydrates have been used extensively in the past but their use now is rare. They will not be discussed here except to note that the symbols used for the two chair conformations of a pyranoid ring in one approach (3) are 1C and C1 and in the other (4) are CA and CE. The conformations of α-D-glucopyranose [2] associated with each of these symbols are shown in Figure 5. A thorough discussion of these older systems of conformational terminology can be found in reference 5.

B. Factors Determining Conformational Energy in Pyranoid Rings

A combination of steric and electronic effects is primarily responsible for determining which of two equilibrating chair conformations will be of lower energy. If a prediction is to be made concerning which chair form is favored, information is first needed

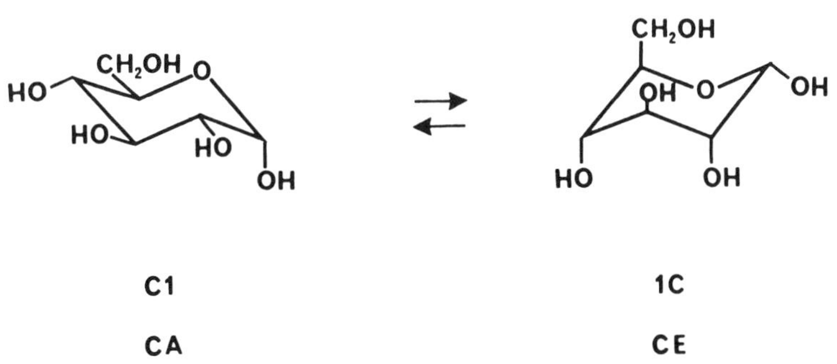

FIG. 5. Conformational designations for the chair forms of α-D-glucopyranose [2] using 1C,C1 and CA,CE terminology.

about the identity and magnitude of these effects. Steric factors can be identified and estimated by determining the percentage of each chair form present in a sufficient number of molecules to establish approximate values for each possible nonbonded interaction. Electronic effects can be estimated by examining the change in conformational equilibria brought about by introducing substituents with various electronic requirements. Both types of information are available. That related to nonbonded interactions will be considered first.

1. Nonbonded Interactions

Six nonbonded interactions of importance in cyclic monosaccharides have been identified and the free energy differences associated with each of these interactions has been estimated (6-8). The identity and magnitude of these destabilizing interactions are given in Figure 6. In this approach it is assumed that

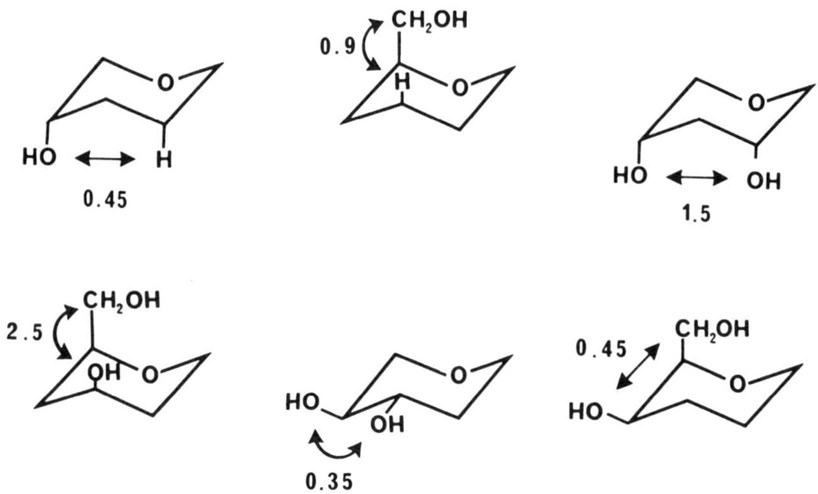

FIG. 6. Estimated values for nonbonded interaction energies (kcal/mole) at room temperature in aqueous solution.

various nonbonded interactions are independent of one another. It also is assumed that interaction is primarily between atoms directly attached to the ring; thus, the same energy value is assigned to the repulsion between two hydroxyl groups as is used for the repulsion between two methoxy groups. Using the same value in such situations is justified on the basis that atoms other than those directly attached to the ring are sufficiently remote that they assume positions which avoid unfavorable repulsions.

If nonbonded interactions were the only factors of importance in determining the free energy difference between two conformations, the values given in Figure 6 could be used by themselves to estimate the energy difference between two conformers. However, a corrective factor must be taken into account before reliable predictions can be made. This corrective factor is necessary because of an electronic phenomenon known as "the anomeric effect" (9).

2. The Anomeric Effect

For a cyclohexane ring bearing a methoxy substituent, the favored conformation has this group in an equatorial orientation (Figure 7). It is initially surprising, therefore, that when one of the hydrogen atoms next to the oxygen atom in tetrahydropyran is replaced by a methoxy group, the favored conformation has this methoxy group in an axial orientation (Figure 7). This unusual conformational preference is an example of the anomeric effect (9). When discussing carbohydrates, therefore, the anomeric effect is a name for the tendency in pyranoid rings for an electronegative substituent to prefer an axial orientation when attached to an anomeric carbon.

The magnitude of the anomeric effect is dependent upon a number of factors (10). These include:

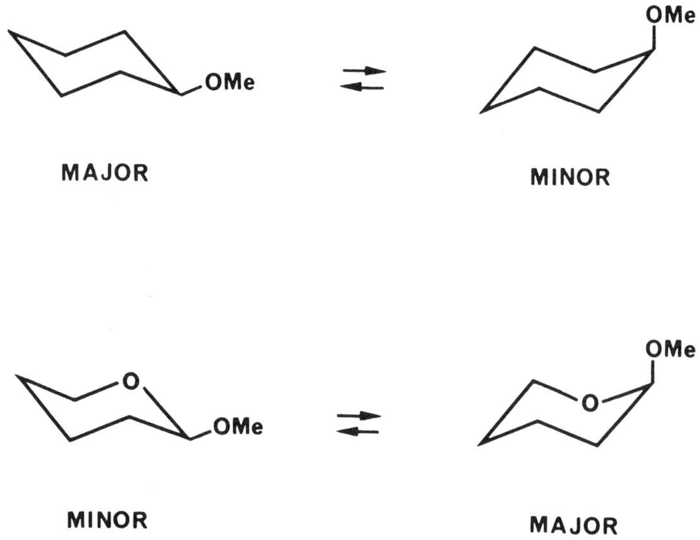

FIG. 7. Chair conformations of cyclohexane and tetrahydropyran, each substituted by a methoxy group.

a. The identity of the electronegative group: Energy differences are in the 0.5–1.5 kcal/mole range for hydroxyl and methoxy groups and tend to cluster around 3 kcal/mole for halogens (11).
b. The nature of the solvent: Generally, the anomeric effect is largest in solvents of low dielectric constant.
c. The location of substituents: The positions of substituents at other than the anomeric carbon atom influence the anomeric effect. For example, the magnitude of the anomeric effect changes if the C-2 hydroxyl group is removed (Figure 8) (6–8).
d. The configuration at positions other than the anomeric carbon atom: The anomeric effect also is a function of the substituents attached to other ring carbon atoms. An example

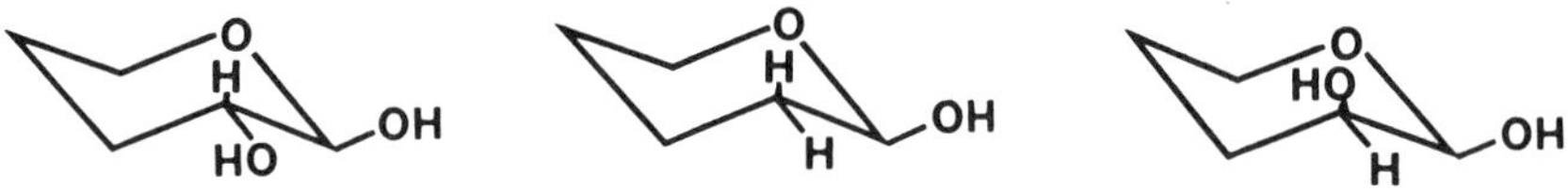

0.55 0.85 1.0

(0.9) (1.20) (1.35)

FIG. 8. Estimated values (6–8) for the anomeric effect (kcal/mole) at room temperature in aqueous solution. The values in parentheses were suggested in reference 11.

of this dependence is found in Figure 8, which shows that the anomeric effect changes when the C-2 configuration is inverted.

There are other factors (e.g., solution temperature), in addition to those listed above, influencing the anomeric effect; however, from the viewpoint of conformational analysis, perhaps the most important conclusion to be drawn relates to the general significance of the anomeric effect. When one compares the magnitude of the energy differences attributed to this effect (see item *a* above and Figure 8) with the energy values for nonbonded interactions (Figure 6), it is clear that the anomeric effect is one of the more important factors in determining favored molecular conformation.

The equilibria between the chair conformations of the α and of the β anomers of methyl 2,3,4-tri-*O*-acetyl-D-xylopyranosides [3 and 1, respectively] in $CDCl_3$ (Figure 9) provide a clear example of the influence of the anomeric effect on conformational preference. In the α anomer (Figure 9), $^4C_1$ is the only detectable conformation, but the β anomer exists in conformational equilibrium between the $^4C_1$ (77%) and the $^1C_4$ (23%) conformers

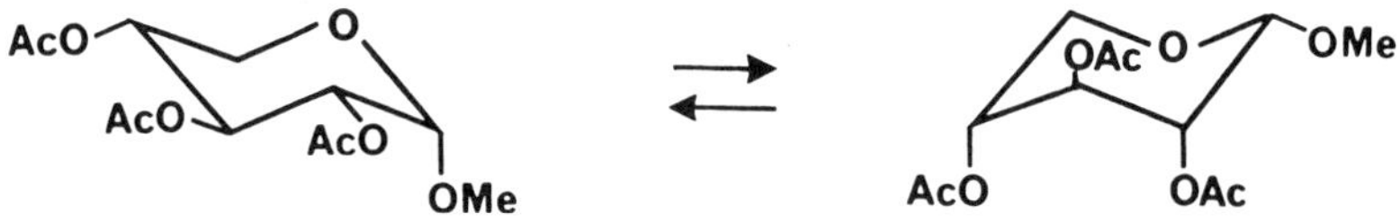

100 % 0 %

3

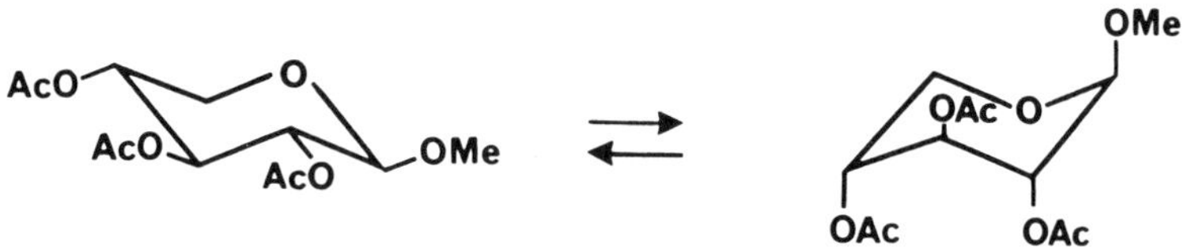

77 % 23 %

1

FIG. 9 Percent of each conformer present in the conformational equilibria for methyl 2,3,4-tri-*O*-acetyl-(α and β)-D-xylopyranosides [3 and 1].

(12). Peracetylated pentopyranosyl halides provide even more striking examples of the influence of the anomeric effect on conformation. In 2,3,4-tri-*O*-benzoyl-β-D-xylopyranosyl choride [4], for example, the need to satisfy the anomeric effect is so great that in $CDCl_3$ the $^1C_4$ conformation constitutes 98% of the equilibrium mixture (13).

The initial explanation offered for the anomeric effect was a proposal based on dipole–dipole interactions (14). This explanation

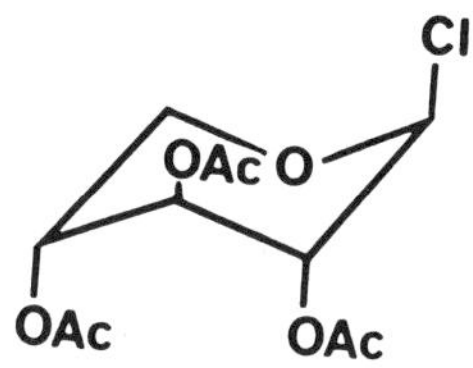

4

suggested that in the acetal linkage in a pyranoid ring, each chair conformation could be viewed as containing two dipoles of importance (Figure 10). The first dipole was parallel to the C-1, O_{Me} bond and was a result of the polarization of this bond (Figure 10). The second dipole bisected the angle between the lone pairs of electrons on the ring oxygen atom and resulted from the concentration of electron density exterior to the ring produced by the electrons in these orbitals. Conformer A (Figure 10) placed the two dipoles nearly parallel to each other (an energetically unfavorable arrangement) while conformer B had the same dipoles oriented in a nearly perpendicular fashion (an energetically more stable arrangement). In the absence of other effects, the existence of these dipoles should favor the conformation with an axial orientation for the electronegative methoxy group.

FIG. 10. Dipoles present in acetal linkages in pyranoid rings. Direction of dipoles shown by arrows.

As experimental evidence accumulated, it became clear that more than simple dipole-dipole interactions were involved and that a more comprehensive explanation was needed for the anomeric effect. Other molecular properties associated with this phenomenon needed to be explained. The most notable of these were the changes in molecular geometry (described below) which were found to be related to the anomeric effect.

Bond distances and bond angles determined by X-ray crystallography for a group of methyl glycosides (15) (mean values for groups of α and β anomers) are given in Figure 11. In the α anomers the C-1, O_{Me} bond is longer than in the β anomers and the C-1, O_{ring} bond is shorter. There is a noticeably greater bond

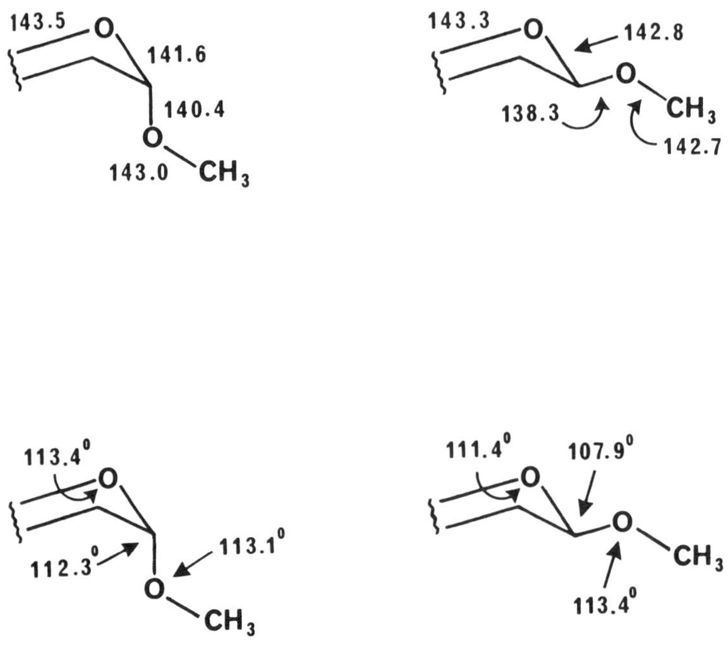

FIG. 11. (top) Bond distances in Angstroms × 10^2 and (bottom) bond angles in degrees (for α and β methyl pyranosides).

angle defined by the atoms O_{ring}, C-1, O_{Me} (Figure 11) in the α anomers (112.3°) than in the β (107.9°). These differences are not adequately explained by dipole-dipole interactions alone. Molecular orbital calculations (16) on model systems provide an explanation for these changes in geometry as well as rationalizing the stability of conformations with an axial orientation of electronegative substituents. It is possible to offer a pictorial interpretation which explains, in a qualitative fashion, the anomeric effect and the most significant of the associated changes in molecular geometry. This interpretation is based upon the orbital interactions shown in Figure 12. Consider the overlap between the orbitals containing the lone pairs of electrons on the ring oxygen atom and the "back lobe" of the orbital associated with the sigma bond between C-1 and the methoxy group (Figure 12). In conformer C the geometry does not allow significant overlap between the orbitals on the oxygen atom containing nonbonded electrons and the back lobe of the sigma orbital; in contrast, the axially oriented orbital on the ring oxygen atom in conformer D is parallel to the back lobe and can interact with it (Figure 12). (This situation has similarity to the parallel p-orbitals which interact to

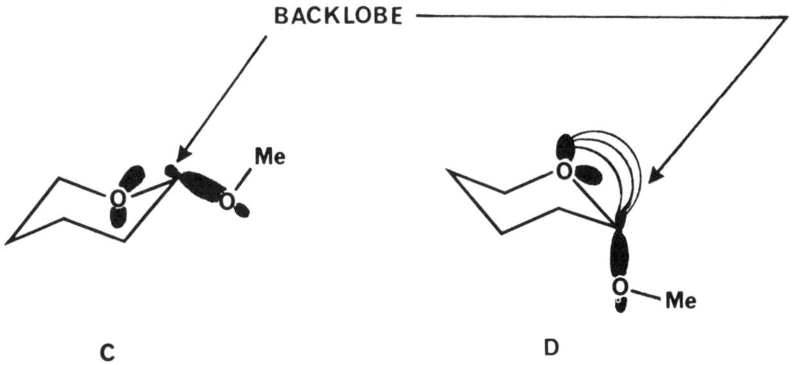

FIG. 12. Orbital interaction in methyl pyranoside.

form a π bond.) Since orbital interaction results in stabilization, only the conformer with the axial substituent (D) is stabilized. Further when comparing α and β anomers, the increased bonding between C-1 and the ring oxygen atom should shorten this bond. Also, the additional bonding between C-1 and the ring oxygen atom uses the orbital on C-1 which is involved in the sigma bond to the methoxy group. This use necessarily must decrease the bonding between C-1 and the methoxy oxygen atom because the combined partial bonding in which this orbital is involved cannot exceed a total of one bond. Therefore, the bond between C-1 and the oxygen atom in the methoxy group is weakened and lengthened. The increased double bond character between the ring oxygen atom and C-1 gives C-1 greater sp^2 character and, as a result, increases the O_{ring}, C-1, O_{Me} bond angle in the α anomer.

3. The Reverse Anomeric Effect

An alkoxycarbonyl group attached to a carbon atom with a neighboring oxygen atom in tetrahydropyran has a larger conformational free energy than does the same group attached to a carbon atom in cyclohexane (17) (Figure 13). Favoring the conformation with the equatorial orientation to a greater extent than predicted on the basis of steric effects alone is known as the "reverse anomeric effect" (18). The existence and magnitude of the reverse anomeric effect appear to be directly related to the amount of positive charge on the atom attached to the anomeric carbon atom; in fact, this phenomenon was first recognized in pyridinium glucopyranosides (18), such as 5, where there is a full positive charge on the nitrogen atom attached to C-1. Less is known about the reverse anomeric effect (when compared to the anomeric effect) because it is encountered relatively infrequently in carbohydrates. (Few carbohydrates have full or partial positive

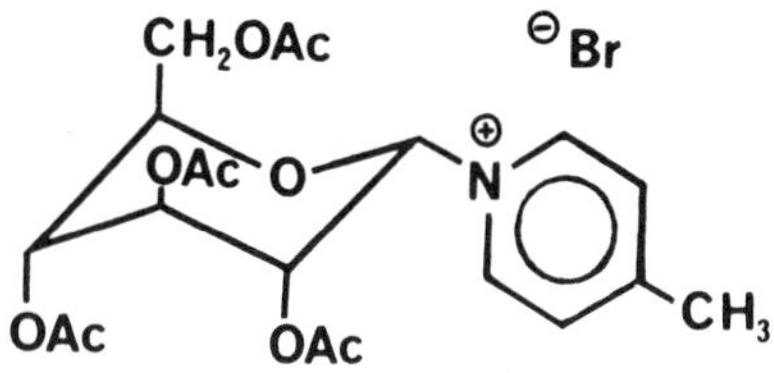

5

charges on atoms attached to C-1). The magnitude of the reverse anomeric effect probably is, like the anomeric effect, dependent upon a variety of factors; for example, it has been shown that the favored conformation for 3,4,5,7-tetra-*O*-acetyl-2,6-anhydro-D-*glycero*-D-*ido*-heptonamide [6] (Figure 14) is solvent dependent. This dependence is rationalized on the basis of the reverse anomeric effect being more important in less polar solvents (19).

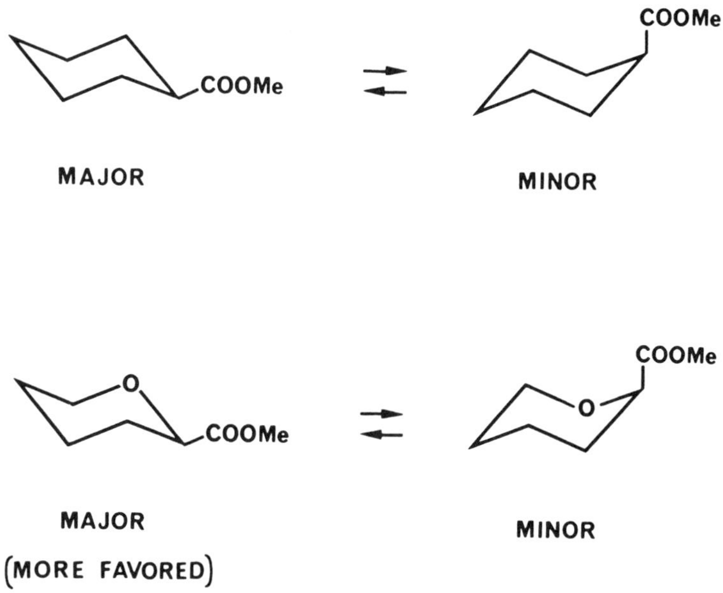

FIG. 13. Chair conformations of cyclohexane and tetrahydropyran, each with a carbomethoxy group attached.

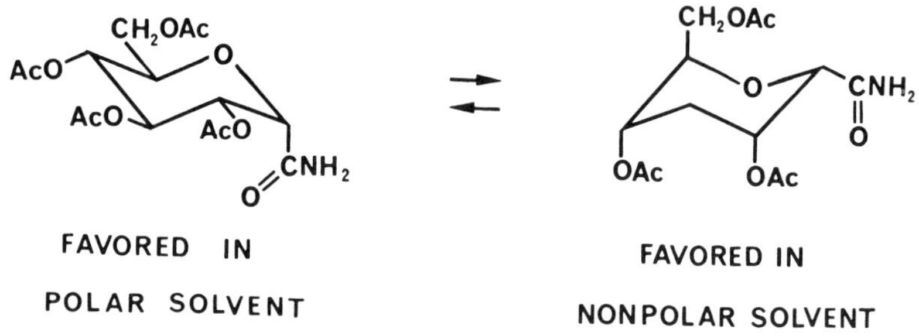

FIG. 14. Equilibrium between the chair conformers of 3,4,5,7-tetra-*O*-acetyl-2,6-anhydro-D-*glycero*-D-*ido*-heptanamide [6].

The reason for the existence of the reverse anomeric effect is probably primarily a dipole-dipole interaction. More complicated explanations are not necessary to explain what is now known about this phenomenon. The C-1 to substituent bond in molecules in which the reverse anomeric effect is of importance appears to be polarized in the opposite direction from that found in most carbohydrates; that is, the negative end of the dipole is at the C-1 end of the bond (Figure 15). The other dipole involved originates at the ring oxygen atom and bisects the angle between the two lone pairs of electrons. Since these two dipoles are oriented in opposite directions, the most stable arrangement for them is to be parallel to each other. This situation exists when the substituent group at C-1 is in an equatorial orientation (Figure 15).

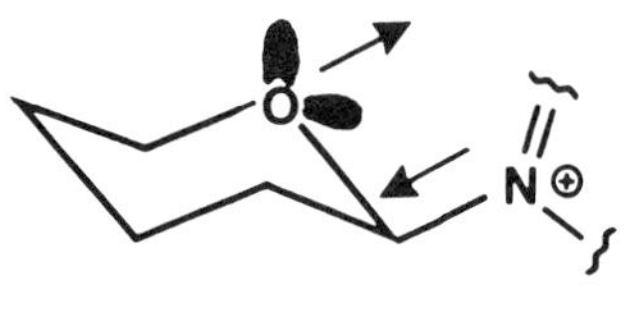

FIG. 15. Dipoles present in a pyranoid ring with a positively charged substituent at C-1. Direction of dipoles shown by arrows.

4. Hydrogen Bonding: Solvent Effects

Hydrogen bonding can play a critical role in determining the favored conformation in a molecule. The most significant factor with respect to hydrogen bonding is whether it is internal or external. In polar media the interaction between the solvent and a hydroxyl group typically is strong and often results in external hydrogen bonding (with the solvent) being favored over internal. In solvents of low dielectric constant, hydrogen bonding, if it is structurally permitted, is internal; therefore, it is possible, as a result of hydrogen bonding, for the preferred conformation in a molecule to be a function of the solvent. Such a situation exists for methyl 3-deoxy-β-L-*erythro*-pentopyranoside [7] (Figure 16) where intramolecular hydrogen bonding and the anomeric effect stabilize the $^4C_1$ conformer in $CDCl_3$ but in the polar solvents D_2O and DMSO-d_6 the $^1C_4$ conformation dominates due to intermolecular hydrogen bonding (20).

Solvent effects in addition to those associated with hydrogen bonding are undoubtedly of importance. Some useful information

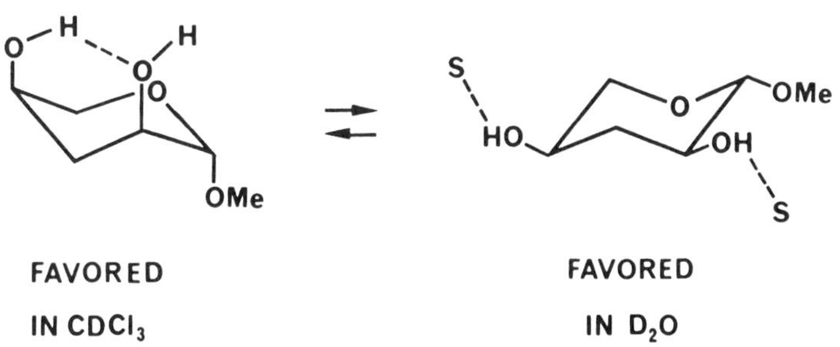

FIG. 16. Effect of solvent on the favored conformation for methyl 3-deoxy-β-L-*erythro*-pentopyranoside [7]. "S" represents the solvent.

about them is available (11,20) but more study is needed before clear generalizations can be made.

5. Calculation of Favored Conformations

a. Empirical Method. It is possible to use the values for nonbonded interactions given in Figure 6 and those for the anomeric effect (Figure 8) to predict the favored conformations for cyclic pyranoses in aqueous solution. An illustration of how this is done for 3-deoxy-α-D-*ribo*-hexopyranose [8] is shown in Figure 17. The $^4C_1$ conformation has destabilizing interactions calculated to be 0.80 (0.45 + 0.35) kcal/mole, while the $^1C_4$ conformation has nonbonded interactions of 3.65 (0.35 + 0.9 + 0.9 + 1.5) kcal/mole. In addition, the anomeric effect favors the $^4C_1$ conformation by 1.0 kcal/mole; thus, the difference in energy between these two conformations is calculated to be [4.65 (3.65 + 1.0) - 0.80] or 3.85 kcal/mole, sufficient to favor greatly the $^4C_1$ conformation.

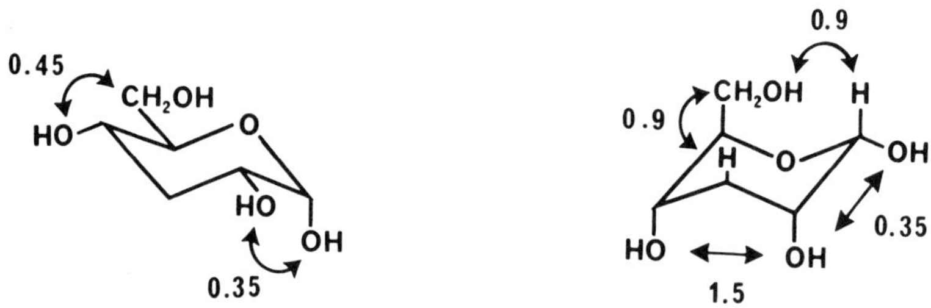

$^4C_1$ $^1C_4$

FIG. 17. Nonbonded interactions in 3-deoxy-α-D-*ribo*-hexopyranose [8].

b. Semi-empirical Method. A semi-empirical system for analyzing the energy differences among various geometries of a molecule has been developed and is known by the term "molecular mechanics". Briefly, in this approach it is assumed that a molecule will adopt a geometry which minimizes total energy. Total potential energy is viewed as consisting of contributions from stretching and compression of bonds, bond angle distortion, torsional interactions, nonbonded interactions, and electrostatic interactions. Some or all of these factors (and even others) can be taken into account in calculating the energy of a molecule. The geometry which leads to an energy minimium is the preferred conformation according to these calculations. The interested reader is referred to reference 21, which contains further information about molecular mechanics and references for further reading. An example of a recent application of this approach to the calculation of the favored conformation for compound 6 (Figure 14) is found in reference 22.

II. UNSATURATED PYRANOID RINGS: THE ALLYLIC EFFECT

A common conformation for unsaturated pyranoid rings is the half-chair, symbolized by an H. While this conformation is a high energy one for saturated systems, it is a preferred conformation in unsaturated ones. The $^4H_5$ and the $^5H_4$ conformations for 3,4,6-tri-*O*-acetyl-1,5-anhydro-2-deoxy-D-*arabino*-hex-1-enitol [9] (tri-*O*-acetyl-D-glucal) are shown in Figure 18.

In attempting to rationalize the conformations adopted by unsaturated pyranoid ring systems, it has been necessary to introduce a factor in addition to nonbonded interactions and the anomeric effect. This factor is known as the "allylic effect" (23) and is an apparent stabilization of *quasi*-axial, polar, allylic

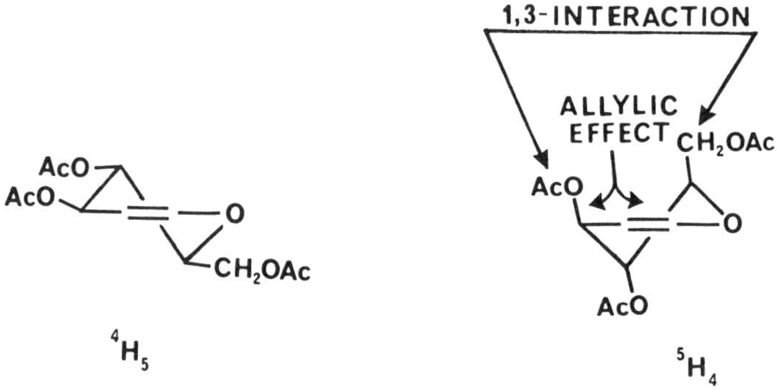

FIG. 18. Conformations of 3,4,6-tri-*O*-acetyl-1,5-anhydro-2-deoxy-D-*arabino*-hex-1-enitol [9].

substituents. The magnitude of the allylic effect has been estimated to be 0.8 kcal/mole for acetoxy groups (23). The allylic effect present in the ${}^5H_4$ (but the ${}^4H_5$) conformation for compound 9 (Figure 18) has been credited with causing the energy of the ${}^5H_4$ and ${}^4H_5$ conformers to be rather close even though the ${}^5H_4$ conformation has a 1,3-interaction (24,25).

III. FURANOID RING SYSTEMS

The situation with respect to five-membered rings is similar to that for six-membered rings in that there is a correspondence between conformations adopted by substituted cyclopentanes and those assumed by compounds containing furanoid rings. The energy barriers between conformations are smaller in five-membered rings than in six-membered rings; thus, equilibration among conformers is more rapid.

The envelope and twist conformations which are assumed by molecules with furanoid rings are given the letters "E" and "T", respectively. Examples of these two types of conformations are pictured in Figure 19. [A total of 10 envelope and 10 twist conformers is possible (26).] In the envelope conformation only one ring atom is not in the plane defined by the other four, while the twist conformation has three atoms defining a plane with one of the remaining ring atoms above the plane and the other below. In Figure 19 the ^1E and $^2T_3$ conformations are shown for methyl β-D-ribofuranoside [10].

In an extensive study of the preferred conformations in methyl furanosides, two prominent features were noted (27). The first is that the C-1, O_{Me} bond favors a *quasi*-axial orientation and the second is that the C-4, C-5 bond prefers a *quasi*-equatorial orientation. The reason for the first of these preferences is attributed to the anomeric effect operative in furanoid systems; while the second is understandable on the basis of the large nonbonded interaction expected from a substituent such as a hydroxymethyl group. The ^1E conformation for compound 10 shown in Figure 19 has both a *quasi*-axial methoxy group and a *quasi*-equatorial hydroxymethyl group; thus, it is not surprising that this conformation was found to be preferred (27).

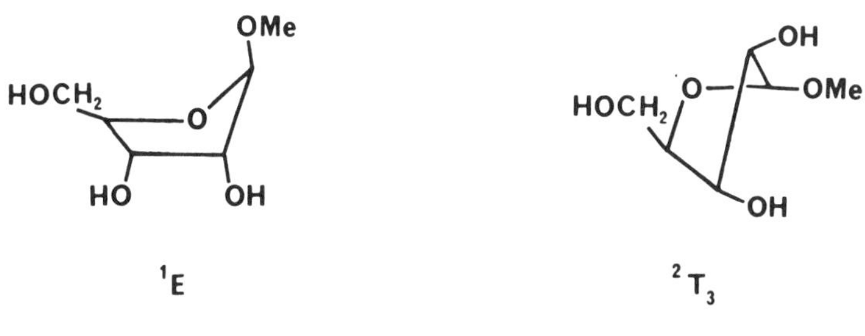

FIG. 19. Envelope and twist conformations for methyl β-D-ribofuranoside [10].

IV. ACYCLIC COMPOUNDS

Conformations of acyclic carbohydrates are usually pictured by using either Newman projections or wedge-slash formulas. These two methods for representing D-threitol [11] are shown in Figure 20. In the Newman projection, D-threitol [11] is viewed along its C-2, C-3 bond. A wedge-slash formula, such as that shown in Figure 20, is sometimes called a "sawhorse" representation.

Investigation of acyclic carbohydrates has led to the generalization that a "zigzag" arrangement of the carbon atom chain, such as that found in 2,3,4,5,6-penta-*O*-acetyl-D-mannononitrile [12],

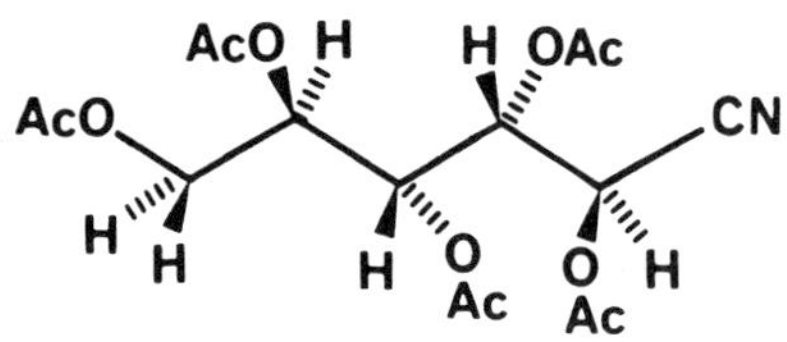

12

is favored unless this results in a 1,3-interaction (28). Since in 2,3,4,5,6-penta-*O*-acetyl-D-gluconononitrile [13] a 1,3-interaction does exist in the zigzag conformation, this compound [13]

FIG. 20. Newman and wedge-slash formulas for D-threitol [11].

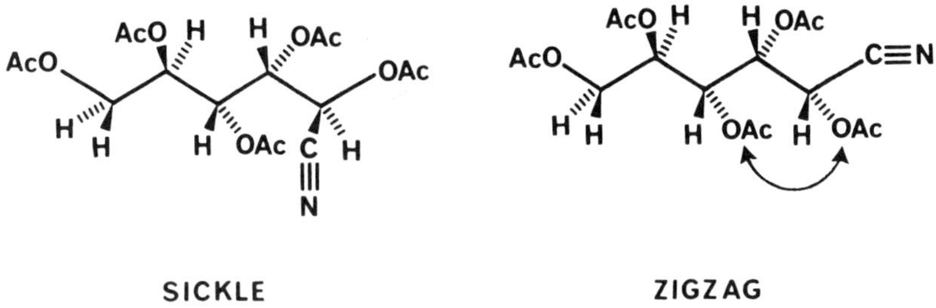

FIG. 21. Zigzag and sickle conformations for 2,3,4,5,6-penta-*O*-acetyl-D-glucononitrile [13]. Arrow indicates 1,3-interaction.

assumes a "sickle" conformation (28) (Figure 21). Study of unprotected sugar alcohols (alditols) has produced similar findings (29).

V. OLIGOSACCHARIDES

A. Torsion Angles

In the conformational analysis of oligosaccharides, it is of importance to know not only the preferred conformation of each monosaccharide residue but also the relative orientation of the ring systems with respect to each other. The overall shape of a molecule will, of course, depend upon this relative orientation. In a molecule such as methyl *O*-(2,6-dideoxy-β-D-*arabino*-hexopyranosyl)-(1→3)-2,6-dideoxy-α-D-*arabino*-hexopyranoside [14] (Figure 22), rotation is possible about the two bonds attached to the bridging oxygen. The two angles of rotation (Figure 22), called torsion angles, are generally designated by the Greek letters ϕ and ψ. The overall shape of the molecule will depend upon the preferred values for these torsion angles.

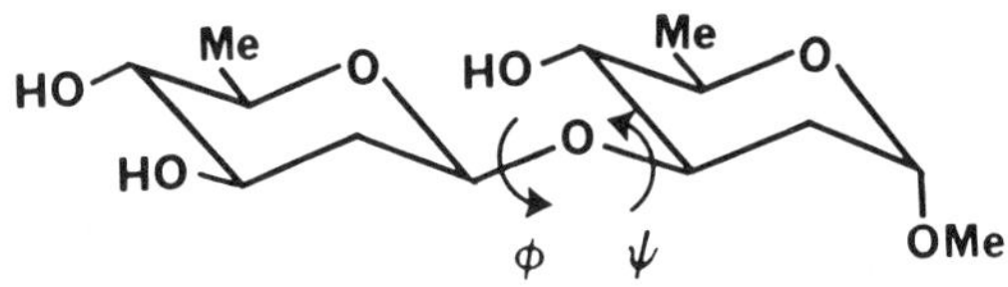

FIG. 22. Torsion angles for methyl *O*-(2,6-dideoxy-β-D-*arabino*-hexopyranosyl)-(1→3)-2,6-dideoxy-α-D-*arabino*-hexopyranoside [14].

It would be desirable to have a graph which showed the change in potential energy as the values of the torsion angles change. By arbitrarily assuming values for these angles and calculating the associated molecular energy using a method such as the molecular mechanics approach (21) (see Section I.B.5.b.), it is possible to obtain the potential energy of the system associated with these arbitrarily selected angles. Systematic alteration of these angles and calculation of new potential energies at each step ultimately will determine which set or sets of angles give the lowest calculated potential energies. These angles then provide the calculated, preferred conformational relationship between the two rings. If potential energy is plotted as a function of the torsion angles, a conformational energy map is obtained. Such a map provides a convenient way for viewing the calculated results. If a molecule with more than two monosaccharide residues is under consideration, the same approach can be used although the complexity is decidedly greater. Some examples of these conformational energy maps are found in reference 30.

B. The Exo-anomeric Effect

Preferred conformations have been found to exist for exocyclic, glycosidic C-O bonds in pyranosides. These preferred conformations are a result of a phenomenon known as the "exo-anomeric effect" (31). This effect results from the same type of orbital interactions which produce the anomeric effect. The difference is that the interactions for the exo-anomeric effect involve the orbitals on the glycosidic oxygen atom (Figure 23), while the important interactions in the anomeric effect involve the ring oxygen atom orbitals (Figure 10).

The three possible gauche arrangements of groups (electron pairs are considered to be groups) around the C-1, O_{Me} bond are shown in Figure 23. Only structures 15 and 16 have an electron pair on the methoxy oxygen atom in a *trans* antiparallel

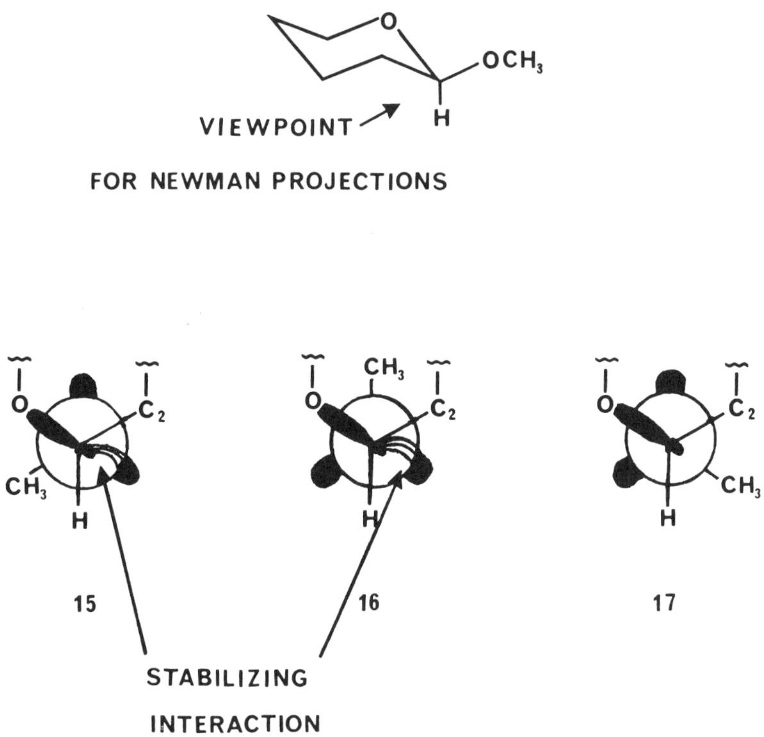

FIG. 23. Possible gauche arrangements of groups around the C-1, O_{Me} bond [15-17].

arrangement to the C-1, O_{ring} bond. Interaction between the back lobe of the orbital involved in the C-1, O_{ring} bond and one of the orbitals on the methoxy oxygen atom containing unshared electrons (Figure 23) leads to the same type of stabilizing interaction observed for the anomeric effect (Figure 12); consequently, conformers 15 and 16 are favored over 17. In comparing 15 and 16, conformation 15 is favored because the steric interaction between the glycosidic methyl group and the atoms attached to C-1 should be less. Thus, the exo-anomeric effect is an important factor in directing the relative positioning between monosaccharide residues.

REFERENCES

1. J. F. Stoddart, *Stereochemistry of Carbohydrates.* Wiley-Interscience, New York, 1971. All boat, skew, and chair forms of the pyranoid ring along with the routes which interconnect them are shown on pages 55–58.
2. This approach to the description of various conformations adopted by pyranoid rings was suggested by J. C. P. Schwartz, *J. Chem. Soc., Chem. Commun.,* 505 (1973).
3. R. E. Reeves, *J. Am. Chem. Soc., 71,* 215 (1949) and *Adv. Carbohydr. Chem., 6,* 123 (1951).
4. H. S. Isbell and R. S. Tipson, *Science, 130,* 793 (1959) and *J. Res. Nat. Bur. Stand., 64A,* 171 (1960).
5. W. Pigman and D. Horton, in *The Carbohydrates* (W. Pigman and D. Horton, eds.), Academic Press, New York, 1972, pp. 59–64.
6. E. L. Eliel, N. L. Allinger, S. J. Angyal, and G. A. Morrison, *Conformational Analysis,* Wiley-Interscience, New York, 1965, p. 356.
7. S. J. Angyal, *Angew. Chem. Intern. Ed., 8,* 157 (1969).

8. S. J. Angyal, *Aust. J. Chem.*, *21*, 2737 (1968).
9. R. U. Lemieux and N. J. Chu, *Am. Chem. Soc.*, *133rd National Meeting, Abstr. Papers, 31N*, (1958) and R. U. Lemieux and S. Koto, *Tetrahedron*, *30*, 1933 (1974).
10. Excellent discussions of the factors important to the anomeric effect are found in ref. 1, pp. 68–74 and in ref. 11, pp. 103–107.
11. P. L. Durette and D. Horton, *Adv. Carbohydr. Chem. Biochem.*, *26*, 49 (1971).
12. A. D. Bruyn, M. Anteunis, R. van Rijsbergen, M. Claeyssens, and P. Kovac, *J. Carbohydr. Chem.*, *1*, 301 (1982).
13. P. L. Durette and D. Horton, *Carbohydr. Res.*, *18*, 57 (1971).
14. J. T. Edward, *Chem. Ind. (London)*, 1102 (1955).
15. G. A. Jeffrey, in *Anomeric Effect (Origin and Consequences)* (W. A. Szarek and D. Horton, eds.), American Chemical Society, Washington, DC, 1979, pp. 54–59.
16. For discussion of molecular orbital calculations relating to the anomeric effect see S. Wolfe, M.-H. Whangbo, and D. J. Mitchell, *Carbohydr. Res.*, *69*, 1 (1979).
17. C. B. Anderson and D. T. Sepp, *J. Org. Chem.*, *33*, 3272 (1968).
18. R. U. Lemieux and A. R. Morgan, *Can. J. Chem.*, *43*, 2205 (1965).
19. M. Chmielewski, J. N. BeMiller, and D. P. Cerretti, *J. Org. Chem.*, *46*, 3903 (1981).
20. R. U. Lemieux, *Pure Appl. Chem.*, *27*, 527 (1971).
21. F. A. Carey and R. J. Sundberg, *Advanced Organic Chemistry*, Plenum, New York, 1977, pp. 72–78.
22. M. Sundaralingam, R. G. Brennan, P. Swaminathan, T. P. Haromy, W. B. Drendel, J. N. Bemiller, M. Chmielewski, and D. P. Cerretti, *J. Carbohydr. Chem.*, *1*, 85 (1982).

23. R. J. Ferrier and G. H. Sankey, *J. Chem. Soc. (C)*, 2345 (1966).
24. A. A. Chalmers and R. H. Hall, *J. Chem. Soc. Perkin Trans. II*, 728 (1974).
25. J. Thiem and P. Ossowski, *J. Carbohydr. Chem.*, *3*, 287 (1984).
26. Ref. 1, pp. 98-99.
27. N. Cyr and A. S. Perlin, *Can. J. Chem.*, *57*, 2504 (1979).
28. L. M. Sweeting, B. Coxon, and R. Varma, *Carbohydr. Res.*, *72*, 43 (1979).
29. G. E. Hawkes and D. Lewis, *J. Chem. Soc. Perkin Trans. II*, 2073 (1984).
30. I. Tvaroska, S. Perez, and R. H. Marchessault, *Carbohydr. Res.*, *61*, 97 (1978).
31. R. U. Lemieux, A. A. Pavia, J. C. Martin, and K. A. Watanabe, *Can. J. Chem.*, *47*, 4427 (1969).

7 Unprotected Sugars

I. INTRODUCTION

Early studies of the chemistry of carbohydrates naturally concentrated on the reactions of those compounds which were readily available. This meant that the chemistry of unprotected sugars was investigated extensively by the first carbohydrate chemists. As the understanding of the chemistry of carbohydrates increased and as synthetic carbohydrate chemistry began to develop, selective reaction in these polyfunctional molecules became increasingly important. Selective reaction usually required the introduction of protecting groups and, in so doing, stimulated an evolutionary movement away from research almost exclusively on unprotected compounds. This movement continued to the point where in recent years the majority of published research related to sugars has involved the use of partially or fully protected compounds.

Even though this change in research emphasis has occurred, there are still findings of considerable importance to be made on unprotected carbohydrates. For example, much is yet to be learned about the factors controlling the complex equilibria of reducing sugars in solution (1). Also, many challenging problems associated with the shape and structure of polysaccharides and higher oligosaccharides are still to be solved (2). The roles played by unprotected sugars and their derivatives in living organisms have only begun to be appreciated and will require further extensive study involving these compounds to be completely understood (3-5).

II. REDUCING SUGARS IN SOLUTION (6)

Unprotected sugars can be divided into two distinct groups, reducing and nonreducing (Figure 1). Nonreducing sugars, when dissolved in a solvent, remain intact while reducing sugars immediately begin to tautomerize upon dissolution. The structural feature of reducing sugars which allows this tautomerism to occur is a ring system containing a hemiacetal or hemiketal linkage (Figure 1).

A. Components Present in an Equilibrated Solution of a Reducing Sugar

Tautomerism of a reducing sugar ultimately produces an equilibrium mixture which includes two pyranoses, two furanoses, an open-chain carbonyl compound, and the hydrated form of the open-chain structure. The equilibrium involving these forms for D-glucose [1] is shown in Figure 2. Trace amounts of other species such as dimers and larger ring forms [e.g., septanoses

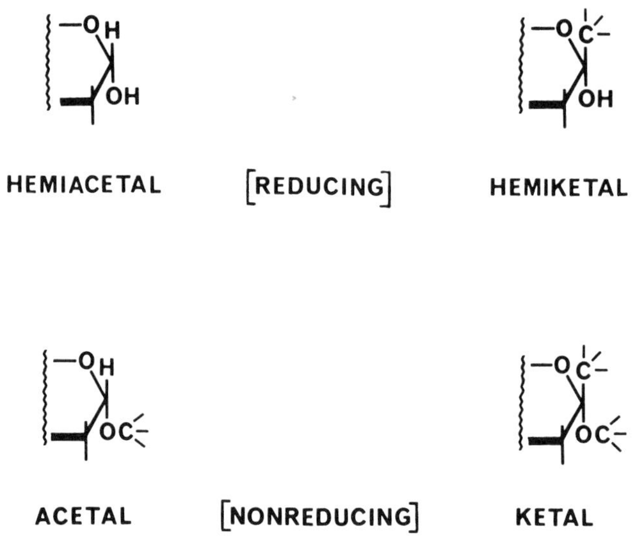

FIG. 1 Essential structural features of reducing and nonreducing sugars.

(seven-membered rings)] also exist. Not every sugar is capable of assuming all forms; for example, the aldotetroses and 2-ketopentoses do not have sufficiently long carbon atom chains to form pyranoid rings.

Identification and quantitation of the components present in an equilibrated solution of a reducing sugar can be done most effectively by nuclear magnetic resonance (NMR). If a component represents greater than 1% of the mixture, it usually can be detected by normal NMR techniques. The distribution among furanose and pyranose forms in equilibrium mixtures of some common reducing sugars is given in Table 1 (7). Advanced NMR techniques and very high field NMR spectrometers have made possible the accurate determination of even the minor components present in solutions of monosaccharides. For example, the

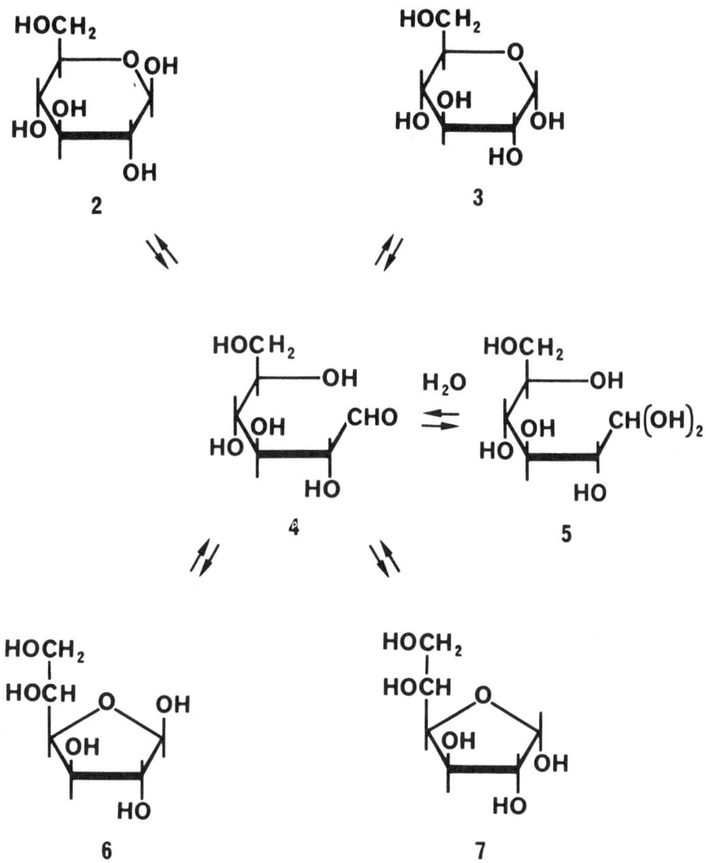

FIG. 2 Various forms of D-glucose [1] present in aqueous solution at equilibrium.

equilibrium composition of D-idose [8] in D_2O has been determined to be: α-furanose, 13.5%; β-furanose, 16.5%; α-pyranose, 35.9%; β-pyranose, 33.4%; open-chain aldehyde, 0.1%; hydrated aldehyde, 0.5% (8). The difficulties which arise in peak assignment and accurate quantitation of the components in these complex mixtures have limited the number of monosaccharides

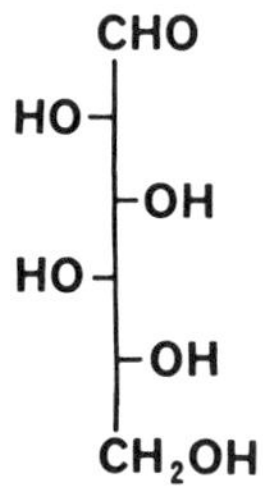

8

Table 1 Composition of Some Reducing Sugars and Their Derivatives in D_2O at 31° C[a]

	Pyranose		Furanose	
Sugar	α	β	α	β
D-fructose	2.5	65	6.5	25
D-galactose	30	64	2.5	3.5
6-deoxy-D-galactose	28	67	5	
D-glucose	38	62	—	—
D-mannose[b]	65.5	34.5	0.6	0.3
6-deoxy-D-mannose[b]	60	40	—	—
3-deoxy-D-*ribo*-hexose	24.5	55	5.	15.5
D-arabinose	60	35.5	2.5	2.
D-lyxose	70	28	1.5	0.5
D-ribose	21.5	58.5	6.5	13.5
D-xylose	36.5	63	<1	

[a]Reference 7, pp. 63-65.
[b]44° C.

which have been analyzed completely by the more sophisticated NMR methods.

Even though it is usually a simple matter to detect by NMR the major cyclic components in an equilibrated solution of a reducing sugar, isolation of each component in a pure form is not necessarily a simple process. For most reducing sugars only the pyranose forms have been obtained as pure compounds and for many sugars only one pyranose form has been isolated as a pure substance.

Pyranoses are generally the most stable forms of reducing sugars and consequently they usually dominate the equilibrium mixtures. Their stability derives, at least in part, from the fully staggered arrangement of groups attached to the pyranoid ring in its chair conformation (Figure 3). Such an arrangement minimizes steric interaction. In contrast, furanose forms have substantial eclipsing interaction (Figure 3); consequently, furanoses tend to be less stable than pyranoses. If a significant eclipsing interaction is removed, the concentration of the furanose forms in an equilibrium mixture increases. For example, the furanose forms [6 and 7] of D-glucose [1] are present at equilibrium in aqueous solution in concentrations of less than 0.2%, but for 3-deoxy-D-*ribo*-hexose [9] (D-glucose [1] in which the C-3 hydroxyl group has been replaced by a hydrogen atom (Figure 3)), the concentration of furanose forms rises to 20% (9). For most aldoses the open-chain carbonyl and its hydrate each represent less than 1% of the equilibrium mixture but for ketoses the concentration of the open-chain structure often is above this level.

B. Mutarotation

Since the various forms of a reducing sugar present in solution are optically active but do not (unless by coincidence) have the same rotatory power, the optical rotation of a solution of a reducing

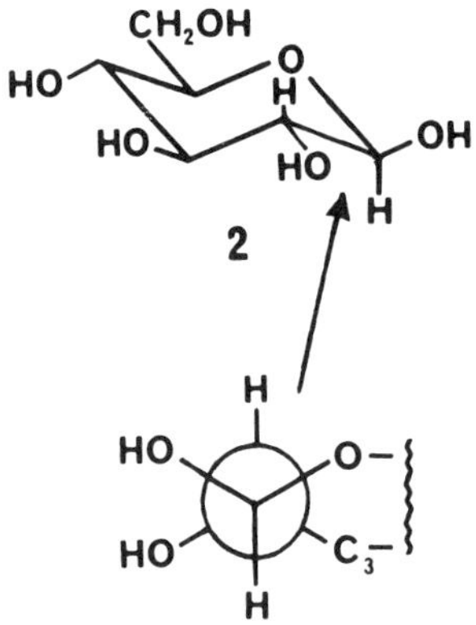

PYRANOSE FORM

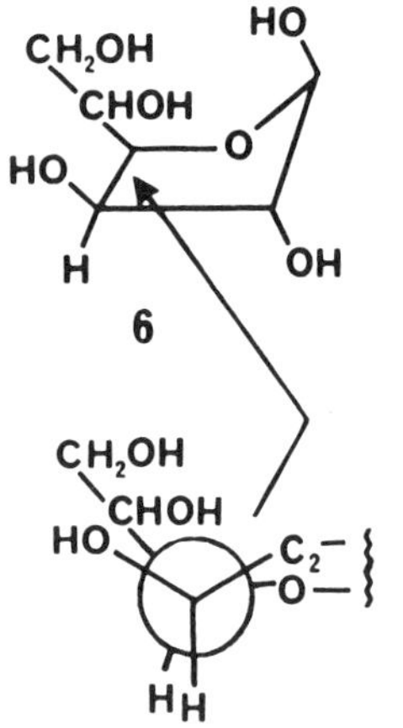

FURANOSE FORM

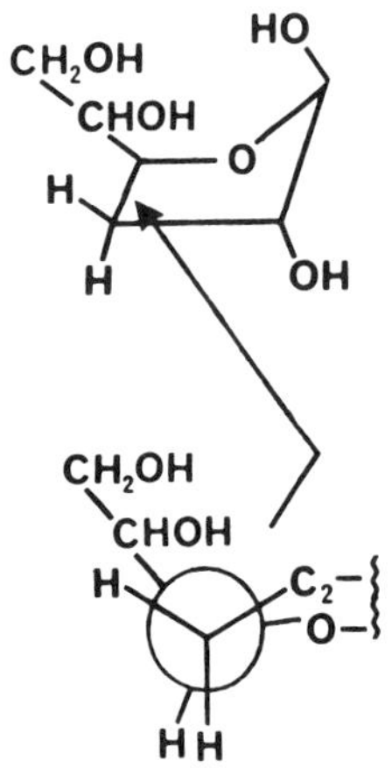

FURANOSE FORM

OF 9

FIG. 3 (top) β-Pyranose [2] and β-furanose [6] forms of D-glucose [1]; (bottom) β-furanose form of 3-deoxy-D-*ribo*-hexose [9]. The arrows point to the bonds, which are also pictured by Newman projections.

sugar will change with time until equilibrium is reached. When a single pure form of a reducing sugar (e.g., α-D-glucopyranose [3], Figure 2) is dissolved in a solvent, it is usually possible to observe a change in rotation as a function of time. This change in rotation is called mutarotation and the reactions which produce it are called the mutarotation reactions (Figure 2).

The central structure in the mutarotation process is an open-chain (acyclic) carbonyl compound. (For D-glucose [1] this carbonyl compound is represented by structure 4 in Figure 2.) Even though the acyclic structure typically is present only in low concentration, detailed studies of ring-opening and ring-closing reactions for several aldo- and ketofuranoses support the long held view that interconversion among the various cyclic forms of a reducing sugar occurs via an acyclic carbonyl intermediate (1).

C. The Effect of Complexation on Equilibration of Reducing Sugars

1. Metal Ions

Complexation with metal ions can alter the equilibrium proportions of the various forms of a reducing sugar present in solution (10). The two situations which structurally favor strong complexation are: an axial-equatorial-axial sequence of hydroxyl groups in a pyranoid ring and a cis-cis hydroxyl sequence in a furanoid ring (11). Compound 10 (α-D-allopyranose) has an axial-equatorial-axial arrangement of hydroxyl groups and compound 11 (α-D-allofuranose) has a cis-cis hydroxyl group arrangement. In both compounds these patterns include the anomeric hydroxyl group. The anomeric hydroxyl groups in the corresponding β-anomers cannot be part of one of these favored arrangements; thus, the β-anomers are not stabilized by complexation. As a result, when a metal ion such as calcium is added to a solution of D-allose,

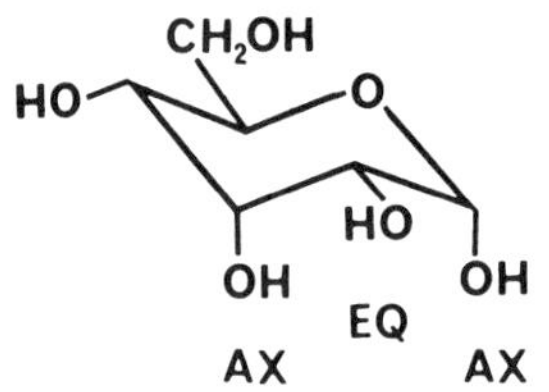

10

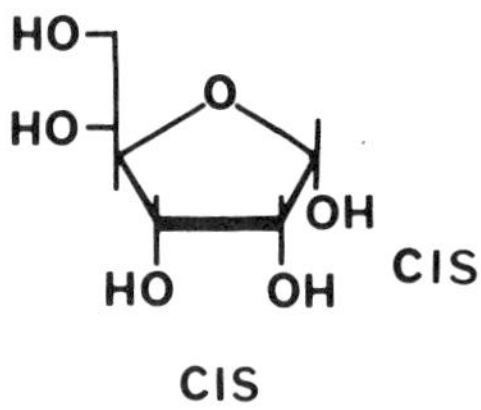

11

the concentrations of α-D-allopyranose [10] and α-D-allofuranose [11] both increase while the concentrations of the corresponding β-anomers decrease (11). Strong complexes with sugars are formed by a number of metal ions, including calcium, copper, strontium, barium, and the lanthanides (10,11,12).

2. Borate Ions

Borate ions and sugars form strong complexes in aqueous solutions (13). This complexation, which is particularly strong for cis 1,2-diols, can cause important equilibrium shifts in solutions of reducing sugars; for example, the equilibrium between D-glucose and D-fructose is shifted almost completely in favor of the latter by formation of D-fructose-borate complexes (14).

III. REACTIONS AT THE ANOMERIC CARBON ATOM

Numerous reactions involving anomeric carbon atoms in unprotected sugars were discovered in the early days of carbohydrate chemistry. Some of these processes remain among the most important reactions of carbohydrates, while others now are primarily of historical interest. Examples of the more common of these reactions are provided below along with comments about each type.

A. Glycoside Formation

When the hydrogen atom of an anomeric hydroxyl group is replaced by a substituted or unsubstituted carbon atom, a glycoside is formed. Although the importance of glycoside formation in biological systems would alone place this process in a select group of vital chemical reactions, the synthetic value of the formation of glycosides should not be overlooked since this process usually is the first step in a laboratory synthesis involving a reducing sugar.

The Fischer method is particularly effective for synthesizing glycosides from unprotected reducing sugars and low molecular weight alcohols. This method normally consists of heating a solution or suspension of a monosaccharide in an alcohol in the presence of an acidic catalyst. Such a procedure produces a mixture of the α and β anomers of both pyranosides and furanosides. Although formation of a mixture may seem to be a poor beginning for a synthetic sequence, some of the compounds (often the furanosides) in a typical equilibrium mixture usually are present in very small quantities. For example, reaction of α-D-xylopyranose [12] in methanolic hydrogen chloride gives an equilibrium mixture consisting of methyl α-D-xylofuranoside [13] (1.9%), methyl β-D-xylofuranoside [14] (3.2%), methyl α-D-xylopyranoside [15] (65.1%), and methyl β-D-xylopyranoside [16] (29.8%) (15).

It is often possible to isolate by crystallization one of the components of a mixture produced by the Fischer method. Reequilibration of the remaining mixture and crystallization to give an additional quantity of the desired component is a simple process and can be repeated to give high yields of a single glycoside. This procedure has been used to obtain crystalline methyl 6-deoxy-α-L-galactopyranoside [17] and 6-deoxy-α-L-mannopyranoside [18] in greater than 90% yields (16).

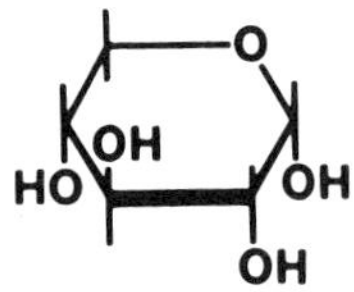

12

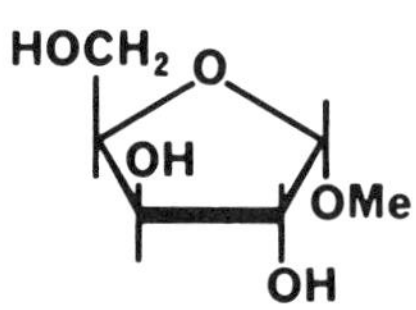

13

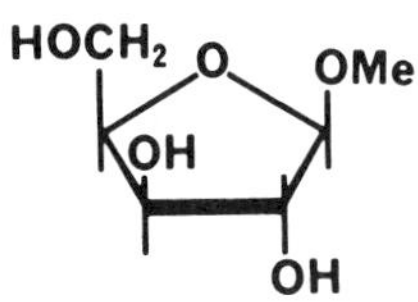

14

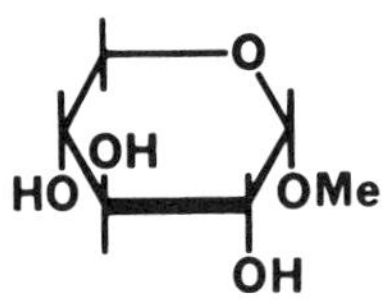

15

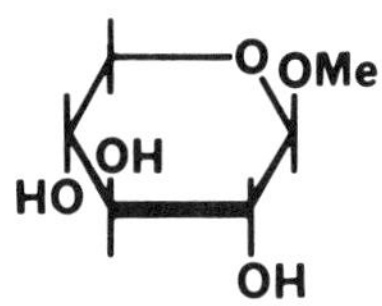

16

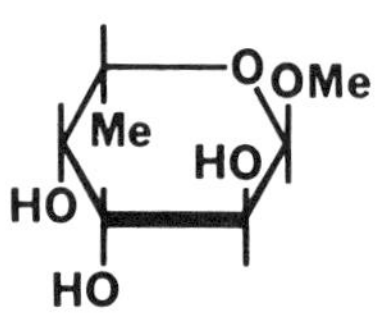

17

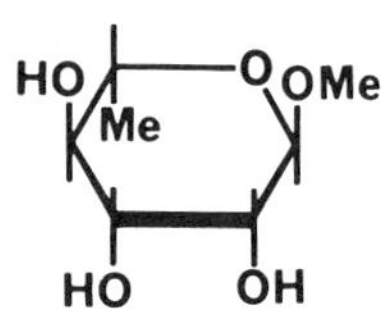

18

An equilibrium mixture produced by the Fischer method does not necessarily reflect the glycoside composition in the early stages of reaction. The pyranosides 15 and 16 are the most stable (i.e., thermodynamically favored) products but the furanosides 13 and 14 (the kinetic products) form more rapidly; thus, in the acid catalyzed methanolysis of D-xylose, the glycosides present in greatest amount early in the reaction are compounds 13 and 14, with a β-furanoside [14] being the major component (17). The equilibration between furanosides and pyranosides occurs more slowly than initial glycoside formation; thus, as equilibrium is established, the concentration of the thermodynamically favored pyranosides 15 and 16 increases.

Even though the xylofuranoses are present in an equilibrium mixture in low concentrations (see Table 1), the interconversion among various forms of D-xylose continuously replenishes the "furanose pool" as the furanoses are removed by reaction. Although at any time the furanose concentration is not high, furanosides form quickly; thus, substantial quantities of furanosides can be produced. The key factor in forming furanosides is stopping the reaction after glycoside formation is nearly complete but before substantial isomerization to pyranosides has occurred.

A generally accepted mechanism for glycoside formation is pictured in Figure 4 using α-D-xylopyranose [12] as the starting material and in Figure 5 using α-D-xylofuranose [19]. The rate-determining step in each of these reactions involves the formation of a resonance-stabilized cation [20 or 21] (Figures 4 and 5). Since furanosides are formed more rapidly than pyranosides, the cation 21 must be generated more easily than 20. One reason for 21 being formed more readily is that there is greater destabilizing interaction in furanoses than in pyranoses (Figure 3). Some of this interaction (e.g., that between groups attached to C-1 and C-2) is eliminated by cation formation. Other things being equal, the

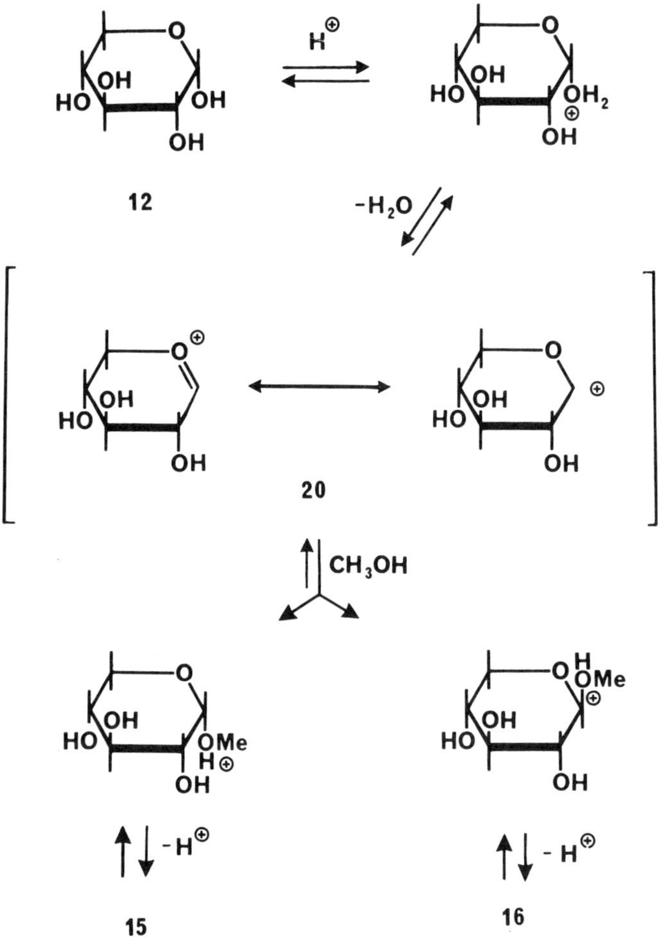

FIG. 4. Proposed mechanism for the formation of methyl D-xylopyranosides [15 and 16] beginning with α-D-xylopyranose [12].

reactant that experiences the greater relief of strain by ionization will form a cation more rapidly.

Acid-catalyzed furanoside-pyranoside interconversion, which also involves the intermediacy of a resonance-stabilized cation

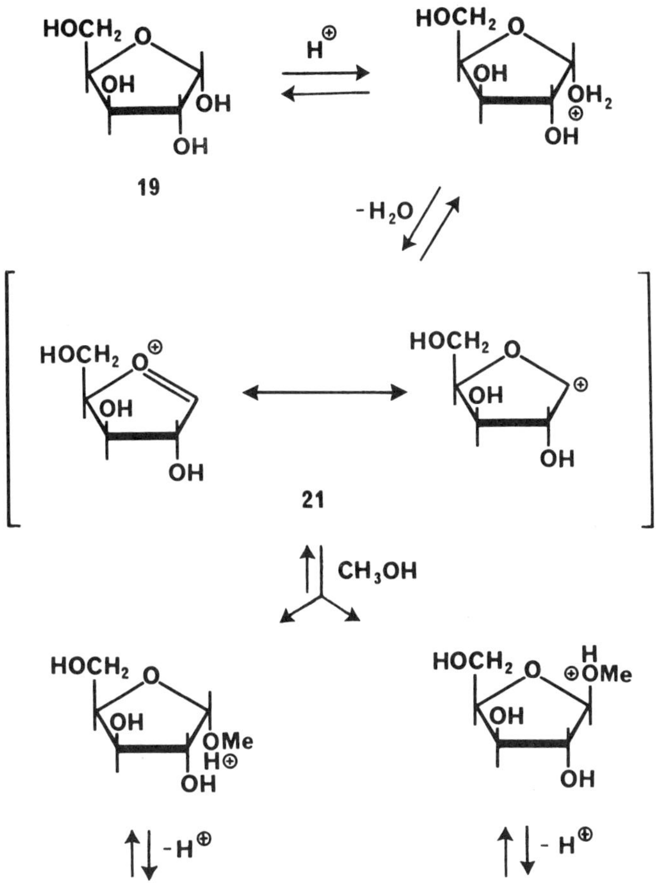

19

13

14

FIG. 5. Proposed mechanism for formation of the methyl xylofuranosides [13 and 14] beginning with α-D-xylofuranose [19].

[22] (Figure 6), takes place at the same time as glycoside formation. Each of the reactions shown in Figure 6 is reversible; therefore, interconversion will continue until the catalyst is removed, even though equilibrium has been reached. (The reactions

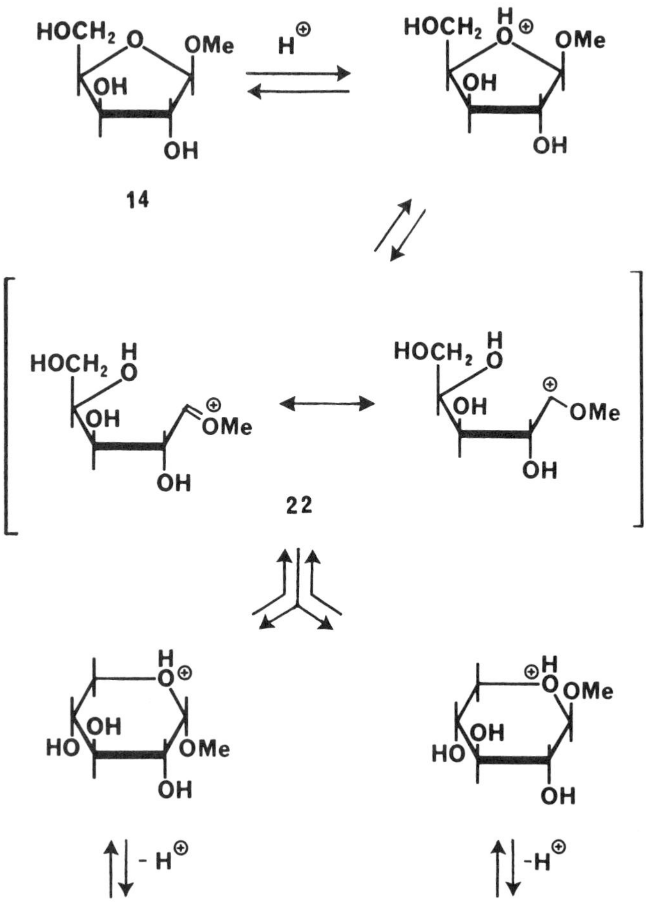

FIG. 6. Proposed mechanism for furanoside-pyranoside interconversion.

shown in Figures 4 and 5 also are reversible but since methanol is the reaction solvent, the cations 20 and 21 are much more likely to react with methanol than water; consequently, at equilibrium there is nearly complete glycoside formation.)

Although a cation (such as 20) is normally considered to be an intermediate in pyranoside interconversion (e.g., between 15 and 16), conditions do exist in which glycosides can be interconverted without formation of a free cation (18). Such interconversion, which requires an incoming alcohol to begin bonding to the anomeric center before a cation is fully formed, has been demonstrated only under rather specialized conditions; thus, the scope of this type of reaction is currently unknown.

B. Formation of Glycosyl Amines and Amadori Products

Unprotected sugars react with primary and secondary amines to give glycosyl amines; that is, compounds in which the anomeric hydroxyl group has been replaced by an amino or substituted amino group. Treatment of either β-D-glucopyranose [2] or α-D-glucopyranose [3] with ammonia results in the crystallization of β-D-glucopyranosylamine [23] from the reaction mixture (Equation 1) (19). Aliphatic and aromatic amines and amino acids

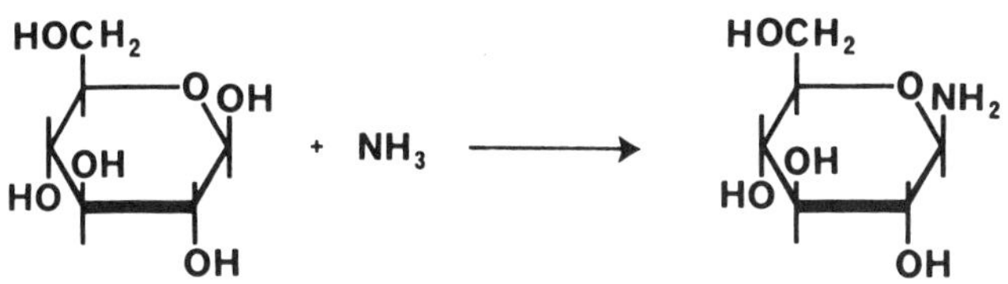

2 23 (1)

react with D-glucose [1] in a similar manner. An indication of how glycosyl amine formation can occur is given in Figure 7.

When an aromatic amine and an aldose are heated together in the presence of an acid, the initially formed glycosyl amine experiences what is known as the "Amadori" rearrangement (20). As an example, when D-glucose [1] and aniline are heated in the

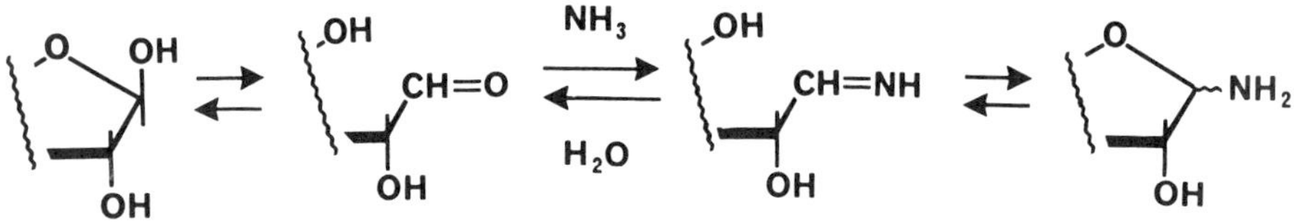

FIG. 7. Proposed scheme for the conversion of a sugar into a glycosyl amine.

presence of an acid, the Amadori compounds 24 and 25 are formed (Equation 2). A mechanism for the Amadori reaction is

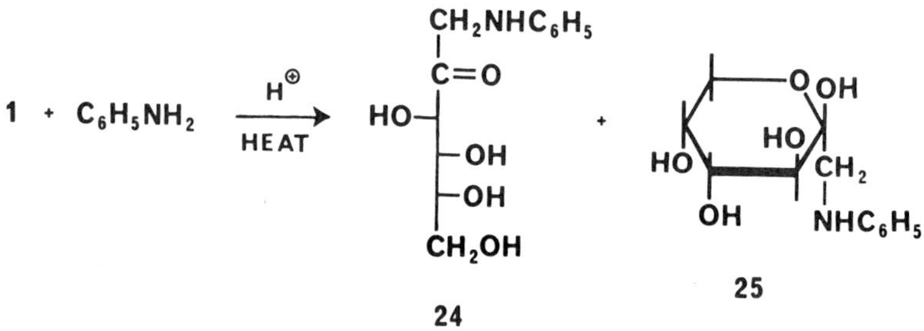

(2)

proposed in Figure 8. A similar type of reaction occurs between ketoses and aromatic amines (21).

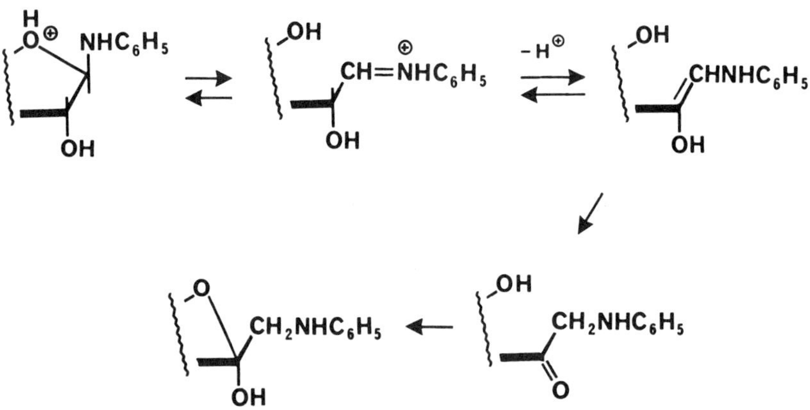

FIG. 8. Proposed mechanism for the Amadori rearrangement.

C. Reduction (22)

Reduction of unprotected aldoses and ketoses to alditols is conveniently done in the laboratory by hydride reduction (23) (Equation 3) or by catalytic hydrogenation (24) (Equation 4). The

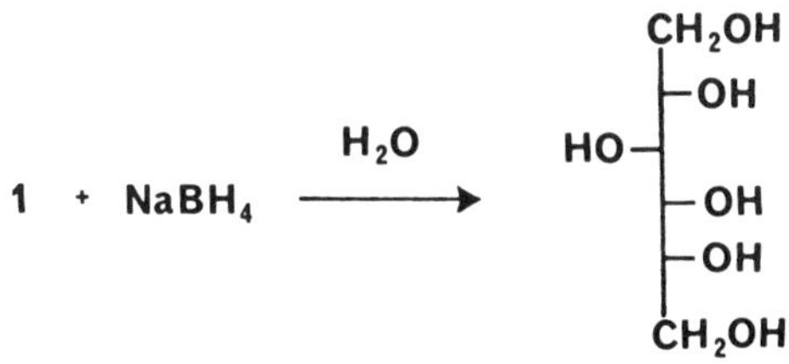

(3)

26

$$1 + H_2 \xrightarrow{\text{RANEY Ni}} 26 \qquad (4)$$

rate of reduction is faster in alkaline solution because ring opening is catalyzed by base and the ring-open form of the sugar is the one being reduced (Figure 9). Sodium borohydride is the hydride reagent most often used since it is relatively stable in water, a typical solvent for reduction of unprotected sugars. Large-scale preparation of alditols such as D-glucitol [26] usually is accomplished by catalytic hydrogenation.

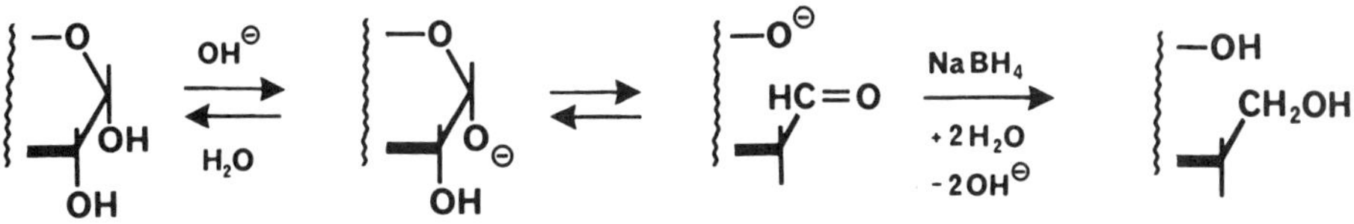

FIG. 9. Proposed mechanism for base-catalyzed ring opening and subsequent sodium borohydride reduction of a sugar.

D. Oxidation (25)

1. Halogens and Hypohalites

Halogens and hypohalites oxidize aldoses to the corresponding aldonic acids; thus, reaction of D-glucose [1] with bromine gives D-gluconic acid [27] (Equation 5) in high yield (26). The

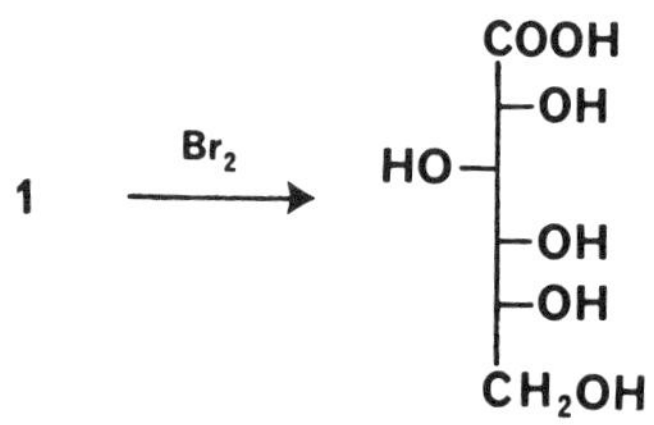

(5)

27

oxidation of unprotected carbohydrates by halogens (chlorine and bromine) and hypohalites (hypochlorite, hypobromite, and hypoiodite) is complicated by the simultaneous presence of several oxidizing agents in solution at concentrations which are a function of reaction conditions. For example, bromine, hypobromous acid, and hypobromite are simultaneously present in an aqueous solution of bromine (Equations 6 and 7). Such a solution becomes

$$Br_2 + H_2O \rightleftharpoons HOBr + HBr \qquad (6)$$

$$Br_2 + 2OH^{\ominus} \rightleftharpoons OBr^{\ominus} + Br^{\ominus} + H_2O \qquad (7)$$

an increasingly powerful oxidizing medium as the pH is raised because hypobromite is the most powerful of these oxidizing agents. In acidic solution, where little hypobromite is present, the halogen is the principal active reagent. Although molecular iodine is too weak to be an effective oxidizing agent, hypoiodite is a powerful one. A summary of information pertinent to the mechanism of this process is given in reference 25; however, oxidation by halogens and hypohalites is complicated and not fully understood.

2. Fehling and Tollens' Solutions

Fehling solution [aqueous copper(II) sulfate and sodium tartrate], which reacts with reducing sugars to give the brick-red precipitate of copper oxide, is used for quantitative determination of reducing sugars (27). Oxidation with Tollens' solution (ammoniacal silver nitrate), used as a test for reducing sugars, produces metallic silver, often in the form of a silver mirror. Although these reactions give carboxylic acid salts, they are complicated by oxidative degradation, which often produces a variety of products.

3. Nitric Acid

Nitric acid is a powerful oxidizing agent which converts aldehydes and primary alcohols into carboxylic acids; thus, oxidation of D-galactose [28] with nitric acid produces galactaric acid [29] (Equation 8) (28). (In Equation 8, from among the forms of

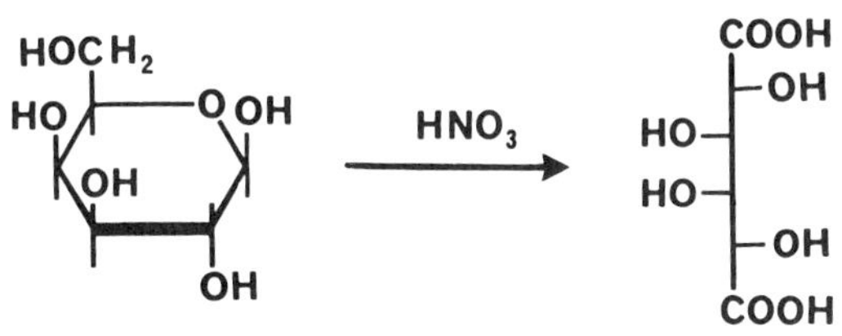

28 29 (8)

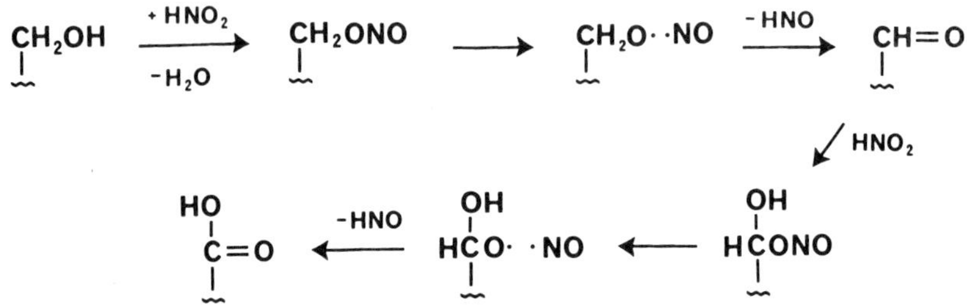

FIG. 10. Proposed mechanism for nitric acid oxidation of a carbon atom bearing a primary hydroxyl group.

D-galactose [28] present in solution, only β-D-galactopyranose is shown.) Oxidation with nitric acid has an induction period which is eliminated if nitrogen oxides or sodium nitrite is introduced into the reaction mixture. This observation has led to the proposal that nitrous acid is the actual oxidizing agent (28). The mechanism shown in Figure 10 includes nitrous acid in the oxidation process.

4. Ruff Degradation (Hydrogen Peroxide and Iron(III))

The oxidative decarboxylation of the sodium salt of an aldonic acid with hydrogen peroxide and iron(III) to give the next lower aldose is known as the Ruff degradation. An example of the Ruff degradation is the reaction of calcium D-galactonate [30] to give D-lyxose [31] (Equation 9) (29).

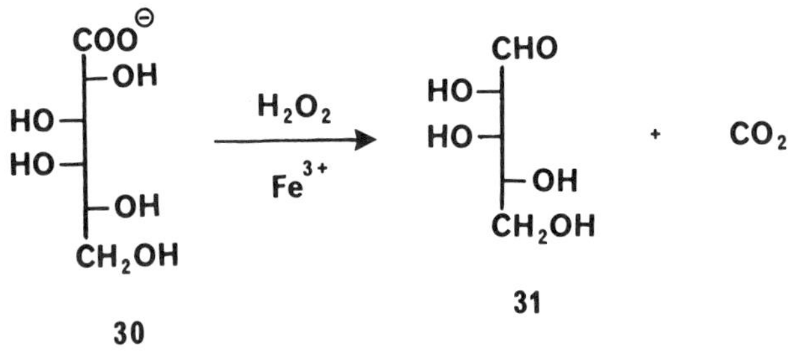

30 31 (9)

E. Cyanohydrin Formation: The Kiliani Reaction (30)

The addition of the elements of hydrogen cyanide to an aldehyde to give a cyanohydrin is a reaction which can be conducted in the presence of acids or bases. When the addition is made to a reducing sugar such as D-glucose [1], a pair of diastereomers is produced, since a new chiral center is formed (Figure 11) (31). Cyanohydrins usually are either hydrolyzed to carboxylic acids (or carboxylates) or reduced to aldoses (Figure 11). In each case the carbon atom chain is extended by one atom. Chain extension by cyanohydrin formation followed by reduction is known as the Kiliani-Fischer synthesis (32). A concise description of the

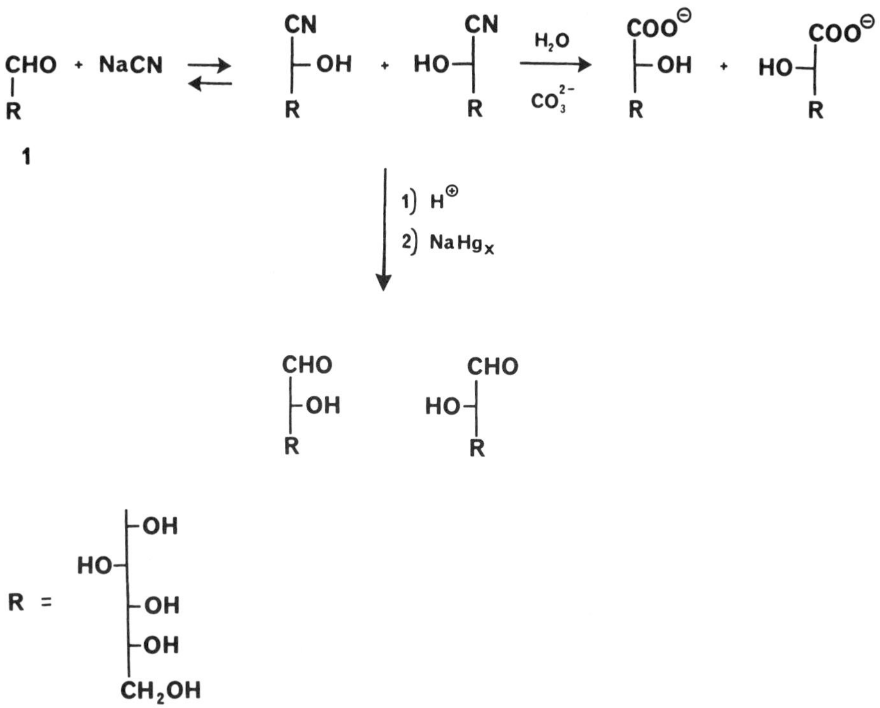

FIG. 11. Scheme for formation and reactions of cyanohydrins.

essential features of the cyanohydrin reaction recently has been published (33).

F. Formation of Phenylhydrazones and Phenylosazones

In neutral solutions, phenylhydrazine reacts with reducing sugars to give the corresponding phenylhydrazones (Equation 10);

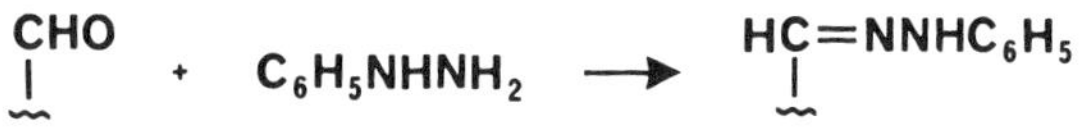

(10)

however, in acidic solutions the protonated form of phenylhydrazine causes oxidation of the phenylhydrazone and ultimately formation of a phenylosazone [32] (Figure 12). This oxidation of the phenylhydrazone and simultaneous reduction of protonated phenylhydrazine is thought to occur via the concerted process shown in Figure 12 (34).

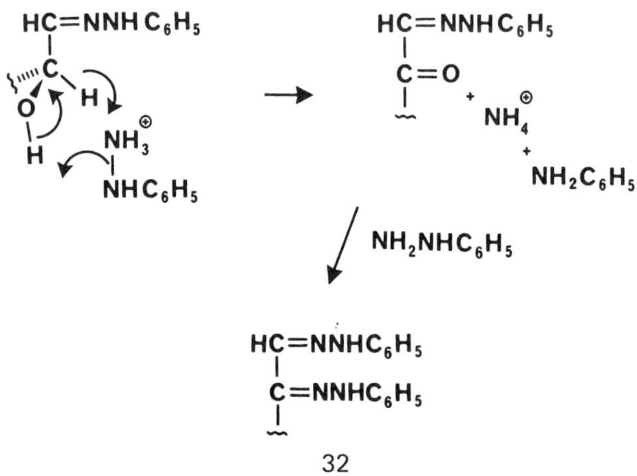

32

FIG. 12. Proposed mechanism for formation of phenylosazones from phenylhydrazones.

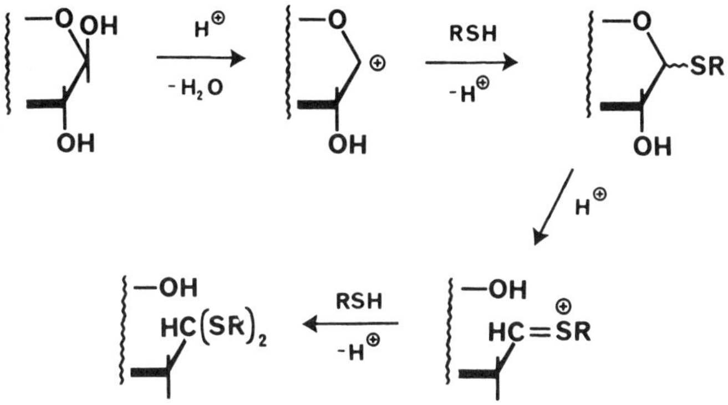

FIG. 13. Proposed mechanism for formation of a dialkyl dithioacetal.

G. Dialkyl Dithioacetal Formation

Treatment of unprotected sugars with alkanethiols in the presence of strong acids produces dialkyl dithioacetals (Figure 13). These compounds often are crystalline materials which are useful synthetic intermediates (35). Reaction of D-glucose [1] with ethanethiol in the presence of hydrochloric acid forms D-glucose diethyl dithioacetal [33] (Equation 11) (36).

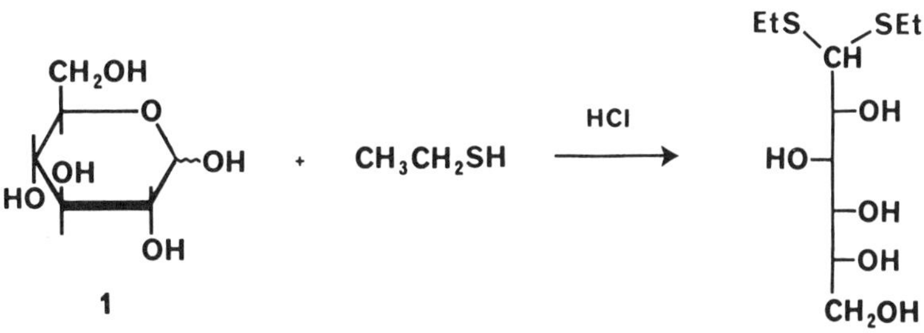

1 33 (11)

IV. REACTIONS INVOLVING NONANOMERIC CARBON ATOMS

Although all of the reactions of unprotected sugars discussed thus far involve the anomeric carbon atom, there are a considerable number of significant processes that lead to changes in structure at other carbon atoms or their attached hydroxyl groups or both. Reactions of this type include those processes which introduce hydroxyl group protection. Group protection, which is the topic of Chapter 8, is the normal prerequisite for regioselective reaction of carbohydrates. The displacement, elimination, condensation, and other reactions discussed in later chapters are typically conducted on partially or fully protected sugars.

In addition to group protection there are a number of other processes in which reaction occurs away from the anomeric center. In most of these cases the presence of the carbonyl-containing, acyclic form of the sugar is the key factor in allowing reaction to take place. The importance of this carbonyl compound is that most reactions involving non-anomeric carbon atoms proceed via an enol (actually an enediol) which is in equilibrium with the acyclic carbonyl form of the sugar (Equation 12).

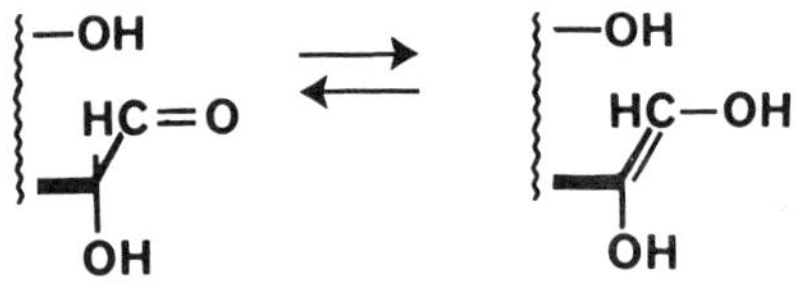

(12)

A. Base-catalyzed Reactions

Treatment of reducing sugars with aqueous base for extended periods of time results in substantial decomposition. The ultimate products from these reactions are primarily hydroxy acids. For example, reaction of D-glucose or D-fructose with aqueous potassium hydroxide for three and one-half months causes the formation of at least seven hydroxy acids, among which lactic acid [34] is the major constituent (37). Clearly, a variety of reactions

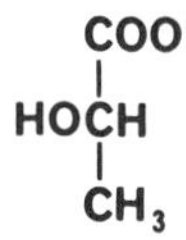

34

is responsible for the extensive degradation and rearrangement which these monosaccharides experience. The basic processes which are thought to be responsible for hydroxy acid formation are outlined below. Excellent reviews of base-catalyzed reactions of sugars are found in references 37 and 38.

1. Lobry de Bruyn-Alberda van Ekenstein Transformation

The initial steps in the base-catalyzed reactions of reducing sugars are a group of isomerizations and epimerizations known as the Lobry de Bruyn-Alberda van Ekenstein transformation (Figure 14) (39). These reactions, which are initiated by formation of the anion of the enediol, produce mixtures of aldoses and ketoses. For example, treatment of D-glucose [1] with aqueous calcium hydroxide for 10 days gives a mixture of D-glucose [1] (63.4%), D-fructose [35] (30.9%), and D-mannose [36] (2.4%) (40).

2. Base-catalyzed Rearrangements and Degradations

The formation of an enediol and the corresponding anion (Figure 14) not only allow the Lobry de Bruyn-Alberda van Ekenstein transformation to take place but the enediol also represents the beginning point for the extensive degradation which

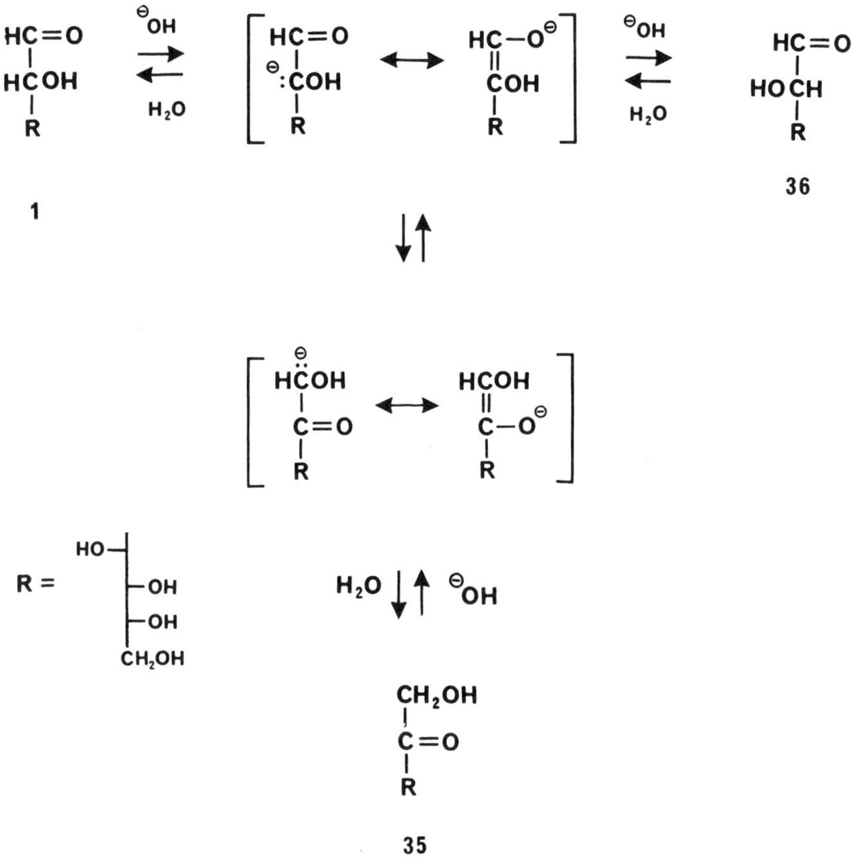

FIG. 14. Proposed mechanism for base catalyzed aldose-ketose isomerization and epimerization reactions (the Lobry de Bruyn-Alberda van Ekenstein transformation).

accompanies the treatment of reducing sugars with base. Included among the reactions which lead to molecular degradation are β-elimination to form α-dicarbonyl compounds, carbon-carbon bond cleavage in these α-dicarbonyl structures, and benzylic acid rearrangement (Figure 15). Also important are the retro-aldol reaction and the related aldol condensation (Figure 16).

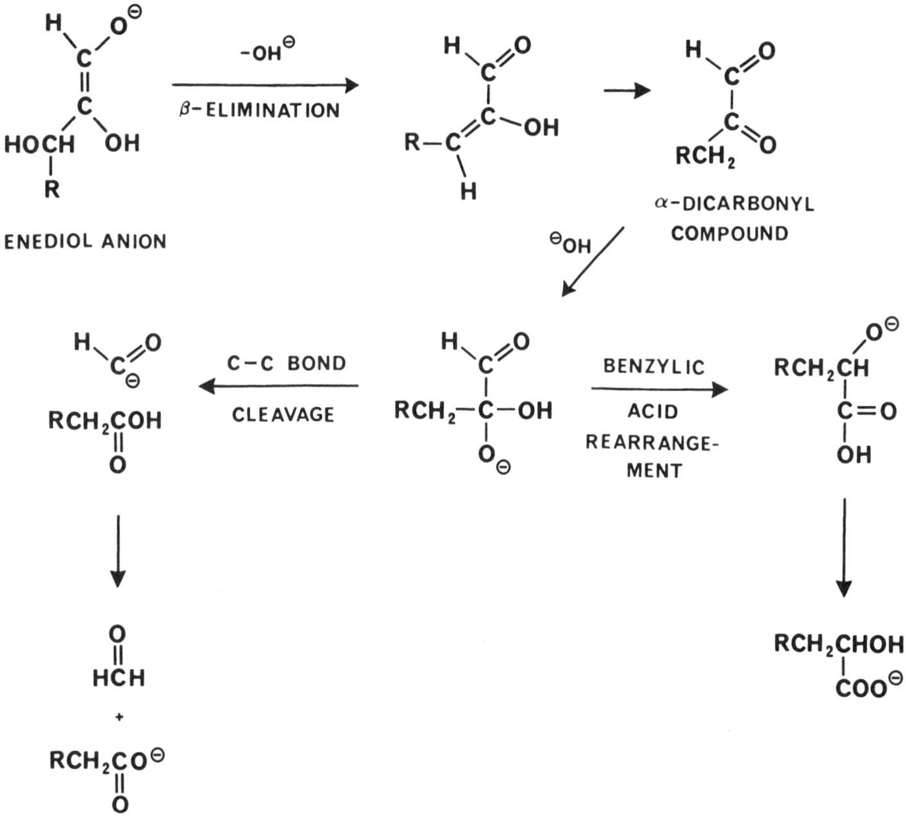

FIG. 15. Proposed mechanisms for formation and reaction of α-dicarbonyl compounds.

B. Dehydration Reactions

1. Acid-catalyzed Processes

From discussion earlier in this chapter of the role of acids in mutarotation and glycoside formation, it is clear that even small amounts of acids can catalyze important changes in carbohydrates. Extensive structural changes can arise from treatment of

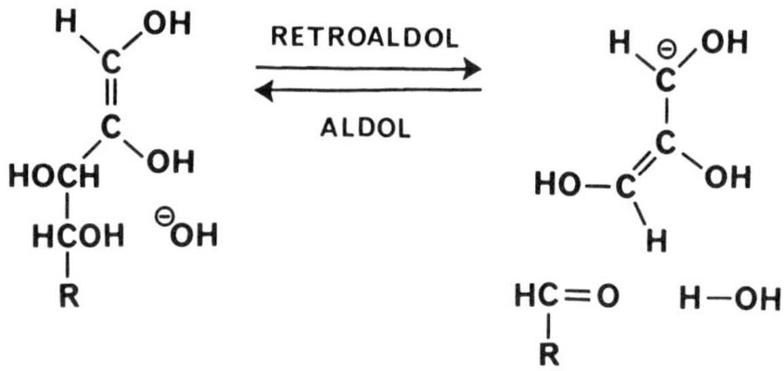

FIG. 16. Aldol and retro-aldol reactions.

carbohydrates with acids under vigorous conditions. These reactions lead to molecular rearrangement and dehydration of the compounds involved; for example, treatment of D-glucose [1] in 15 M sulfuric acid at 250° C for 1.5 minutes produces 5-(hydroxymethyl)-2-furaldehyde [37] (Equation 13) (41). Reactions such

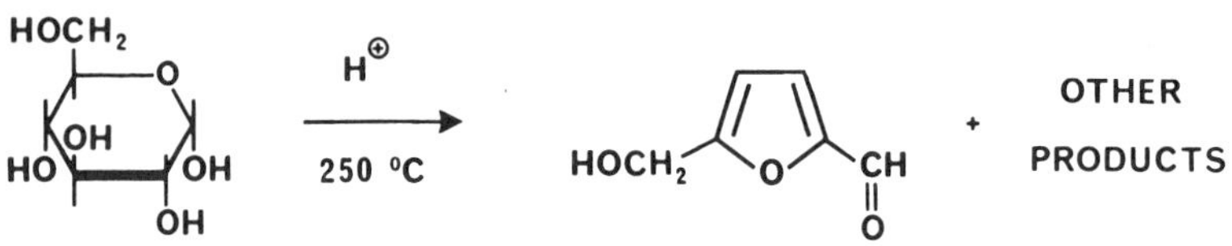

1

37

(13)

as that shown in Equation 13 are mechanistically complex and result not only in formation of compound 37 and other partially deoxygenated compounds but also oligosaccharides (known as reversion products) and complex polymeric compounds known as humines (42). Reviews on acid-catalyzed reactions of carbohydrates are found in references 42 and 43.

2. Pyrolysis

Over one hundred compounds have been identified from the pyrolysis of various mono- and oligosaccharides (44,45). The

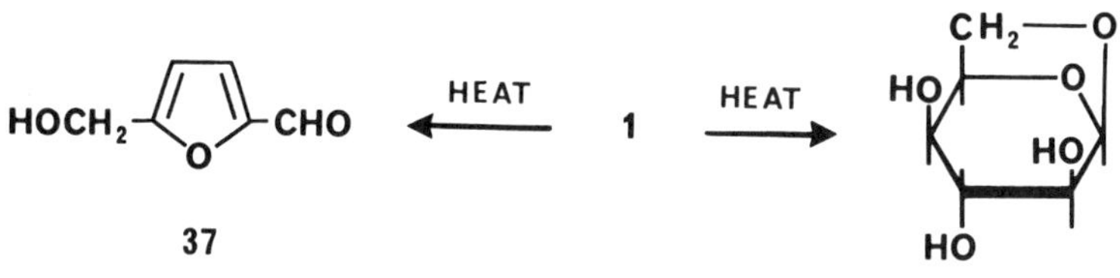

FIG. 17. Pyrolysis reactions of D-glucose [1].

principal reactions are loss of water to form anhydro and unsaturated sugars. For example, the pyrolysis of D-glucose [1] is thought to proceed along several pathways, one leading to 1,6-anhydro-β-D-glucopyranose [38] by loss of water and a second to the unsaturated 5-(hydroxymethyl)-2-furaldehyde [37] by a series of processes including elimination of the elements of water (Figure 17) (45).

C. Autoxidation

Carbohydrates react with oxygen (autoxidation) to produce a variety of oxidation products. Autoxidation begins with formation of an enediol which, in the presence of oxygen, is converted into an α-ketoaldehyde and hydrogen peroxide (Figure 18) (46). Once hydrogen peroxide has been formed, further reactions producing hydroxyl radicals and hydroxyalkyl radicals (both of which have been observed by electron spin resonance) take place (46). A change in reaction mechanism occurs at pH values greater

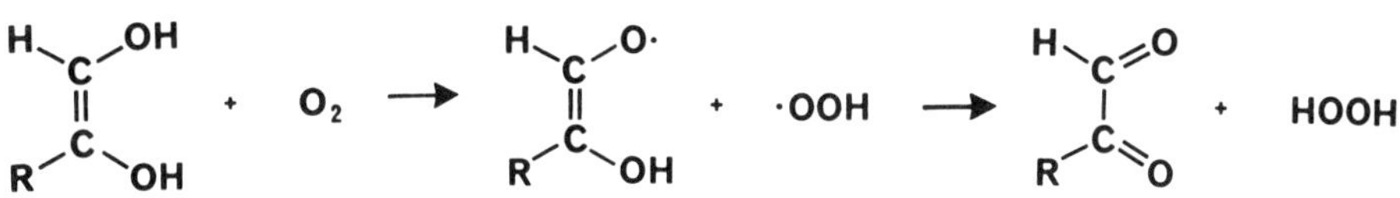

FIG. 18. Proposed mechanism for autoxidation reactions.

than 8.2. The greater reactivity at higher pH values is attributed to deprotonation of the enediol to give the more reactive enediolate anion.

V. THE FORMOSE REACTION

When formaldehyde is treated with base, a complex set of reactions takes place which produces a sugar-like syrup known as formose (47). Chromatography of formose reveals it to be a mixture which contains aldoses and ketoses including glucose, mannose, and fructose as well as all of the aldopentoses, the four 2-ketopentoses, and a variety of branched-chain compounds (48). The primary process in the formose reaction is the aldol condensation, although once sugar or sugar-like products have formed, all the base-catalyzed reactions shown in Figures 15 and 16 can occur. An example of an aldol condensation taking place during formose formation is the reaction leading from glycoaldehyde [39] and 1,3-dihydroxyacetone [40] to the four 2-ketopentoses [41-44] (Figure 19). The alkaline degradation of "formose sugar" is minimized by conducting the formose reaction for relatively short periods of time; nevertheless, dozens of products and intermediates have been isolated or proposed as being formed during this complex process (37,38).

There is a long induction period for the formose reaction which can be eliminated by the addition of compounds 39, 40, or glyceraldehyde [45] to the reaction mixture. This situation has led to the proposal that traces of aldehydes and ketones containing two and three carbon atoms, present as impurities in formaldehyde, actually start the formose reaction (37). This proposal is supported by the observation that ultrapure formaldehyde, from which these impurities have been removed, does not form formose (49). The formose reaction is much faster in the

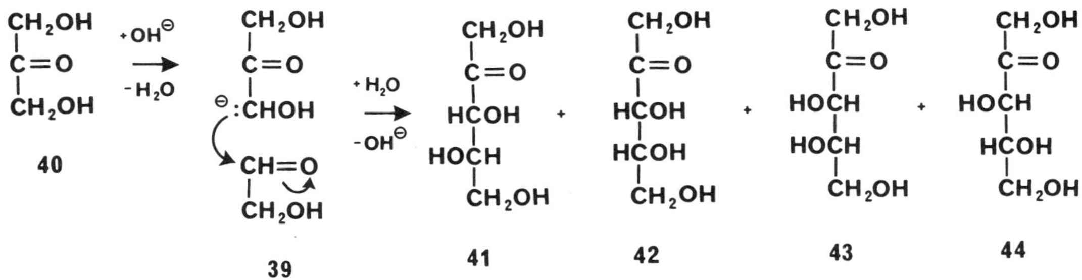

FIG. 19. Proposed mechanism for formation of the four 2-ketopentoses from glycoaldehyde [39] and dihydroxyacetone [40].

presence of bases, such as thallium, lead, and calcium hydroxides, compounds with cations which are effective complexing agents (50).

REFERENCES

1. J. Pierce, A. S. Serianni, and R. Barker, *J. Am. Chem. Soc.*, *107*, 2448 (1985).
2. D. A. Rees, *Polysaccharide Shapes*, John Wiley and Sons, New York, NY, 1977.
3. Y.-T. Li and S. C. Li, *Adv. Carbohydr. Chem. Biochem*, *40*, 235 (1982).
4. T. Feizi and R. A. Childs, *Trends Biochem. Sci.*, *10*, 24 (1985).
5. T. Feizi, *Nature*, (London), *314*, 53 (1985).
6. An excellent review of the composition of reducing sugars in solution has been prepared by S. J. Angyal, *Adv. Carbohydr. Chem. Biochem.*, *42*, 15-58 (1984).
7. Reference 6, p. 63.
8. J. R. Snyder and A. S. Serianni, *J. Org. Chem.*, *51*, 2694 (1986).
9. S. J. Angyal and V. A. Pickles, *Aust. J. Chem.*, *25*, 1696 and 1711 (1972).
10. S. J. Angyal, *Chem. Soc. Rev.*, *9*, 415 (1980).
11. S. J. Angyal, *Pure Appl. Chem.*, *35*, 131 (1973).
12. K. Araki and S. Shiraishi, *Carbohydr. Res.*, *148*, 121 (1986).
13. J. Boeseken, *Adv. Carbohydr. Chem.*, *4*, 189 (1949).
14. H. Pelmore and M. C. R. Symons, *Carbohydr. Res.*, *155*, 206 (1986).
15. C. T. Bishop and F. P. Cooper, *Can. J. Chem.*, *41*, 2743 (1963).

16. R. W. Binkley, G. S. Goewey, and J. C. Johnston, *J. Org. Chem.*, *49*, 992 (1984).
17. R. J. Ferrier and L. R. Hatton, *Carbohydr. Res.*, *6*, 75 (1968).
18. J. L. Jensen, S.-C. Tsuang, and A. H. Uslan, *J. Org. Chem.*, *51*, 816 (1986).
19. J. E. Hodge and B. F. Moy, *J. Org. Chem.*, *28*, 2784 (1963).
20. The Amadori rearrangement has been reviewed by J. E. Hodge, *Adv. Carbohydr. Chem.*, *10*, 169-205 (1955).
21. K. Heyns, H. Paulsen, R. Eichstedt, and M. Rolle, *Ber.*, *90*, 2039 (1957).
22. A thorough review of the reduction of carbohydrates has been written by J. W. Green, in *The Carbohydrates* (W. Pigman and D. Horton, eds.), Academic Press, New York, Vol. 1B, 1980, pp. 989-1011.
23. M. L. Wolfrom and H. B. Wood, Jr., *J. Am. Chem. Soc.*, *73*, 2933 (1951).
24. L. Hough and R. S. Theobald, *Methods Carbohydr. Chem.*, *1*, 94 (1962).
25. J. W. Green also has written an excellent review of the oxidation of carbohydrates in *The Carbohydrates* (W. Pigman and D. Horton, eds.), Academic Press, New York, Vol. 1B, 1980, pp. 1101-1166.
26. C. S. Hudson and H. S. Isbell, *J. Am. Chem. Soc.*, *51*, 2225 (1929).
27. W. W. Binkley in *The Carbohydrates* (W. Pigman and D. Horton, eds.), Academic Press, New York, Vol. IIB, 1970, p. 761.
28. W. W. Pigman, B. L. Browning, W. H. McPherson, C. R. Calkins, and R. L. Leaf, Jr., *J. Am. Chem. Soc.*, *71*, 2200 (1949).

29. H. G. Fletcher, Jr., H. H. Diehl, and C. S. Hudson, *J. Am. Chem. Soc.*, *72*, 4546 (1950).
30. H. Kiliani, *Ber. Dtsch. Chem. Ges.*, *18*, 3066 (1885).
31. N. K. Richtmyer, *Methods Carbohydr. Chem.*, *1*, 160 (1962).
32. E. Fischer, *Ber. Dtsch. Chem. Ges.*, *22*, 2204 (1889).
33. R. Barker and A. S. Serianni, *Acc. Chem. Res.*, *19*, 307 (1986).
34. H. El Khadem, *Adv. Carbohydr. Chem.*, *20*, 139 (1965).
35. J. D. Wander and D. Horton, *Adv. Carbohydr. Chem. Biochem.*, *32*, 16 (1976).
36. M. L. Wolfrom and A. Thompson, *Methods Carbohydr. Chem.*, *2*, 427 (1963).
37. A. P. G. Kieboom and H. van Bekkum, *Recl. Trav. Chim. Pays-Bas*, *103*, 1 (1984).
38. J. M. de Bruijn, A. P. G. Kieboom, and H. van Bekkum, *Sugar Tech. Rev.*, *13*, 21 (1986).
39. C. A. Lobry de Bruyn and W. Alberda van Ekenstein, *Recl. Trav. Chim. Pays-Bas*, *14*, 203 (1895).
40. M. L. Wolfrom and W. L. Lewis, *J. Am. Chem. Soc.*, *50*, 837 (1928).
41. M. S. Feather and J. F. Harris, *Carbohydr. Res.*, *15*, 304 (1970).
42. H. E. van Dam, A. P. G. Kieboom, and H. van Bekkum, *Starch/Starke*, *38*, 95 (1986).
43. M. S. Feather and J. F. Harris, *Adv. Carbohydr. Chem. Biochem.*, *28*, 161 (1973).
44. T. Urashima, K. Suyama, and S. Adachi, *Carbohydr. Res.*, *135*, 324 (1985).
45. I. S. Fagerson, *J. Agr. Food Chem.*, *17*, 4 (1969).
46. P. J. Thornalley and A. Stern, *Carbohydr. Res.*, *134*, 191 (1984).

47. A. Butlerow, *Justus Liebigs Ann. Chem.*, *120*, 295 (1861).
48. L. Hough and J. K. N. Jones, *J. Chem. Soc.*, 1122 and 3191 (1951).
49. R. F. Socha, A. H. Weiss, and M. M. Sakharov, *React. Kinet. Catal. Lett.*, *14*, 119 (1980).
50. E. Pfeil and G. Schroth, *Ber.*, *85*, 293 (1952).

8 Protecting Groups

I. GENERAL COMMENTS ON PROTECTING GROUPS

A. The Purpose of Group Protection

B. Characteristics of Effective Group Protection

II. GENERAL COMMENTS ON PROTECTION OF CARBOHYDRATES

III. ACETALS AND KETALS

A. Glycosides

1. The Fischer Method

a. Methyl and Benzyl Glycosides

b. Allyl Glycosides

2. Glycal Addition

3. Koenigs-Knorr Reaction

B. Isopropylidene Ketals and Benzylidene Acetals

1. Thermodynamically Controlled Reactions

2. Kinetically Controlled Reactions

3. Reagents for Acetonation (Isopropylidenation) and Benzylidenation

        4. Deprotection
            a. Hydrolysis
            b. Reductive Cleavage of Benzylidene Acetals
            c. Oxidative Cleavage of Benzylidene Acetals
    C. Cyclohexylidene Ketals and Ethylidene and Methylene Acetals
    D. (2-Methoxyethoxy)methyl (Mem) and Tetrahydropyranyl (THP) Derivatives
    E. Dialkyl Dithioacetals
IV. ESTERS
    A. Acetates and Benzoates
        1. Acetylation
        2. Benzoylation
        3. Regioselective Protection
        4. Regioselective Deprotection
        5. Acyl Migration
        6. Deprotection of Base-Sensitive Compounds
    B. Alkyl- and Arylsulfonates
    C. Nitrates
    D. Boronates
    E. Orthoesters
V. ETHERS
    A. Benzyl Ethers
        1. Reagents for Benzylation
        2. Regioselective Benzylation
            a. Direct Methods
            b. Indirect Methods

I. GENERAL COMMENTS ON PROTECTING GROUPS

A. The Purpose of Group Protection

When a polyfunctional molecule is used in a synthetic reaction, usually transformation of a single functional group is desired. Achieving such reaction requires either (a) a reagent which can select a single functional group for reaction or (b) group protection which leaves unprotected only the group for which reaction is desired. In practice a combination of these two strategies often is adopted. In a given situation selective reaction usually is preferred over group protection because selective reaction is simpler; that is, it requires no protection and deprotection steps. However, selective reaction is easiest to conduct when there is a considerable difference in reactivity between functional groups. When many functional groups in a molecule are of the same type (e.g., hydroxyl groups in carbohydrates), group protection assumes an essential role in a reaction sequence; therefore, it is not surprising that considerable effort has been directed toward the development of effective protecting groups (1).

B. Characteristics of Effective Group Protection

There are a number of desirable characteristics for a protecting group. The most important of these are that a group should be:

a) introduced easily in high yield;
b) stable under a variety of reaction conditions;
c) removed easily and in high yield at the appropriate time;
d) readily separated after deprotection from the deprotected molecule.

Other considerations such as cost and toxicity of the compounds used in the protection-deprotection process can become significant factors at times. No single protecting group is satisfactory in all situations. Many can be used in particular instances. Some are effective under a wide variety of conditions.

II. GENERAL COMMENTS ON PROTECTION OF CARBOHYDRATES

Carbohydrates are most frequently protected by conversion into acetals, ketals, esters, or ethers. A combination of these protecting groups usually is needed. Within each of these four classes of compounds, only one or two structural types usually are encountered. For instance, the ketal which most often provides group protection is that derived from acetone (Equation 1) (2). The esters

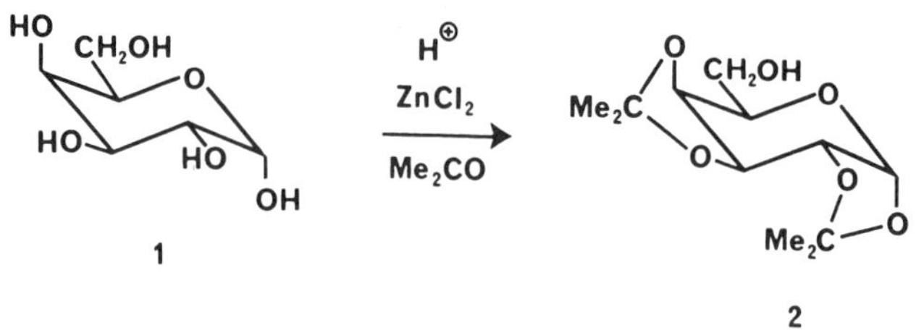

(1)

used to protect carbohydrates usually are acetates (Equation 2) (3) or benzoates (Equation 3) (4), while two types of

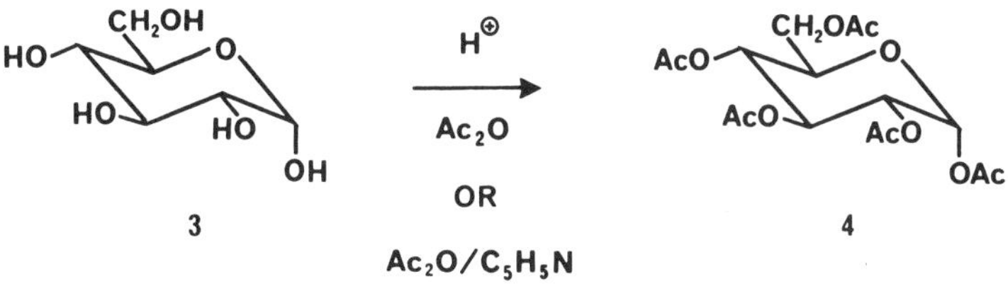

(2)

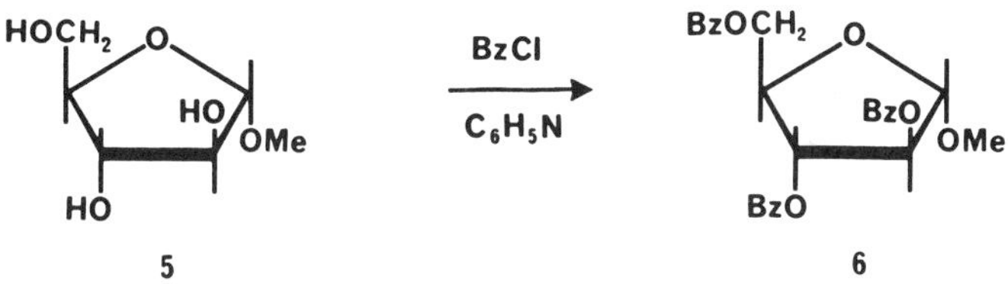

(3)

acetal protection are important. The first is that resulting from conversion of the hemiacetal linkage in a sugar into an acetal (Equation 4) (5). Such a reaction (glycoside formation) is the

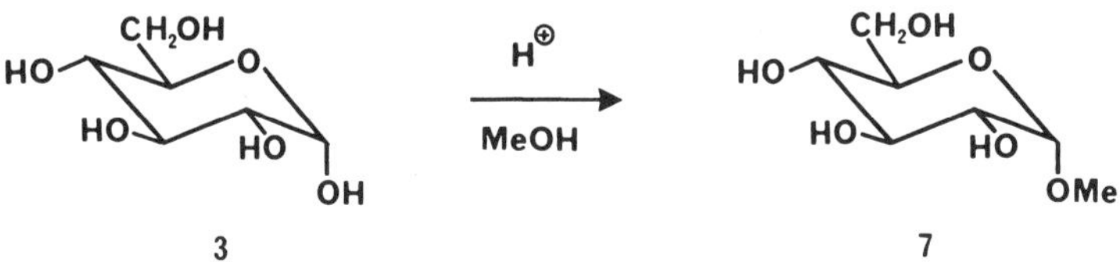

(4)

most common method for protection of an anomeric center and usually involves synthesis of a methyl or benzyl glycoside. Acetal protection also results when two hydroxyl groups in a sugar or sugar derivative react with benzaldehyde (Equation 5) (6).

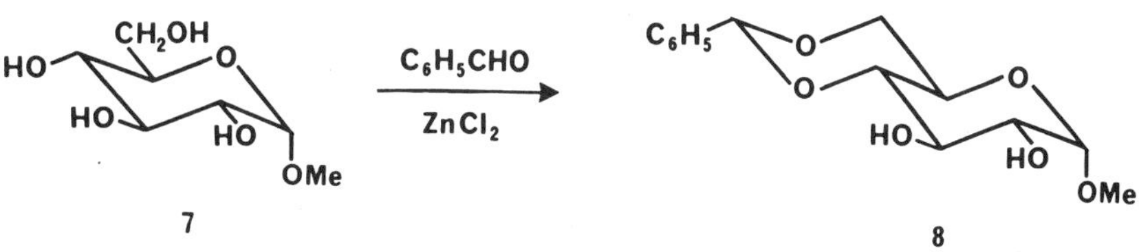

(5)

Ether protection generally takes the form of a benzyl ether (Equation 6) (7).

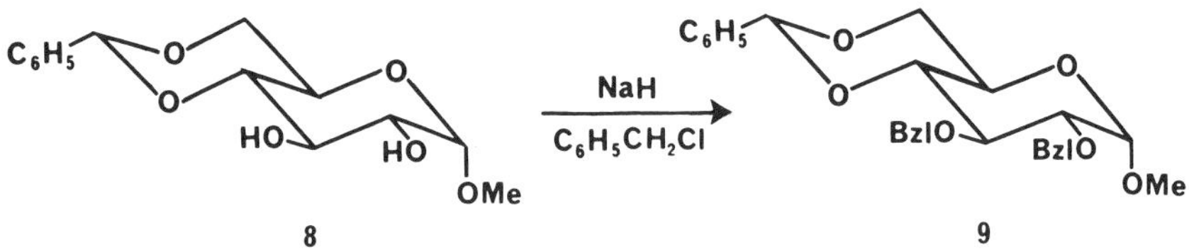

(6)

Acetal and ketal protection to a large extent complements esterification. Esters are moderately stable in the presence of acids but cyclic acetal and ketal protecting groups are not (Equation 7)

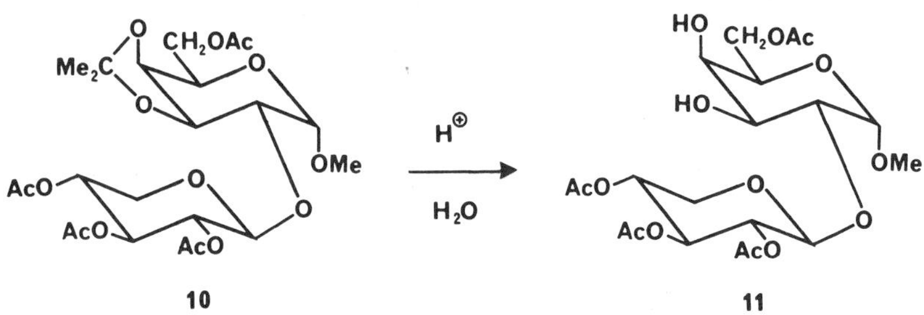

(7)

(8). Acetals and ketals, in contrast, are stable under basic reaction conditions, conditions which cause ester deprotection (Equation 8)

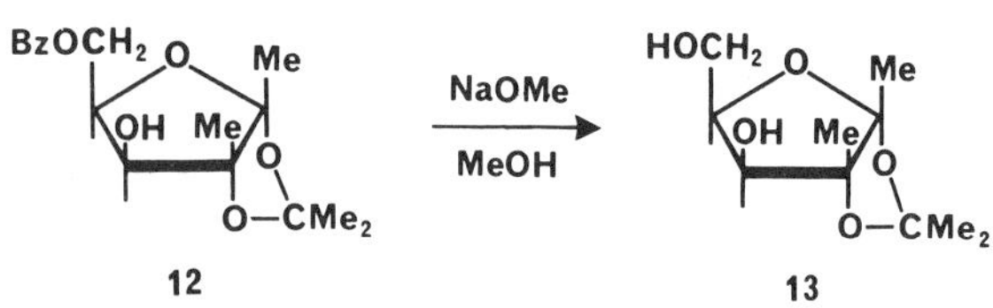

(8)

(9). Thus, appropriate group selection can provide a broad range of protection. Careful planning is needed, however, to ensure that the protecting groups present at each step in a reaction sequence will be stable under the reaction conditions.

A number of standard abbreviations are found in Equations 1-8. All of these are ones in common use and, with the exception of the notation for the benzyl and benzoyl groups, they have created no misunderstanding. The symbols for the benzyl and benzoyl groups, however, have at times been confused. To avoid misunderstanding, the practice adopted here (and now in most journals and books) is for the benzoyl group to be represented by "Bz" and the benzyl group by "Bzl".

In this chapter acetal, ketal, ester, ether, and other useful forms of group protection for carbohydrates will be discussed. This discussion will be centered around the introduction and removal of protecting groups, their relative advantages and disadvantages, and mechanistic understanding important to these protection-deprotection processes.

Finally, attention should be drawn to several review articles which contain valuable information on group protection. These articles, which have appeared in the past ten years, include general discussions on the reactivity of hydroxyl groups (10) as well as more specialized treatments of cyclic acetals (11-13), selective deprotection (14), and photochemical deprotection (15). These reviews have the added value that they provide references to earlier writings on their respective topics.

III. ACETALS AND KETALS

A. Glycosides

In the laboratory, glycosides typically are formed either for group protection or as a part of a synthesis of a larger molecule, usually an oligosaccharide. Group protection is, of course, under consideration here. Oligosaccharide synthesis will be discussed in Chapter 14.

1. The Fischer Method

a. Methyl and Benzyl Glycosides. If, as is often the case, a synthetic sequence begins with an unprotected sugar, the first step generally is to convert this compound into a glycoside. Such conversion protects the most reactive functional group in the molecule. Methyl glycosides are easily prepared by the acid-catalyzed reaction of an unprotected sugar dissolved or suspended in methanol (Equation 4). (The mechanism for glycoside formation was discussed in Chapter 7, Section III.A.) Methyl glycosides satisfy well the criteria for effective group protection. Their primary disadvantage is that removal requires acidic reaction conditions. If a target molecule in a synthetic sequence is acid labile and protection by glycoside formation is to be used, a methyl glycoside is not a good choice. The traditional alternative has been to form a benzyl glycoside (Equation 9) (16). (The wavy

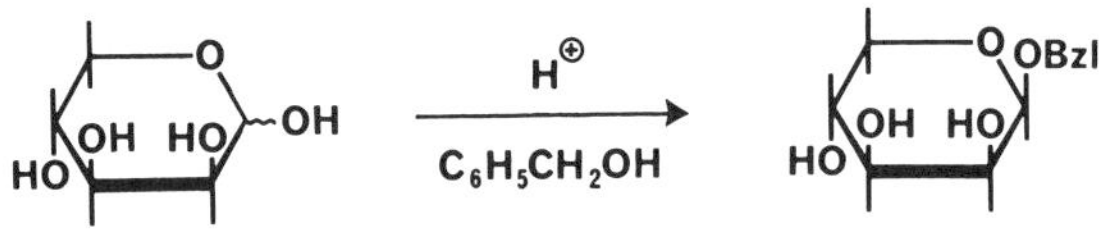

(9)

line connecting C-1 and O-1 in structure 14 indicates that a mixture of stereoisomers was used for this reaction. Whenever a wavy line connects two atoms, it will indicate a mixture of stereoisomers.) Benzyl glycoside synthesis is simple, particularly if the product crystallizes from the reaction mixture. If crystallization does not occur and excess benzyl alcohol has to be removed by distillation or chromatography, glycoside formation is somewhat less convenient. The important advantage of benzyl glycosides, however, is that they can be deprotected by hydrogenolysis under neutral conditions (Equation 10) (16).

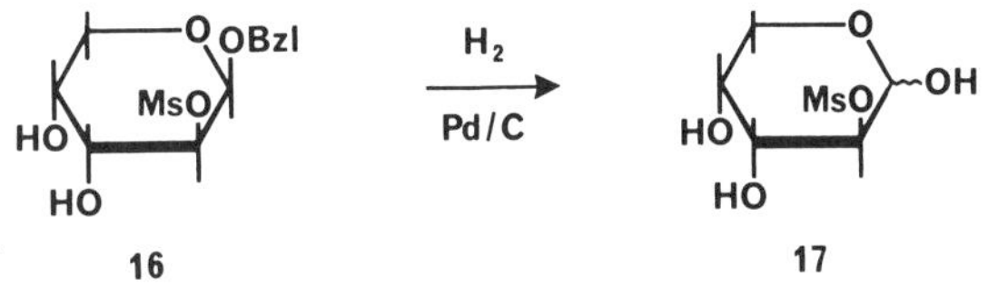

(10)

b. **Allyl Glycosides.** Another alternative to methanol in glycoside synthesis is allyl alcohol. The allyl group is easily introduced (Equation 11) (17), is stable under a variety of conditions,

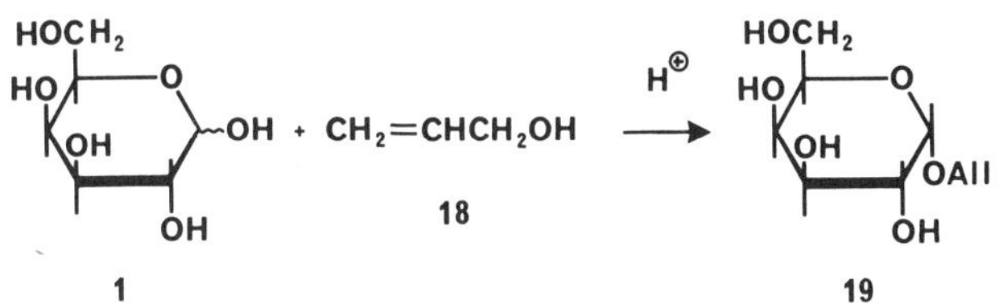

(11)

and can be removed in a mild, two-step reaction sequence. The first deprotection step is the catalytic isomerization of the allyl double bond into conjugation with the oxygen atom using tris(triphenylphosphine)rhodium(I) chloride (Figure 1).

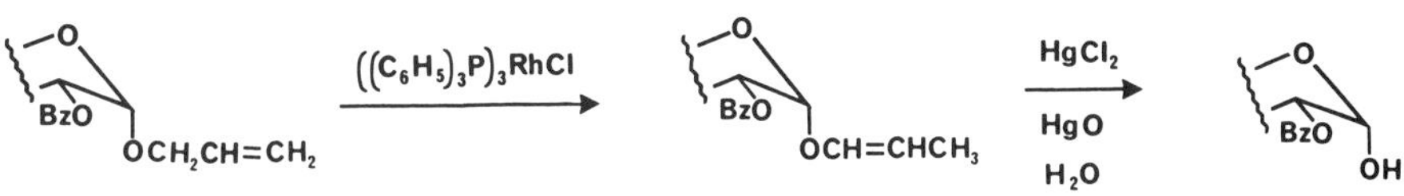

FIG. 1. Deprotection of an allyl glycoside.

Following isomerization of the double bond, the 1-propenyl group can be removed by treatment of the glycoside with mercury(II) chloride and mercury(II) oxide (or zinc(II) chloride and zinc(II) oxide) in aqueous acetone. The reactions shown in Figure 1 effectively deprotect the allyl glycoside 20 without disturbing the glycosidic linkage between the two rings or the other protecting groups (18).

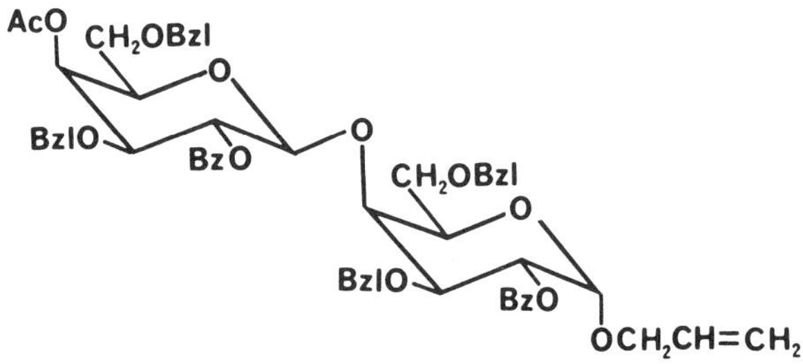

20

2. Glycal Addition

Another method for glycoside formation is an addition reaction involving a glycal. This process requires the reaction of an electrophilic reagent with a glycal in the presence of an alcohol (Figure 2). Glycal addition is particularly useful in the synthesis of 2-deoxy-2-halogeno sugars; for example, the reaction of *N*-iodosuccinimide (a source of molecular iodine) with 3,4,6-tri-*O*-acetyl-1,5-anhydro-2-deoxy-D-*arabino*-hex-1-enitol (tri-*O*-acetyl-D-glucal [21]) in cyclohexanol produces a pair of cyclohexyl glycosides [22 and 23] (Equation 12) (19). Since hydrogenolysis readily replaces

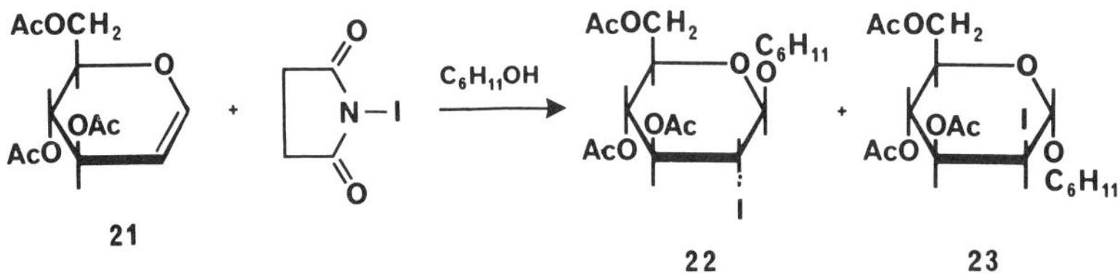

(12)

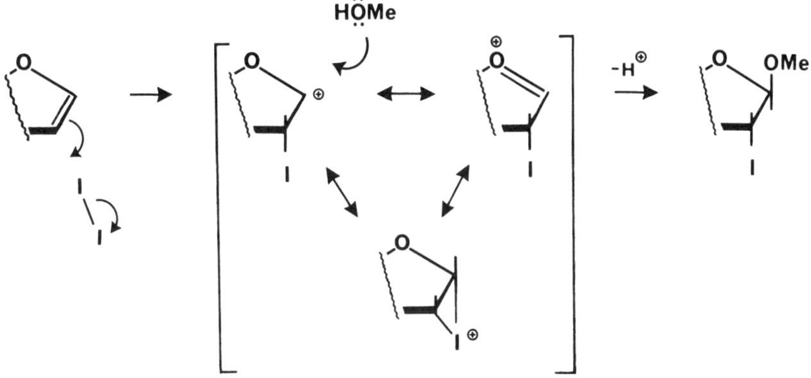

FIG. 2. Proposed mechanism for the reaction of iodine with a glycal in the presence of methanol.

iodine and bromine atoms with hydrogen atoms, this addition reaction is an attractive method for forming compounds destined to become 2-deoxy sugars. Electrophilic reagents other than halogens (e.g., HCl, HBr, H_2SO_4, $HgCl_2$) also can add to glycals in the presence of alcohols to produce glycosides. Halogen additions are, however, much more common.

3. Koenigs-Knorr Reaction

One of the classical methods for glycoside synthesis is the Koenigs-Knorr reaction, which is the reaction of a glycosyl halide with a compound containing an unprotected hydroxyl group. While this reaction can be used to synthesize simple glycosides (Equation 13) (20), its greater value is in oligosaccharide synthesis

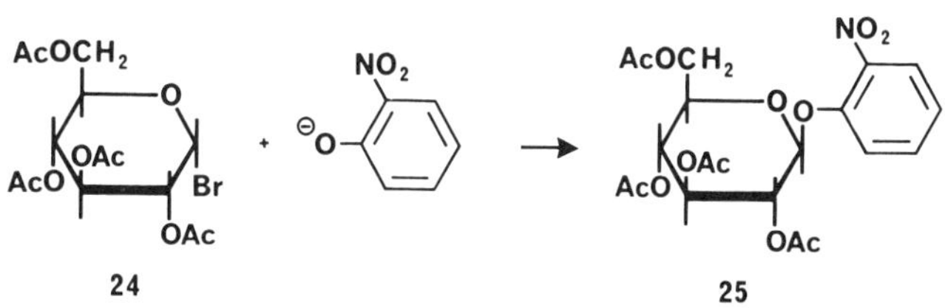

(13)

(Chapter 14); therefore, further discussion of the Koenigs-Knorr reaction will be delayed until Chapter 14.

B. Isopropylidene Ketals and Benzylidene Acetals

1. Thermodynamically Controlled Reactions

Although isopropylidene and benzylidene groups often are introduced into a molecule under similar reaction conditions (Equations 1 and 5), the location of the protection can be quite different. The reason for this difference is directly related to the stability of each protected molecule. Since protection normally occurs under conditions which allow reversibility, reaction proceeds until equilibrium is reached. The distribution of the products at equilibrium is determined by their relative thermodynamic stabilities; in other words, these reactions are thermodynamically controlled.

Benzylidene groups prefer to be part of six-membered ring acetals, while the ketals resulting from acetonation (also called isopropylidenation) generally are five-membered rings. This difference is attributed to the effect of the methyl and phenyl substituents on the stability of the particular ring systems. Consider the five- and six-membered rings shown in Figure 3. Each of these can be formed by benzylidenation. The most stable

FIG. 3. 1,3-Dioxane (left) and 1,3-dioxolane (right) rings formed by benzylidenation. The double-headed arrow points to carbon atoms bearing groups experiencing eclipsing interactions.

arrangement is the six-membered (1,3-dioxane) ring in which the chair conformation has the phenyl substituent in an equatorial position. This arrangement is more stable than the five-membered (1,3-dioxolane) ring in which the ring substituents necessarily experience eclipsing interactions (Figure 3).

The stabilities of five- and six-membered rings formed by acetonation present a different picture. Acetonation produces a six-membered ring in which each chair conformation necessarily has the destabilizing effect of an axial methyl group (Figure 4). In order to avoid such structures, isopropylidene protection generally results in formation of a five-membered ring. Even with the eclipsing interactions present in such a ring system, the five-membered ring is thermodynamically more stable than the six-membered one.

The generalizations which have been made about ring size and stability provide guidelines for predicting the products from benzylidenation and acetonation. These guidelines are not inviolate. If the most stable structure possible from benzylidenation, for instance, contains a five-membered ring, then the five-membered ring system will form. For example, benzylidenation of methyl 6-deoxy-α-L-mannopyranoside [26] could, in principle, produce either a five-membered (2,3-O-benzylidene) or a six-membered (2,4-O-benzylidene) ring. In this case the five-membered

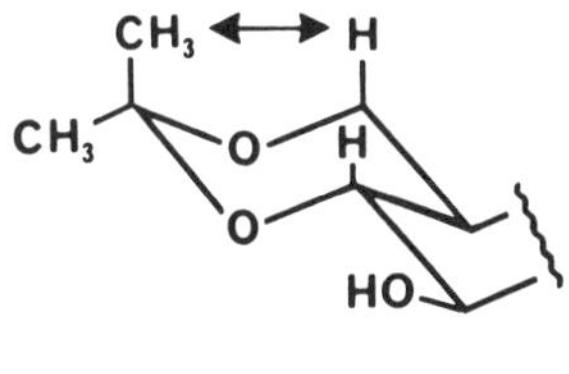

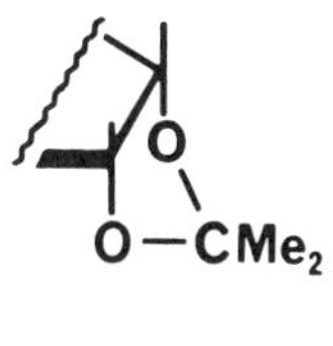

FIG. 4. 1,3-Dioxane (left and 1,3-dioxolane (right) rings formed by acetonation. The double-headed arrow points to one destabilizing interaction caused by the axial methyl group.

ring is formed because it introduces less ring strain into the molecule (Equation 14) (21). Thermodynamic stability also can

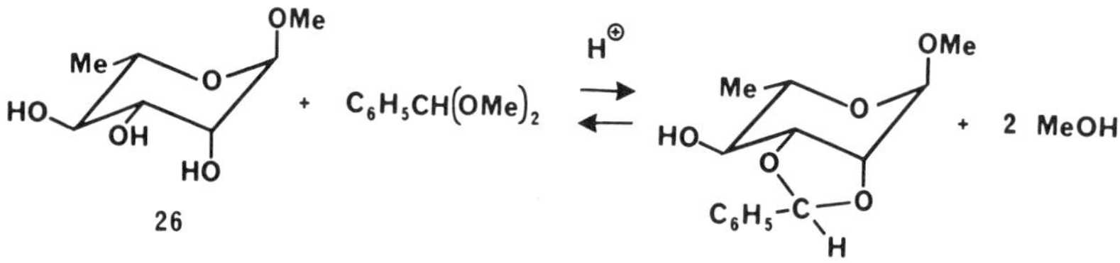

26 27 (14)

dictate the formation of a six-member ring from acetonation. In the reaction of allyl 2-acetamido-2-deoxy-β-D-glucopyranoside [28], 3,4-*O*-isopropylidenation is possible but would produce a ketal [30] with a *trans*-fused ring system. Such a compound

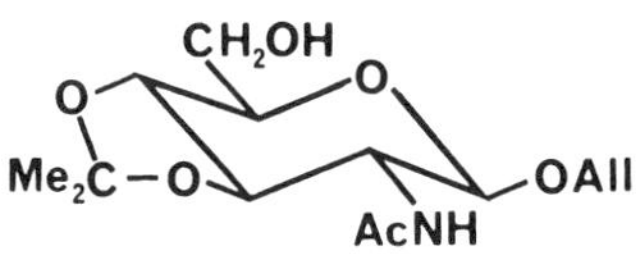

30

[30] is less stable than the product of 4,6-*O*-isopropylidenation (Equation 15) (22).

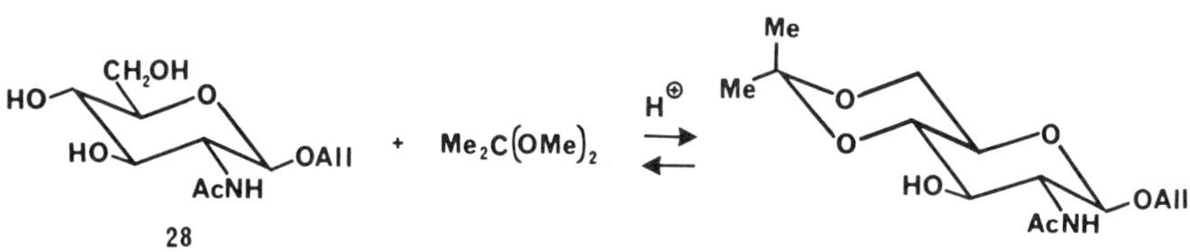

28 29 (15)

2. Kinetically Controlled Reactions

One of the characteristics of an irreversible reaction is that the product formed most rapidly is the major product. Such a reaction is described as being kinetically controlled. Consider the interaction of protonated benzaldehyde with methyl α-D-glucopyranoside [7]. The least hindered hydroxyl group in 7 (i.e., the primary one at C-6) is, on the average, approached within bonding distance more easily (and, thus, more frequently) than the other hydroxyl groups in the molecule. This favored approach to C-6 leads to preferential formation of methyl 4,6-*O*-benzylidene-α-D-glucopyranoside [8], the product of kinetic control.

Even though typical benzylidenation conditions allow isomerization of the first-formed product [8] to a different benzylidene derivative (Figure 5), isomerization of 8 is not observed since it is not only the most rapidly formed benzylidene acetal but it is also the most stable one. Any other acetal is a minor product, regardless of whether the reaction is reversible or not; therefore,

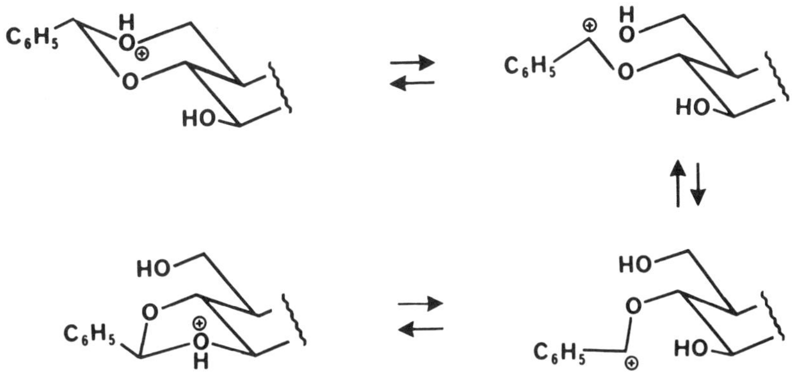

FIG. 5. Proposed mechanism for acid-catalyzed isomerization of benzylidene acetals.

the 4,6-*O*-benzylidene acetal 8 is both the kinetically and thermodynamically favored reaction product.

Although kinetics and thermodynamics often favor the same reaction product, this is not always the case. Consider the approach of protonated acetone to α-D-galactopyranose [1]. Here also the primary hydroxyl group is the least hindered one and, therefore, the most rapid reaction is that leading to a 4,6-*O*-isopropylidene ketal. If reaction conditions permit reversal of ketal formation, no 4,6-*O*-isopropylidene protected product will be isolated; rather, 1,2:3,4-di-*O*-isopropylidene-α-D-galactopyranose [2], the most stable derivative, will be obtained (Equation 1). In this reaction the product of kinetic control is different from the thermodynamically controlled product.

If acetonation is conducted under conditions in which the protecting group remains attached at its initial point of contact, a 4,6-*O*-isopropylidene acetal will be formed. Reaction of α-D-galactopyranose [1] with 2-methoxypropene in the presence of an acid catalyst produces 4,6-*O*-isopropylidene-α-D-galactopyranose [31] as the major product (Equation 16) (23). Compound 31 is,

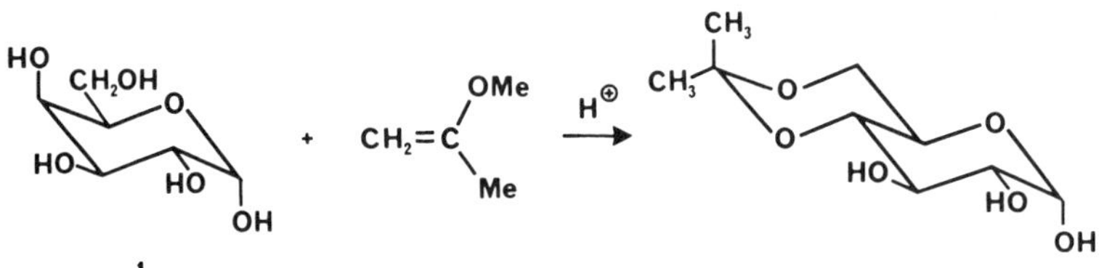

1 31

(16)

therefore, the product of kinetic control. Similar reactions to produce 4,6-*O*-isopropylidene derivatives take place under kinetically controlled conditions with other sugars (13,24). It is interesting to note that kinetically controlled reaction not only produces 4,6-*O*-isopropylidene protection but also can lead to formation of other relatively strained ring systems. For example, reaction of methyl α-D-glucopyranoside [7] with excess 2-methoxypropene gives methyl 2,3:4,6-di-*O*-isopropylidene-α-D-glucopyranoside [32]. The latter compound [32] contains a strained, *trans*-fused, five-membered ring system (Equation 17) (23).

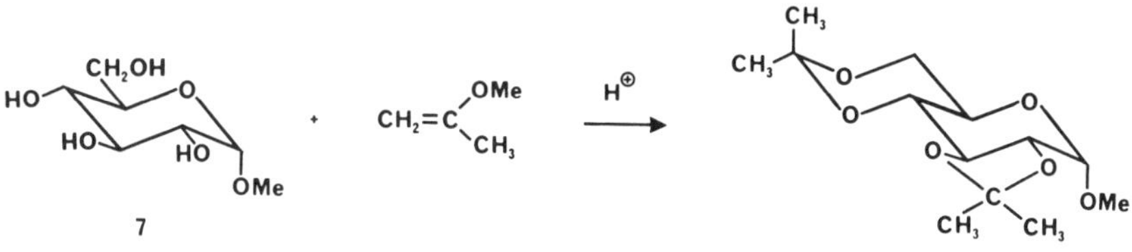

7 32

(17)

3. Reagents for Acetonation (Isopropylidenation) and Benzylidenation

In addition to the reagents already mentioned (acetone, benzaldehyde, and 2-methoxypropene), other compounds are useful for acetonation and benzylidenation. Among these, 2,2-dimethoxypropane and α,α-dimethoxytoluene are particularly valuable. Examples of the use of these reagents are given in the reactions shown in Equations 14 and 15. These reactions are acetal exchange processes which can be "driven to completion" by the removal of methanol. For example, the procedure for the synthesis of methyl 2,3-*O*-benzylidene-6-deoxy-α-D-mannopyranoside [27] (Equation 14) calls for distillation of methanol from the reaction mixture as the reaction proceeds (25).

Another reagent for benzylidenation is α,α-dibromotoluene (Equation 18) (26). This reagent is different from the others

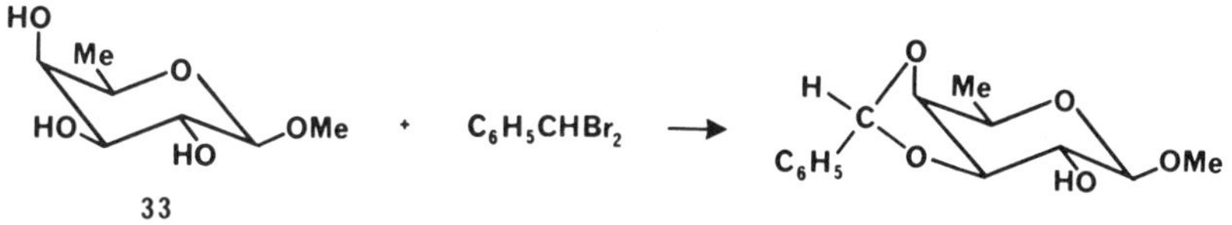

33 34

(18)

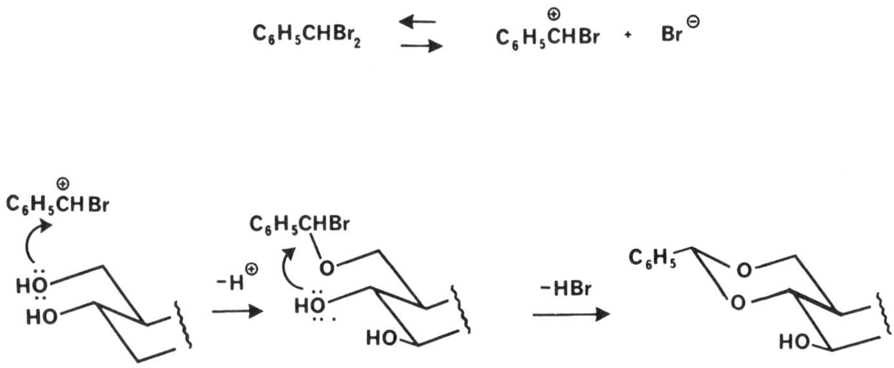

FIG. 6. Proposed mechanism for benzylidene acetal formation using α,α-dibromotoluene.

mentioned thus far in that it does not require acid catalysis for benzylidenation (Figure 6); in fact, benzylidene protection can be introduced with pyridine as the reaction solvent.

4. Deprotection

a. Hydrolysis. Removal of benzylidene and isopropylidene groups can be accomplished by treatment of protected compounds with aqueous acids; thus, the benzylidene acetal 35 is hydrolyzed by aqueous acetic acid (Equation 19) (27). If more than one acetal

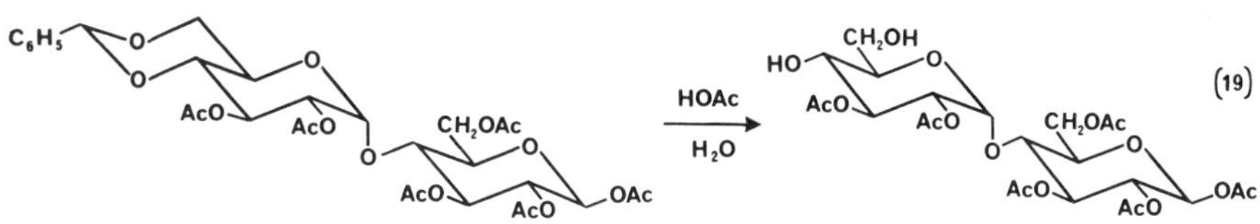

35 36 (19)

or ketal protecting group is present in a molecule, deprotection often can be selective; for example, reaction of 3-*O*-acetyl-1,2:5,6-di-*O*-isopropylidene-α-D-allofuranose [37] with aqueous acid at room temperature gives 3-*O*-acetyl-1,2-*O*-isopropylidene-α-D-allofuranose [38] (Equation 20) (28). More vigorous treatment removes both isopropylidene groups.

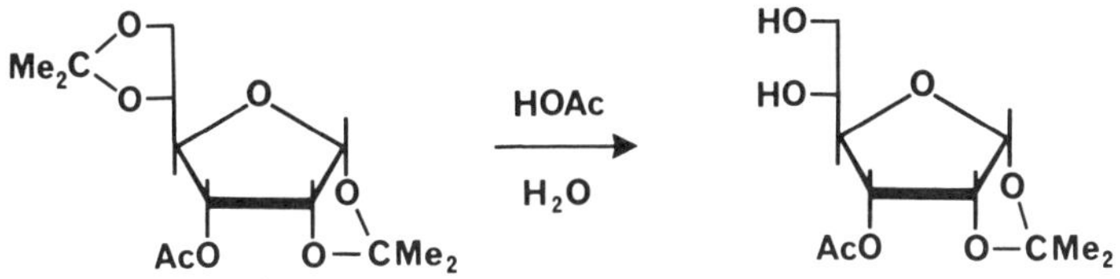

37 38 (20)

b. **Reductive Cleavage of Benzylidene Acetals.** Considerable flexibility exists in the deprotection of benzylidene acetals. In addition to complete hydrolysis, reaction with certain reagents produces molecules which are still partially protected. The most useful partial deprotection results from reductive cleavage of a benzylidene acetal to give a free hydroxyl group and a benzyl ether (Equations 21 and 22). The course of the reductive, ring-opening process is dependent upon the reagent used and the

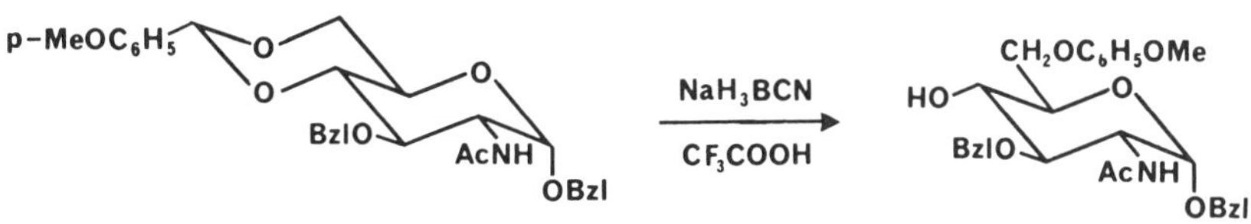

39 40 (21)

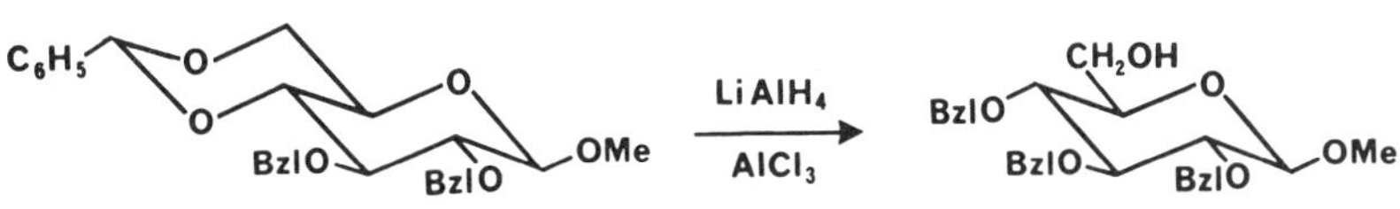

41 42 (22)

FIG. 7. Proposed intermediates in the reductive ring opening of a 4,6-*O*-benzylidene acetal.

reaction conditions. Each reagent consists of a combination of an acid and a metal hydride; two pairs which have complimentary reactivity are lithium aluminum hydride-aluminum chloride and sodium cyanoborohydride-trifluoroacetic acid. Reaction of a 4,6-*O*-benzylidene acetal with sodium cyanoborohydride-trifluoroacetic acid leads to the formation of a 6-*O*-benzyl ether (Equation 21) (29). This reaction appears to be governed by preferential protonation at O-4 rather than O-6 to produce an intermediate which reacts with sodium cyanoborohydride to give the 6-*O*-benzyl derivative (Equation 21, Figure 7). The sterically larger Lewis acid, aluminum chloride, favors coordination with the less hindered O-6 (Figure 7) to give a complex which reacts with lithium aluminum hydride to produce a 4-*O*-benzyl derivative (Equation 22, Figure 7) (30).

c. **Oxidative Cleavage of Benzylidene Acetals.** Ring opening of benzylidene acetals to give partially protected benzoates can be accomplished by light-catalyzed reaction of the acetal with *N*-bromosuccinimide in the presence of water (Equation 23) (31).

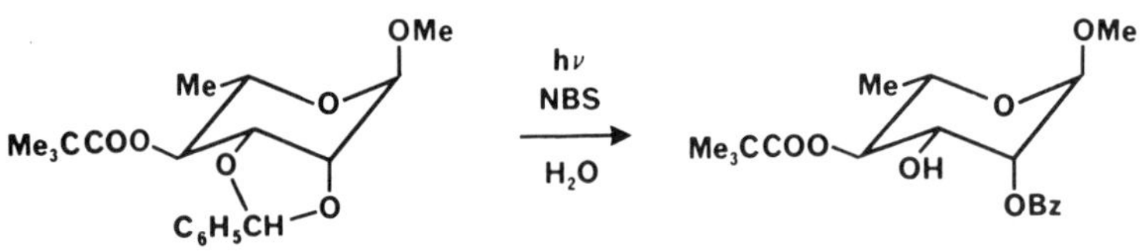

43 44 (23)

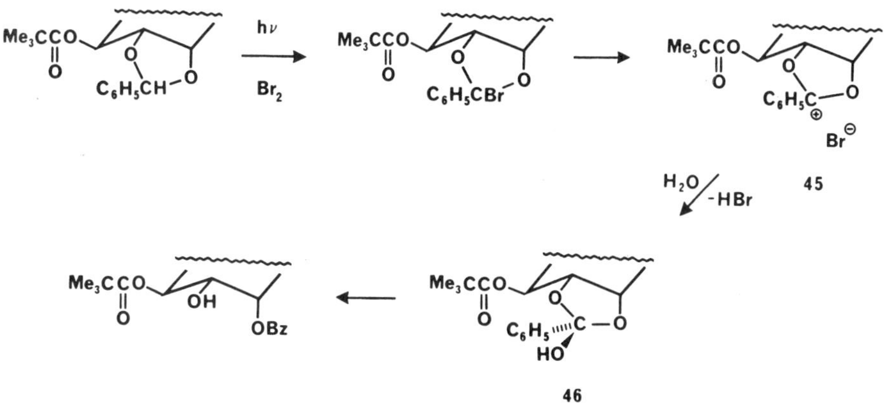

FIG. 8. Proposed mechanism for oxidative ring opening of a 2,3-*O*-benzylidene acetal.

The "key" intermediate in this reaction is the cation 45 which reacts with water to give the unstable ortho acid 46 (Figure 8). Ortho acids such as 46 ring open to produce benzoates with the benzoyloxy group in an axial position (32,33).

C. Cyclohexylidene Ketals and Ethylidene and Methylene Acetals

Cyclohexylidene ketals are sometimes used in place of isopropylidene protection while ethylidene acetals can provide an alternative to benzylidenation. An example of the introduction of cyclohexylidene protection is provided by the reaction shown in Equation 24 (34). Also, by way of example, ethylidenation of methyl α-D-glucopyranoside [7] is seen in the reaction described by Equation 25 (35) to proceed in essentially the same fashion as benzylidenation of 7 (Equation 5). The reactivity of the cyclohexylidene and ethylidene groups is not greatly different from that of isopropylidene and benzylidene groups, respectively. Methylene acetals, however, are distinctly more stable toward hydrolysis than other acetals. Complete hydrolysis of a methylene

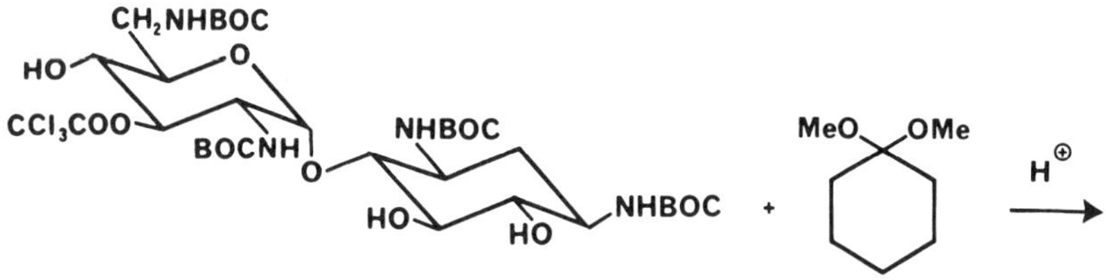

47

$$BOC = C_6H_5CH_2\underset{\underset{O}{\|}}{C}-$$

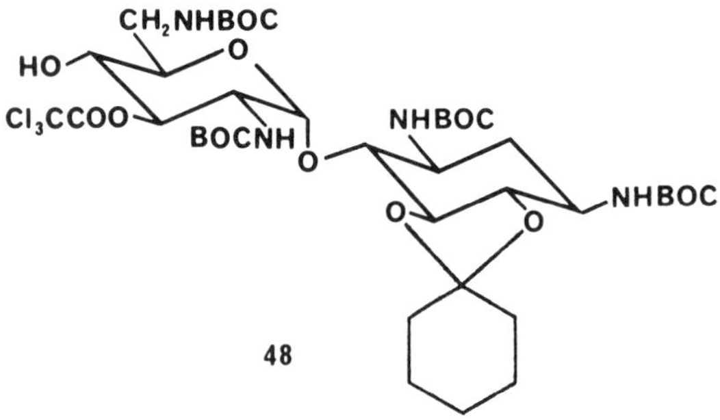

48 (24)

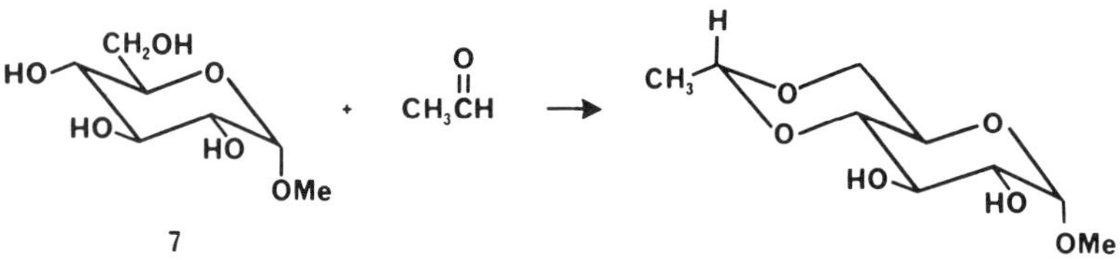

7 49 (25)

acetal can require prolonged heating in the presence of strong acids at temperatures above 100° C. In contrast, isopropylidene ketals generally are hydrolyzed in a similar manner at room temperature. Benzylidene acetals hydrolyze even more easily. The ease of hydrolysis of acetals and ketals parallels the ease of formation of the cation produced by ring opening (Figure 9). The reason for this association is that cation formation (Figure 9) is generally the rate-determining step in ketal and acetal hydrolysis.

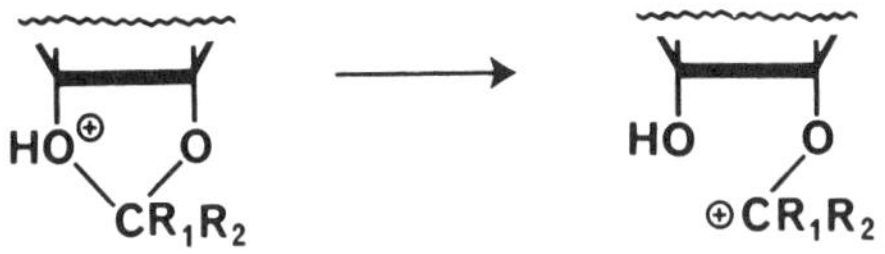

R_1	R_2	STABILITY
H	H	LEAST
CH_3	CH_3	
C_6H_5	H	MOST

FIG. 9. Stability of cationic intermediates generated during acetal and ketal hydrolysis.

D. (2-Methoxyethoxy)methyl (Mem) and Tetrahydropyranyl (THP) Derivatives

Acetals sometimes are used to protect single hydroxyl groups other than anomeric hydroxyls. Tetrahydropyranyl (THP) (Equation 26) (36) and (2-methoxyethoxy)methyl (Mem) (Equation 27) (18) groups afford this type of protection. These

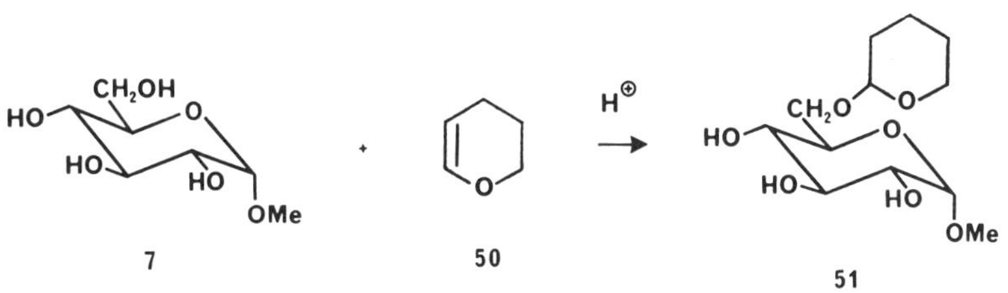

(26)

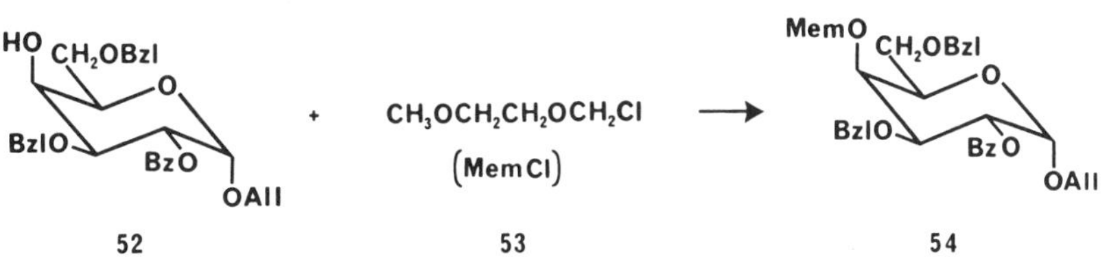

(27)

groups have reactivity similar to other acetals, including stability in the presence of bases but not acids. They are not widely used to protect carbohydrates, however, because protection generally can be provided more effectively by other groups. Benzyl ethers (Section V.A.), for instance, provide superior protection because they are stable over a broader range of reaction conditions. THP and Mem derivatives are subject to inadvertent deprotection; for example, attempted conversion of allyl 2-*O*-benzoyl-3,6-di-*O*-benzyl-4-*O*-(2-methoxyethoxy)methyl-α-D-galactopyranoside [54] into the glycosyl chloride 55 always resulted in loss of the Mem group (Equation 28) (18). The benzyl and benzoyl groups remained in place during these reactions.

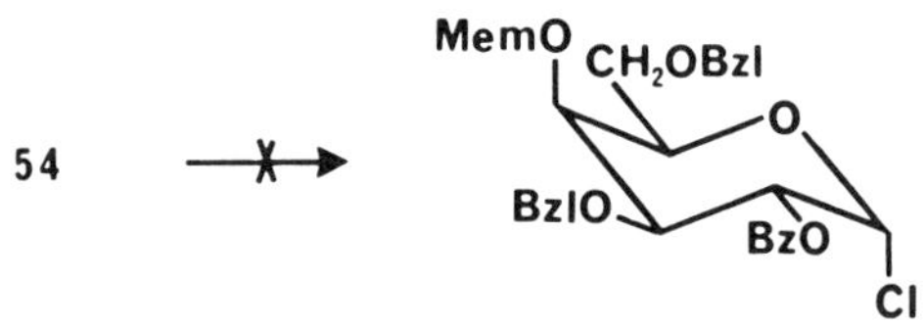

55

(28)

E. Dialkyl Dithioacetals

The reaction of unprotected sugars with alkanethiols to form dialkyl dithioacetals is one of the reactions described in Chapter 7 (Section III.G.). Dialkyl dithioacetal protection is particularly useful in situations where protection of an aldehydo group is necessary but a cyclic structure is not desired. Reaction of a sugar with an alkanethiol produces an "open-chain" structure (37). Reaction of D-arabinose [56] with ethanethiol, for example, gives the corresponding diethyl dithioacetal [57] (Equation 29) (37). Diethyl dithioacetal protection can be removed by treatment with mercury(II) salts. For instance, if compound 57 is further protected by formation of the corresponding 2,3:4,5-di-*O*-isopropylidene acetal prior to deprotection (Figure 10), the

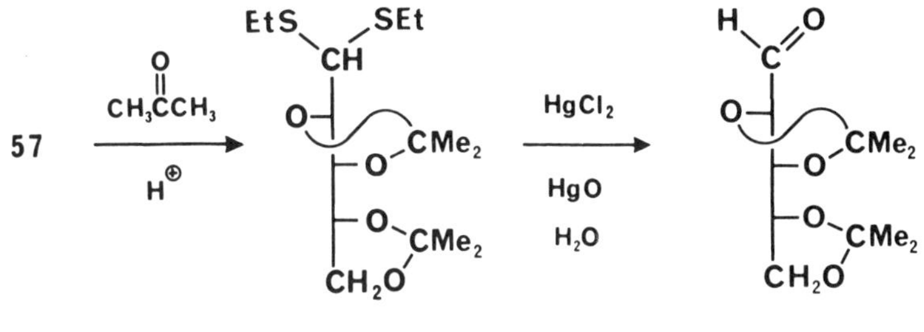

58

FIG. 10. Scheme for removal of diethyl dithioacetal protection to give a protected aldehydo sugar.

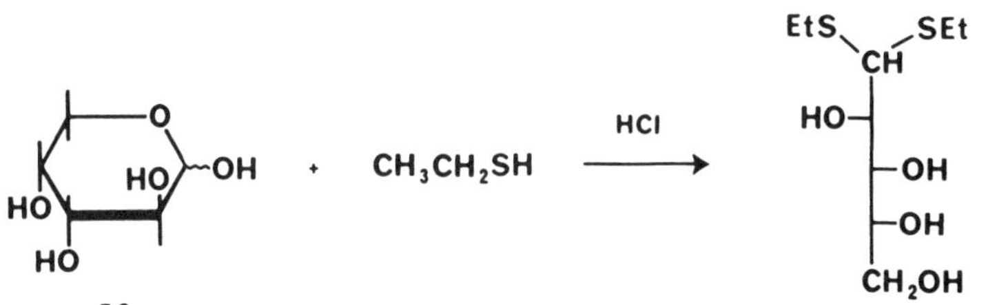

56 57 (29)

open-chain 2,3:4,5-di-*O*-isopropylidene-aldehydo-D-arabinose [58] is formed by reaction with mercury(II) chloride, mercury(II) oxide, and water (38).

IV. ESTERS

A. Acetates and Benzoates

1. Acetylation

It is usually possible to esterify all the hydroxyl groups in a sugar in a single reaction; for example, stirring a suspension of α-D-glucopyranose [3] in a mixture of acetic anhydride and pyridine at low temperature produces 1,2,3,4,6-penta-*O*-acetyl-α-D-glucopyranose [4] in 88% yield (Equation 2) (2). Acylation under these conditions occurs without anomeric change. At higher temperatures equilibration between pyranose and furanose ring forms and α and β anomers produces mixtures of products; nevertheless, fair to good yields of a single product often can be obtained from reaction at elevated temperature. Addition of α-D-glucopyranose [3] to boiling acetic anhydride containing sodium acetate produces a mixture of pentaacetates from which the β anomer can be isolated in 73% yield (Equation 30) (3).

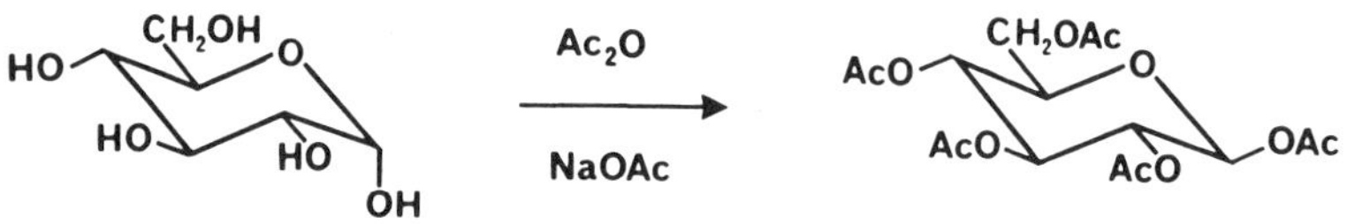

3 59 (30)

Acetylation also can be conducted under acidic conditions with compounds such as zinc chloride or sulfuric acid serving as catalysts (Equation 2) (3).

2. Benzoylation

Benzoylation generally is conducted in the presence of pyridine using benzoyl chloride as the acylating agent (Equation 3). Since benzoyl groups tend to be less susceptible to deprotection under acidic conditions than acetyl groups, benzoyl protection is sometimes chosen for the increased stability of the protected compounds. Benzoates also tend to crystallize readily.

3. Regioselective Protection

Regioselective reaction is possible in the esterification of carbohydrates. Primary hydroxyl groups react more rapidly than secondary ones; thus, selective formation of primary esters generally is easily accomplished (Equation 31) (39). Even secondary

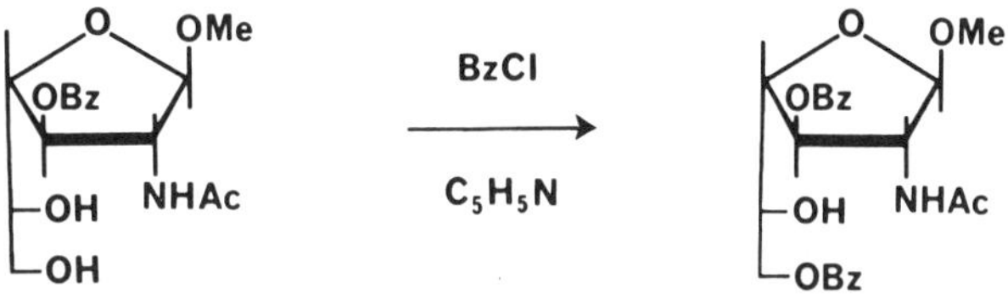

60 61 (31)

hydroxyl groups exhibit a considerable range of reactivity. In molecules which exist predominantly in a single chair conformation, equatorial groups often can be esterified with little reaction of the sterically more hindered axial ones (Equation 32) (17).

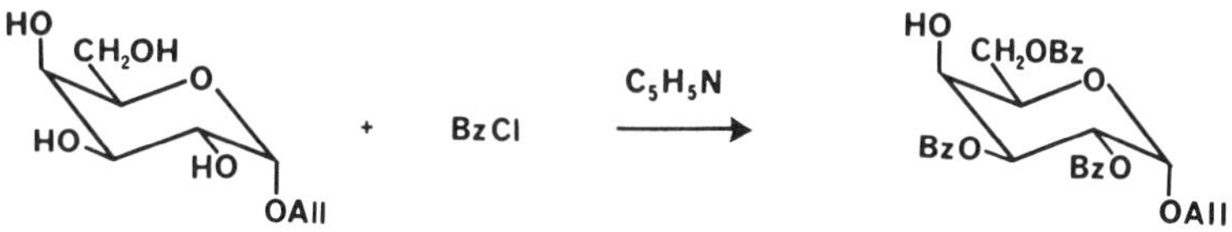

62 63 (32)

Some secondary hydroxyl groups are sufficiently sterically hindered that they remain unprotected while all other hydroxyl groups in the molecule are esterified (Equation 33) (40). Tertiary

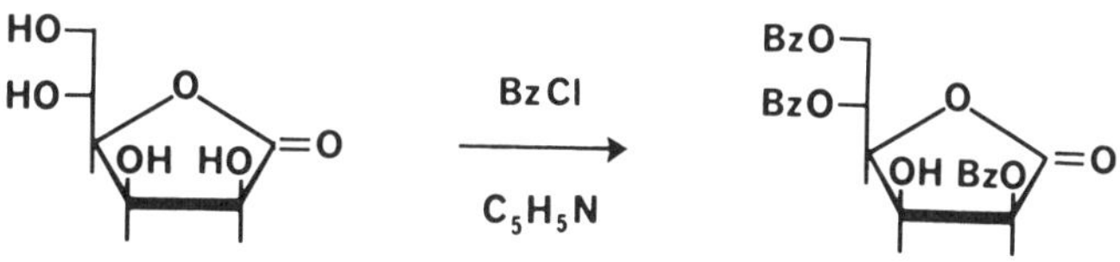

64 65 (33)

hydroxyl groups tend to be so unreactive that they may require conversion into the corresponding alkoxide ion in order for esterification to occur (Equation 34) (41). (If the tertiary alkoxide

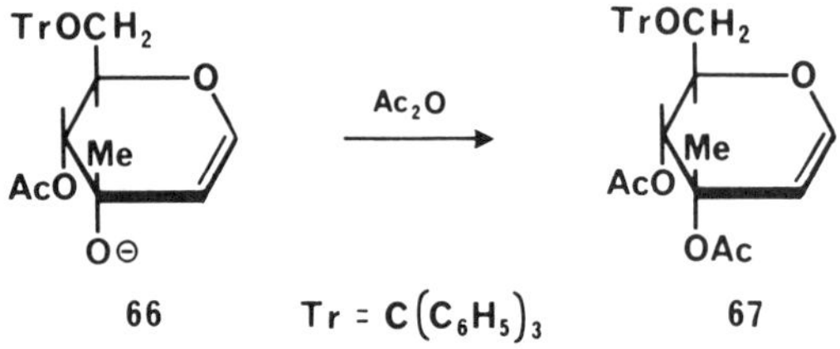

66 67

(34)

shown in Equation 34 is not formed prior to the addition of acetic anhydride, esterification does not take place.)

Several methods for improving selective protection have been developed. One of these depends upon less reactive esterifying agents, such as *N*-acetylimidazole (42) [68, R = Ac], *N*-benzoyl-imidazole (43) [69, R = Bz], and 1-(benzoyloxy)benztriazole (44) [69], to show greater selectivity. Such reagents are capable, in

68 69

some instances, of exhibiting considerable regioselectivity even when two secondary, equatorial hydroxyl groups are involved. For example, reaction of methyl 4,6-*O*-benzylidene-α-D-glucopyranoside [8] with 1-(benzoyloxy)benztriazole [69] gives the 2-*O*-benzoyl derivative 71 in 90% yield (Equation 35) (44). The combination of steric and electronic factors involved in selective esterification can change dramatically with a simple change in molecular structure. When methyl 4,6-*O*-benzylidene-β-D-glucopyranoside [72] is benzoylated with 1-(benzoyloxy) benztriazole [69] very little regioselectivity is observed (Equation 35) (44).

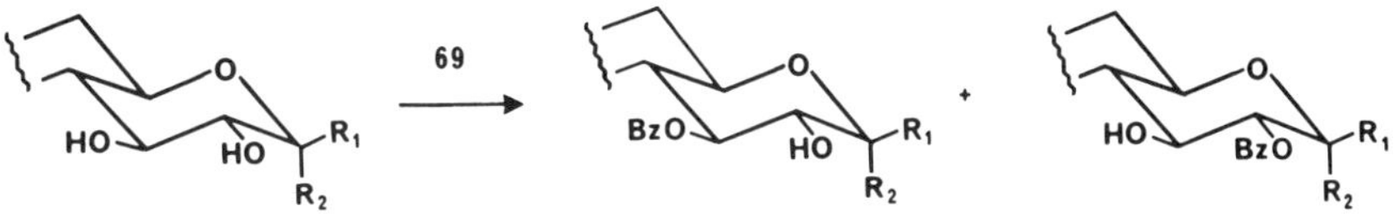

(35)

8 R_1 = H, R_2 = OMe 70 (4 %) 71 (90 %)

72 R_1 = OMe, R_2 = H 73 (43 %) 74 (50 %)

4. Regioselective Deprotection

Another approach to selective ester formation is one which focuses attention on deprotection, that is, selective removal of ester groups. When one considers that even structurally similar esters react with sodium hydroxide at considerably different rates (45), selective deprotection is recognized as an attractive possibility. One example of selective deprotection is found in the reaction shown in Equation 36 where the benzoyl group is removed

75 76

(36)

without causing reaction of the more hindered pivaloyl ester (21). The special reactivity of chloroacetyl groups offers a general method for selective de-esterification. Chloroacetyl protection can be removed in the presence of acetyl and benzoyl groups by treatment with thiourea (Equation 37) (46). A proposed mechanism for this deprotection process is shown in Figure 11. The yields for de-chloroacetylation, however, are only fair (43% for compound 78) and this process can result in decomposition of sensitive disaccharides (17).

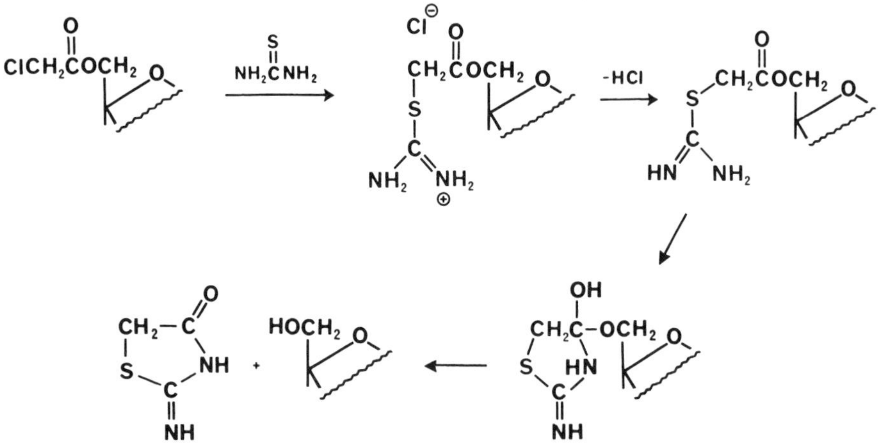

FIG. 11. Proposed mechanism for the deprotection of α-chloroacetates using thiourea.

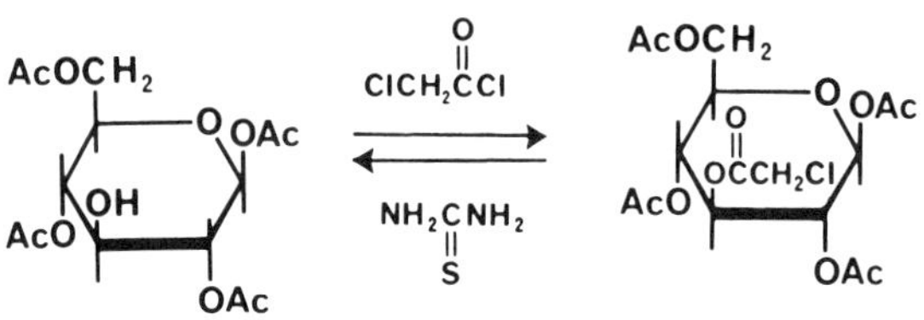

77 78 (37)

Hydrazine selectively deprotects acetylated carbohydrates. The O-1 acetyl group in 1,2,3-tri-*O*-acetyl-4,6-di-*O*-benzyl-α-D-galactopyranose [79] is removed by hydrazine without reaction at O-2 or O-3 (Equation 38) (47). Also, treatment of methyl

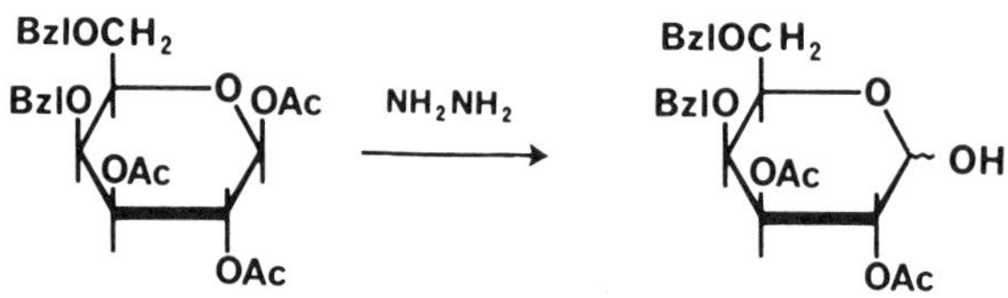

79 80 (38)

2,3-di-*O*-benzoyl-4,6-*O*-benzylidene-α-D-glucopyranoside [81] with hydrazine regioselectively removes the 2-*O*-benzoyl group (Equation 39) (48).

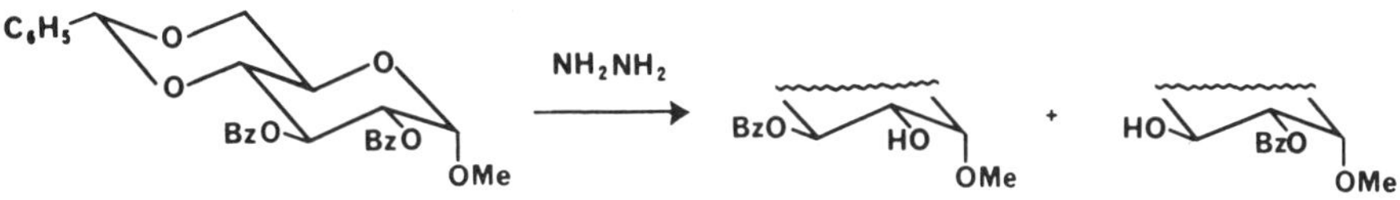

(39)

81 82 (65 %) 83 (18 %)

5. Acyl Migration

One of the potential dangers in partial deprotection of esters under basic conditions is acyl migration. This danger can be turned to an advantage, if deliberate use is made of the migration process. An acyl group migrating between two adjacent oxygen atoms will favor attachment to the atom which produces an axial, acyloxy substituent; thus, methyl 4-*O*-benzoyl-2,6-dideoxy-α-L-*lyxo*-hexopyranoside [85] can be synthesized from the corresponding 3-*O*-benzoyl compound [84] by benzoyl migration (Equation 40)

(40)

84 85

(49). The migratory ability of acyl groups also can be used to establish other protection patterns. Irreversible migration of acetyl groups from O-4 to O-6 is a process observed for a number of compounds (50). Reaction of benzyl 3,4-di-*O*-acetyl-2-benzamido-2-deoxy-β-D-glucopyranoside [86] with aqueous pyridine, for example, causes migration of the acetyl group to O-6 (Equation 41) (50). In addition to the generalizations noted here, study of

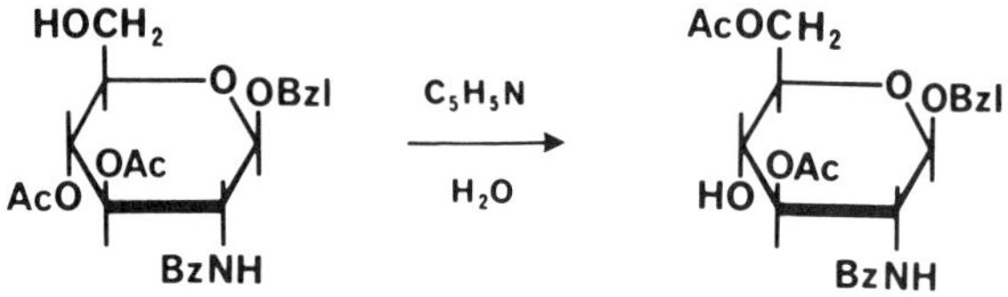

86 87 (41)

acetyl migration has enabled other observations to be made concerning the type of migration reactions that can be expected (50).

6. Deprotection of Base-Sensitive Compounds

Most esters are deprotected without difficulty using sodium methoxide in methanol (Equation 8). Other reagents such as basic ion exchange resins (Equation 36) or mixtures of ammonia and methanol also are effective. A method which has been used successfully for deprotection when strong bases cause molecular decomposition is treatment of the esterified compound with sodium cyanide in 95% ethanol; thus, the disaccharide 88, which

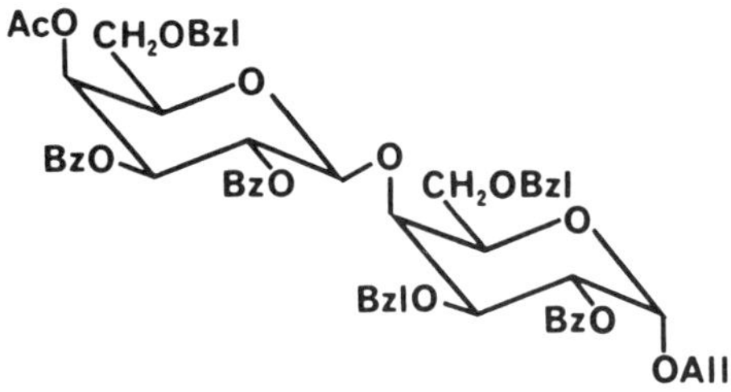

88

was decomposed both by thiourea and by alkoxides, was deprotected effectively with sodium cyanide (17,51).

B. Alkyl- and Arylsulfonates

Esters of sulfonic acids, particularly *p*-toluenesulfonic and methanesulfonic acids, are among the most common functional groups in carbohydrate chemistry. They usually are introduced not specifically as protecting groups but to serve as leaving groups in substitution or elimination reactions. Often, however, they provide valuable, temporary protection. These esters are easily formed (Equation 42) and they are stable under a variety of

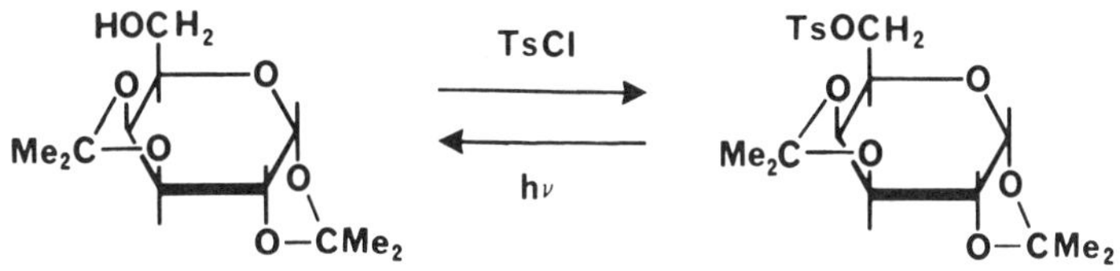

2 89 (42)

conditions, including the presence of acids and bases. They, of course, are displaced by effective nucleophiles and eliminated when heated with strong bases. A difficulty with the methylsulfonyl protecting group is that, unless it is to be displaced later, there is no mild method for deprotection. For *p*-tolylsulfonyl protected compounds, however, deprotection can be achieved under mild conditions by photochemical reaction (Equation 42) (52).

C. Nitrates

Esters formed by reaction of carbohydrates with nitrating agents also are capable of hydroxyl group protection (Equation 43) (53).

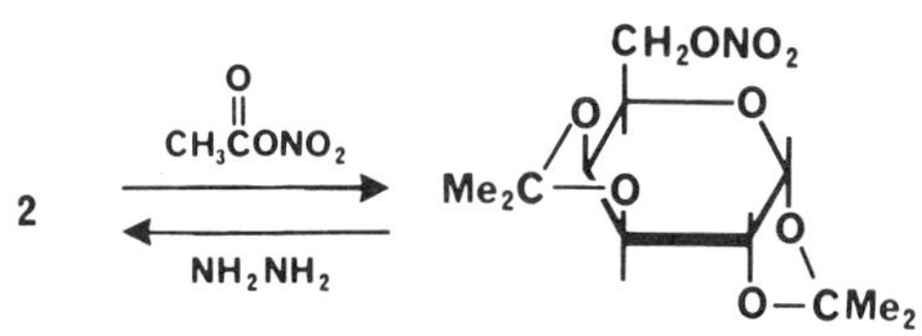

(43)

Utilization of these esters as protecting groups is fairly rare because they offer no real advantage over the more convenient carboxylic acid esters. One disadvantage in the use of nitrates is the complex set of reactions which takes place when they are treated with hydroxide or alkoxides (54). Another disadvantage of nitrates when compared to acetates or benzoates is that deprotection requires more forcing conditions. Refluxing a nitrate in a mixture of hydrazine (a powerful nucleophile) and ethanol is the best deprotection process (Equation 43).

D. Boronates

Among the boron-containing esters of carbohydrates, those derived from phenylboronic acid [$PhB(OH)_2$] have been used most effectively as protecting groups. Reaction of phenylboronic acid with methyl α-D-xylopyranoside [90] forms methyl α-D-xylopyranoside 2,4-phenylboronate [91] (Equation 44) (55). The C-3

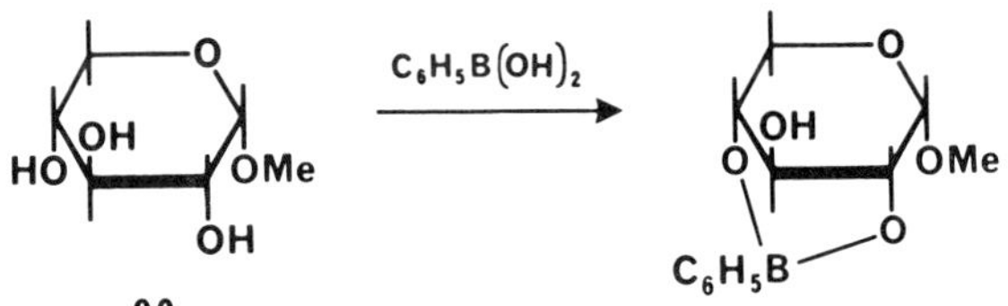

90 91 (44)

hydroxyl group in 91 can be oxidized or esterified prior to deprotection but utilization of boronate protection is limited by the susceptibility of these esters to hydrolysis and alcohololysis; in fact, boronates should be handled under anhydrous, alcohol-free conditions (56).

E. Orthoesters

A protecting group with structural similarity to acetals and ketals is the orthoester. The orthoester can be formed by reaction of a diol with ethyl orthoformate in the presence of an acid catalyst (Equation 45) (57). This type of protection generally is used only

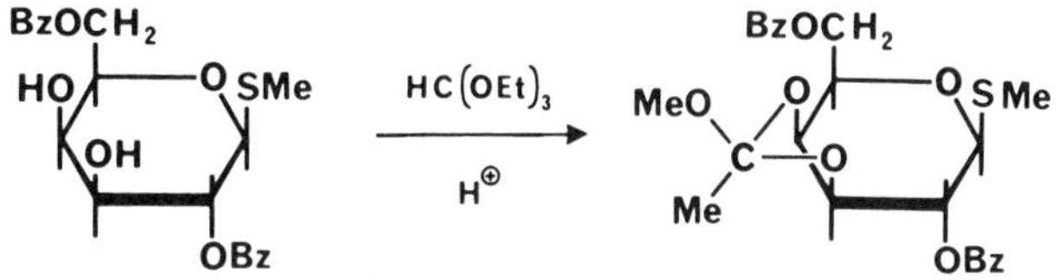

92 93 (45)

in situations where a ketal is unsatisfactory because the orthoester is much more acid labile and, consequently, deprotection occurs very easily. Controlled hydrolysis of the orthoester (Figure 12), however, can produce a partially protected compound (Equation 46)

93 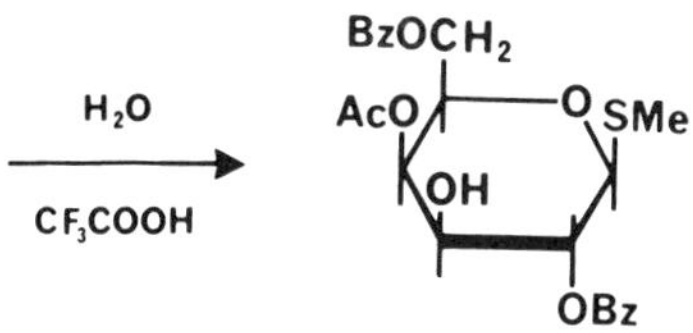 (46)

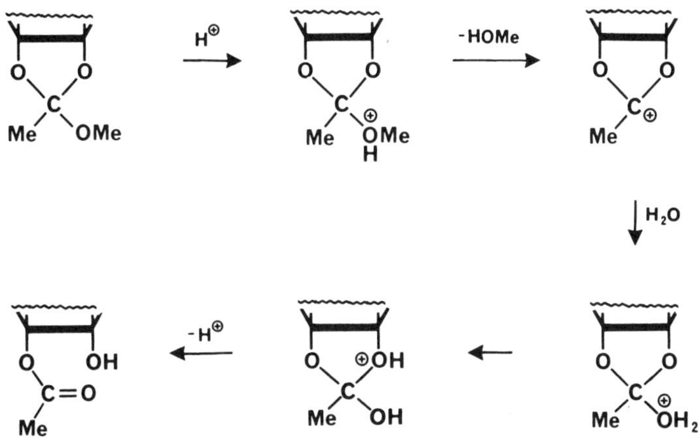

FIG. 12. Proposed mechanism for partial hydrolysis of an orthoester.

(57). Orthoesters are valuable intermediates in oligosaccharide synthesis (Chapter 14).

V. ETHERS

A. Benzyl Ethers

1. Reagents for Benzylation

Among common protecting groups the benzyl ether holds a position of considerable importance. It is stable over a broader range of conditions, including both acidic and basic media, than any other common protecting group; in addition, it is easily removed. The primary disadvantage of benzyl ethers is that strong bases generally are used in their introduction. A typical procedure for forming a benzyl ether is to react the compound being protected with sodium hydride in *N,N*-dimethylformamide (DMF) to form an alkoxide ion which then is treated with benzyl bromide (Equation 47) (58). Base-sensitive groups, such as esters, are

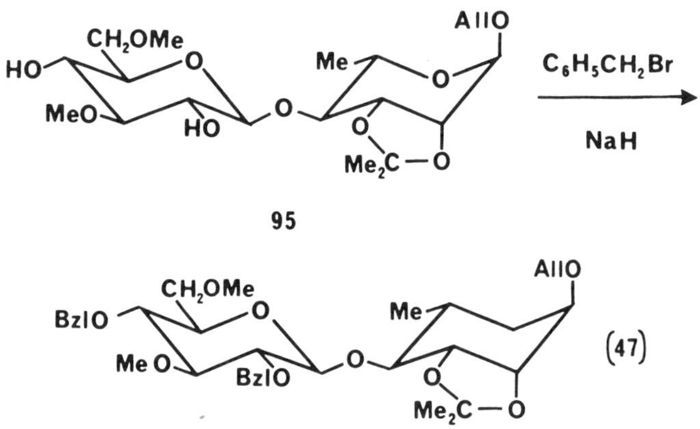

95

96

(47)

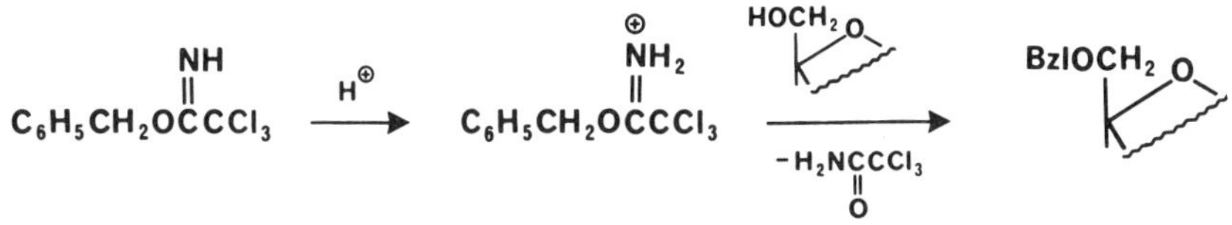

FIG. 13. Benzylation using benzyl trichloroacetimidate.

usually introduced into a molecule after benzylation to avoid deprotection.

If base-sensitive groups must be present during benzylation, several reagents can be used. Benzyl protection can be introduced using benzyl trichloroacetimidate in the presence of an acidic catalyst (Figure 13) (59). In addition, if benzylation must be conducted under essentially neutral conditions, the combination of a benzyl halide and silver triflate will introduce a benzyl group (Figure 14) (60).

2. Regioselective Benzylation

a. Direct Methods. Selectivity in the direct formation of benzyl ethers is generally more difficult to achieve than is selectivity in esterification reactions. Benzylation in the presence of phase transfer agents, such as tetrabutylammonium hydroxide, can lead to regioselective reaction, but the selectivity is not necessarily high. It is not surprising that regioselectivity is difficult to achieve

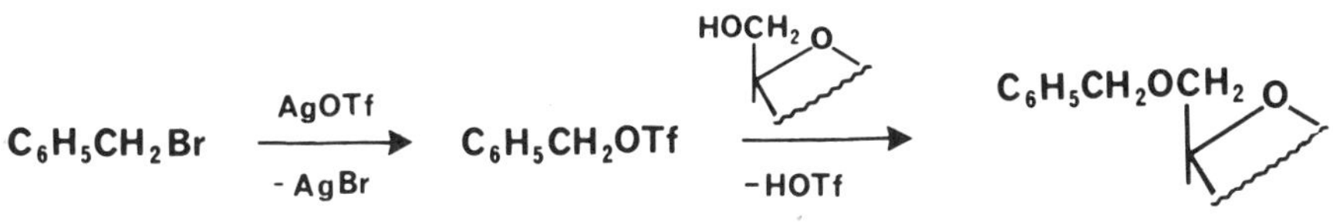

FIG. 14. Benzylation using silver triflate and benzyl bromide.

in such reactions because so many factors contribute to determining where reaction will occur. For instance, since alkoxide formation is the first step in benzylation, reaction of the hydroxyl group best able to form an anion should be favored (61). The favored anion may not be the one best suited for participation in the second reaction step, bromide displacement. These and other factors, such as the relative solubility of the various tetrabutylammonium alkoxides in the organic and aqueous phases, may favor reaction of different hydroxyl groups. Treatment of benzyl 2-acetamido-3-*O*-benzyl-2-deoxy-β-D-glucopyranoside [97] with benzyl bromide and tetrabutylammonium hydroxide gives a mixture of benzyl ethers from which the desired compound [98] is isolated in modest (43%) yield (Equation 48) (62).

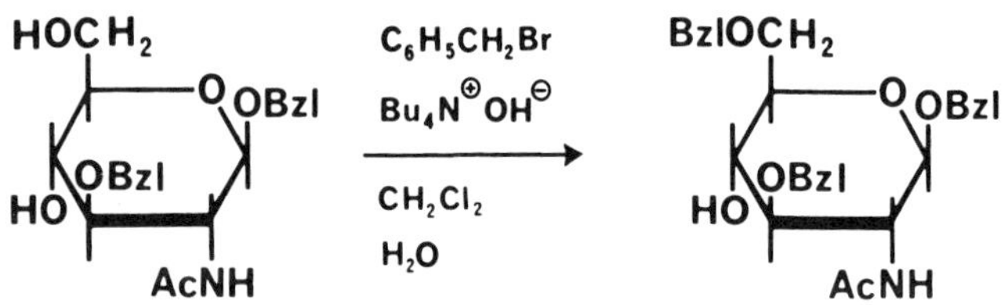

97 98 (48)

b. Indirect Methods. Regioselectivity in benzylation can be achieved by indirect methods. One of these, the reductive deprotection of benzylidene acetals (Equations 21 and 22), already has been discussed (Section III.C.). When reductive debenzylidenation was used to synthesize benzyl 2-acetamido-3,6-di-*O*-benzyl-2-deoxy-β-D-glucopyranoside [98] (Figure 15) (62), the yield (77%) was considerably better than that obtained from the phase transfer reaction (Equation 48). A second indirect method for regioselective benzylation requires temporary protection by reaction with an oxide of tin; thus, treatment of benzyl 2-acetamido-3-*O*-benzyl-2-deoxy-β-D-glucopyranoside [97] with

97  98

99

FIG. 15. Scheme for selective benzylation of benzyl 2-acetamido-3-*O*-benzyl-2-deoxy-β-D-glucopyranoside [97] by benzylidene acetal reduction.

dibutyltin oxide converted it into the corresponding cyclic stannyl derivative [100] (Figure 16) (62). Subsequent reaction of this compound [100] with benzyl bromide gave the 6-*O*-benzyl derivative [98] (Figure 16) in 65% yield.

3. Debenzylation

Deprotection of benzyl ethers, as with benzyl glycosides (Section II.B.), is nearly always accomplished by hydrogenolysis (Equation 49) (63). Deprotection thus occurs under mild

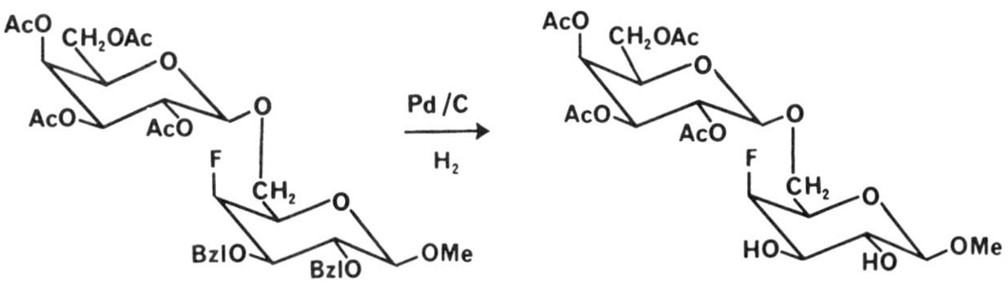 (49)

101 102

conditions. The disadvantage to this deprotection process is that compounds such as unsaturated carbohydrates can be hydrogenated and deoxyhalogeno sugars can experience hydrogenolysis

97 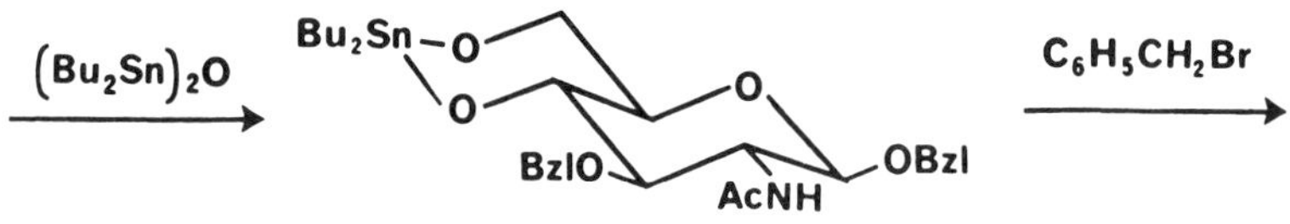 98

100

FIG. 16. Scheme for selective benzylation of benzyl 2-acetamido-3-*O*-benzyl-2-deoxy-β-D-glucopyranoside by reaction with dibutyltin oxide.

under these conditions. Such reactions normally present no serious limitation to the use of benzyl ethers as protecting groups, however, since these types of functional groups usually have undergone desired transformation prior to deprotection.

B. *p*-Methoxybenzyl Ethers

With respect to formation and most reactions, *p*-methoxybenzyl ethers differ little from benzyl ethers. There is, however, a significant difference between the two in possible reactions for deprotection. Both can be deprotected by hydrogenolysis but only the *p*-methoxybenzyl ether can also be removed by oxidation with 2,3-dichloro-5,6-dicyanobenzoquinone (DDQ) (Equation 50) (64).

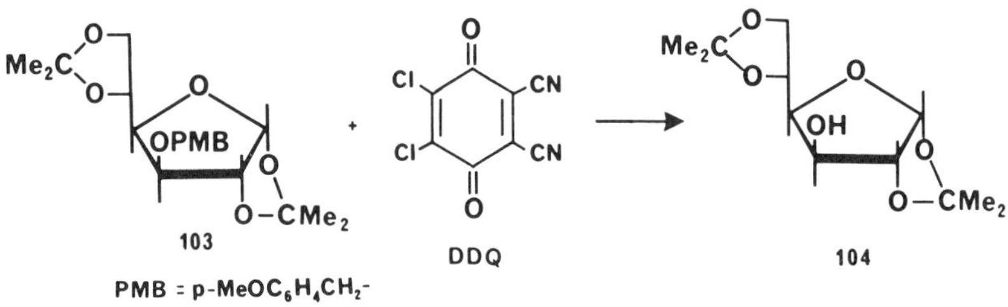

(50)

A proposed mechanism for DDQ oxidative deprotection is shown in Figure 17 (64). Cerium(IV) ammonium nitrate (CAN) is an

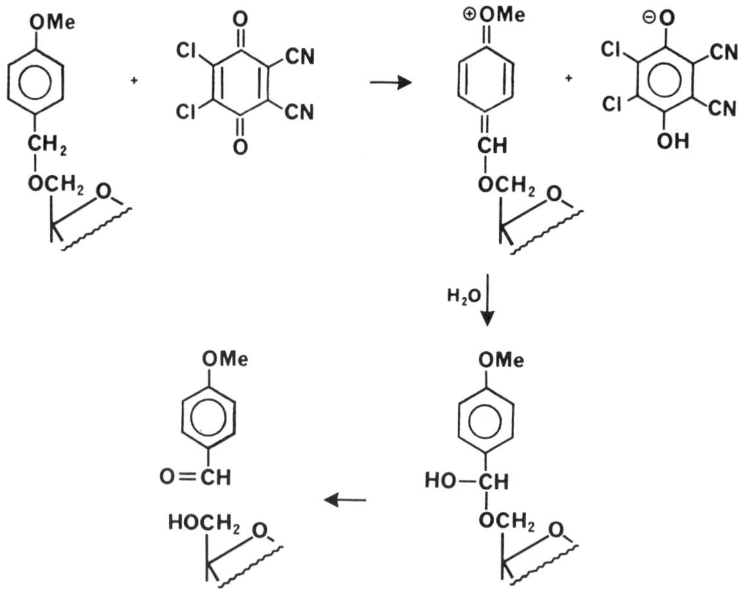

FIG. 17. Proposed mechanism for oxidative deprotection of a *p*-methoxybenzyl ether using 2,3-dichloro-5,6-dicyanobenzoquinone (DDQ).

oxidizing agent also capable of selectively reacting with *p*-methoxybenzyl ethers. The *p*-methoxybenzyl group is removed in 95% yield when methyl 2,3-di-*O*-benzyl-4-*O*-(*p*-methoxybenzyl)-α-D-glucopyranoside [105] is treated with CAN (Equation 51) (29).

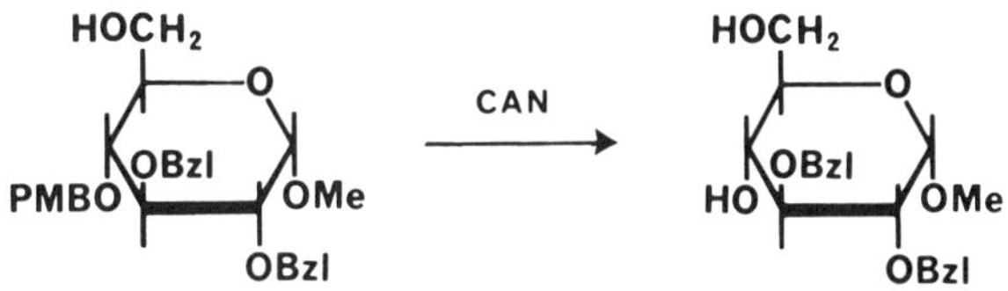

105 106 (51)

C. Trityl Ethers

Regioselective ether protection of primary hydroxyl groups in the presence of secondary or tertiary ones can be easily accomplished by reaction of the compound to be protected with chlorotriphenylmethane (trityl chloride) (Equation 52) (65). Ionization of trityl

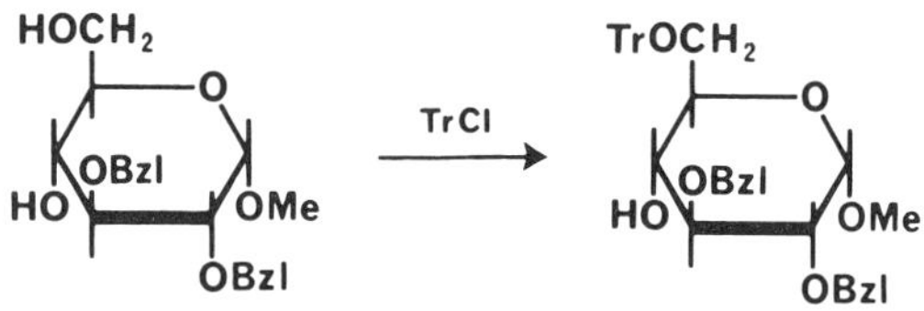

107 $Tr = C(C_6H_5)_3$ **108** (52)

chloride forms a resonance-stabilized cation which is sufficiently sterically hindered at the methyl carbon atom, so that reaction occurs exclusively with primary hydroxyl groups (Equation 53),

(53)

unless forcing conditions are used. Trityl ethers are readily deprotected by heating in aqueous acetic acid or by treatment with stronger acids without heating.

D. Allyl Ethers

Allyl ethers afford similar protection to that provided by benzyl ethers, but the presence of nonaromatic unsaturation imparts to them a reactivity which benzyl ethers do not have. Allyl ethers can experience simple electrophilic addition reactions. This represents a disadvantage in that the protecting group reacts with electrophilic reagents which do not attack the benzyl group. The advantage enjoyed by allyl protection over benzyl protection also stems from the presence of the nonaromatic unsaturation, which enables the allyl group to be removed under mild conditions that do not involve hydrogenolysis. Allyl ethers are formed by generation of alkoxide anions which then react with allyl bromide (Equation 54) (66). Deallylation can be achieved by the reaction

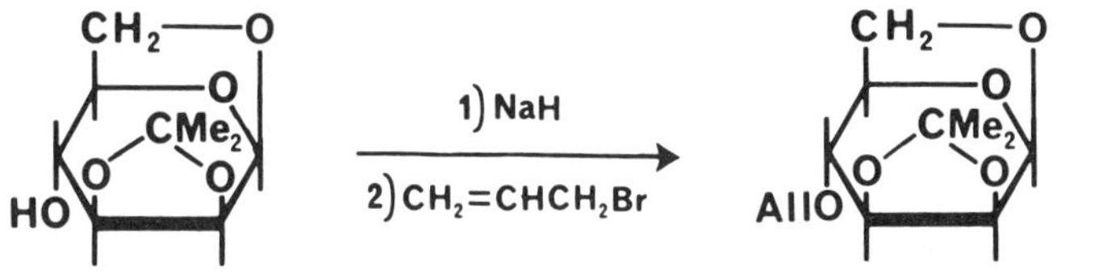

109 110 (54)

sequence discussed in Section II.B. (and pictured in Figure 1) for allyl glycosides. If the allyl protected compound is stable in the presence of strong bases, the isomerization of the double bond can be accomplished with potassium *tert*-butoxide in dimethyl sulfoxide (Equation 55) (58). Another significant factor in the use

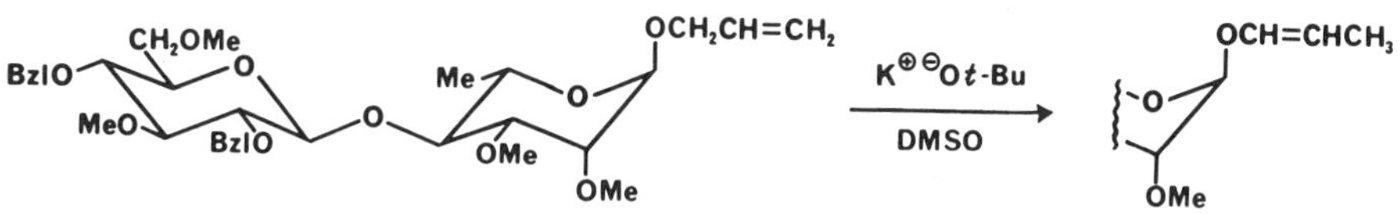

111 112 (55)

of allyl protection is that, like benzyl protection, it can be regioselectively introduced by treatment of a cyclic stannyl derivative with allyl bromide (Equation 56) (67).

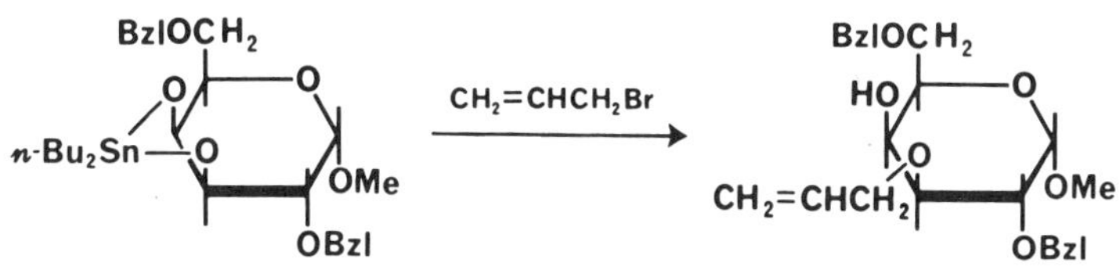

113 114 (56)

E. Methyl Ethers

Methylation is rarely used as a means of group protection in synthetic reactions (except, of course, formation of methyl glycosides) because of the difficulty associated with demethylation. Methyl ethers, however, are useful protecting groups in situations where deprotection is not necessary. These situations arise when, for example, a group of sugars is to be analyzed for identification and quantitation of the compounds present. Methylation of the entire mixture converts the sugars into derivatives which are more thermally stable and volatile than the free sugars themselves and, thus, are much better suited for analysis techniques such as gas chromatography/mass spectrometry.

The almost universally used method for methylation is the Hakomori procedure (68). In this procedure methyl sulfoxide is both a participant in and a solvent for the reaction. Methyl sulfoxide is first converted into the methylsulfinyl carbanion by reaction with a strong base such as potassium hydride or butyl lithium (Figure 18). This carbanion then removes a proton from an unprotected hydroxyl group. S_N2 displacement of iodide ion from methyl iodide completes the reaction.

F. Trialkylsilyl Ethers

Trialkylsilyl ether formation is a simple method for protecting hydroxyl groups. The most common silylating agent is

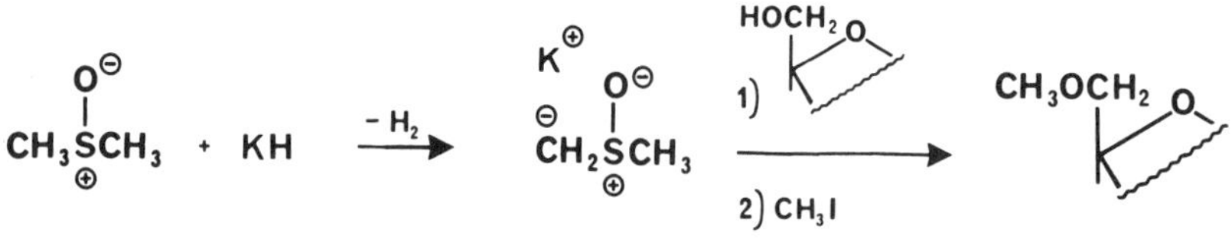

FIG. 18. Proposed mechanism for Hakomori methylation.

chlorotrimethylsilane, which reacts with alcohols (usually in the presence of a tertiary amine to neutralize the hydrochloric acid produced) to give the corresponding trimethylsilyl ethers (Figure 19) (62). Trimethylsilyl ethers are stable under neutral and basic conditions but undergo hydrolysis in the presence of aqueous acids (Figure 19) (62). A number of trialkylsilylating agents have been used in an effort to increase the stability of the ethers formed and to find silylating agents which react with greater regioselectivity. Selective reaction of primary hydroxyl groups can be achieved by the hindered silylating agent *t*-butylchlorodimethylsilane [119]. Reaction of 3-*O*-benzoyl-1,2-*O*-isopropylidene-α-D-glucofuranose [120] with 119 results in exclusive protection of

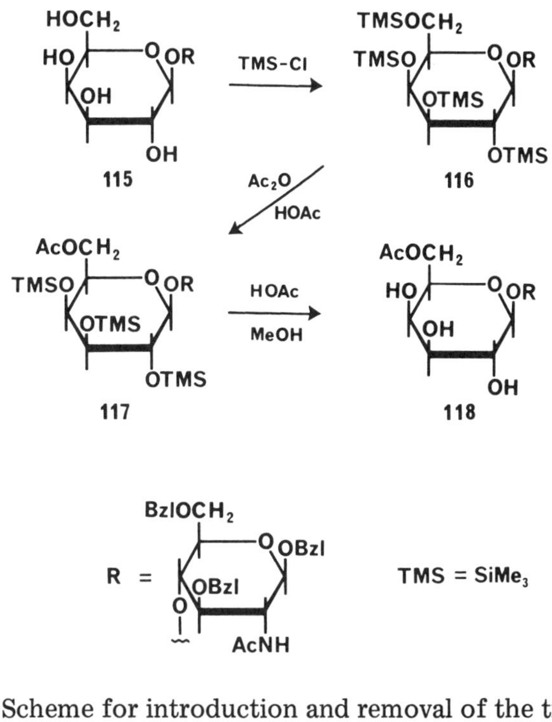

FIG. 19. Scheme for introduction and removal of the trimethylsilyl protecting group.

the C-6 hydroxyl group (Figure 20) (69). An effective and specific method for desilylation under basic conditions consists of reacting the silyl ether with tetrabutylammonium fluoride (Figure 20) (69).

Silylation greatly increases the volatility of compounds containing hydroxyl groups. Formation of trialkylsilyl ethers is a standard reaction prior to analysis of a mixture of sugars by gas chromatography/mass spectrometry.

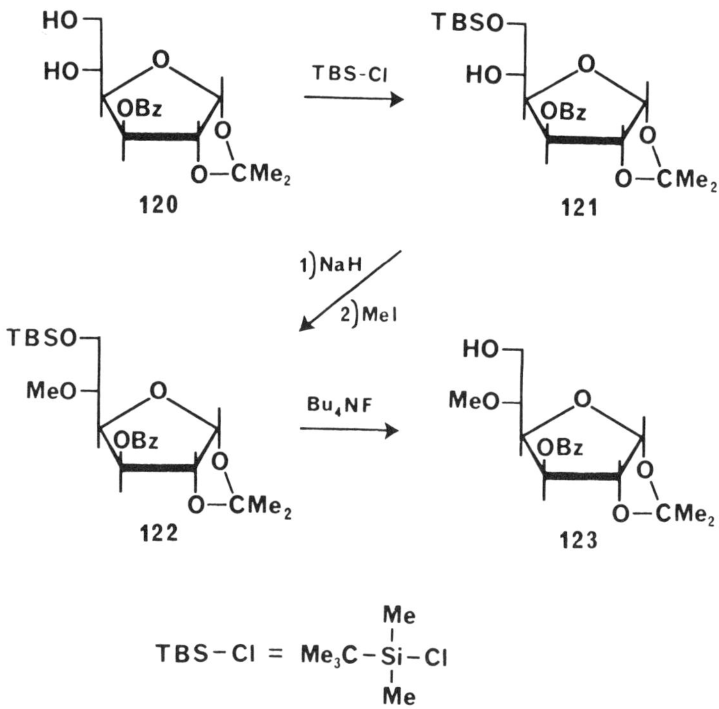

119

FIG. 20. Scheme for introduction and removal of the *t*-butyldimethylsilyl protecting group.

VI. AMIDE AND CARBAMATE PROTECTION

The reactions which have been discussed thus far have been for protecting hydroxyl groups. A second type of functional group present in many carbohydrates and requiring group protection is the amino group. Conversion of an amine into an amide (or related derivative such as a carbamate) protects it from reaction with most reagents. Regioselective reaction of an amine in the presence of hydroxyl groups can be realized because amines are more reactive toward acylating agents (i.e., amines are more nucleophilic than alcohols); thus, reaction of methyl 3-amino-2,3,6-trideoxy-α-D-*arabino*-hexopyranoside [124] with benzyl chloroformate [125] regioselectively protects the amino group (Equation 57) (70).

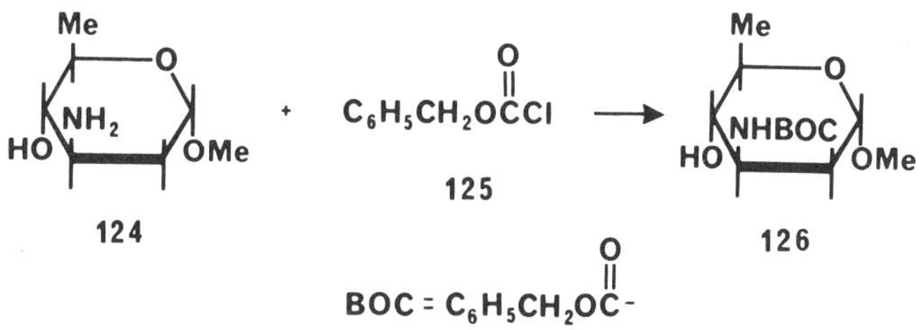

(57)

Esters are hydrolyzed by bases under much milder conditions than amides; consequently, treatment of a compound with sodium methoxide in methanol rapidly deprotects esters but not amides. The difficulty with protection of aminodeoxy sugars normally does not arise with the introduction of protecting groups but rather with their removal. Amide hydrolysis requires either heating in the presence of strong bases or prolonged heating in the presence of strong acids. The problems associated with subjecting carbohydrates to these harsh reaction conditions can be circumvented to a large extent by using the benzyloxycarbonyl (BOC) protecting group. This protecting group, which contains a benzyl ether, can be removed by hydrogenolysis (Equation 58) (71).

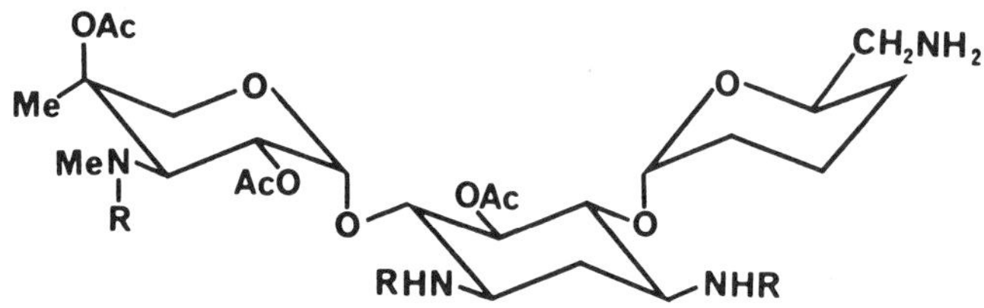

127 R = BOC ($C_6H_5CH_2OCO$)

128 R = H

$$127 \xrightarrow[H_2]{Pd/C} 128 \qquad (58)$$

VII. ANHYDRO COMPOUNDS

Hydroxyl groups can be protected by the formation of anhydro compounds. Even though the formal elimination of the elements of water from a molecule technically does not constitute the introduction of a protecting group, formation of an anhydro compound can be considered to be equivalent to providing hydroxyl group protection. Structures in which the anhydro linkage is between adjacent carbon atoms (e.g., a 2,3-anhydro sugar) are generally too reactive to afford adequate protection, but formation of larger, less strained rings can be effective. In practice the 1,6-anhydro linkage, such as that found in compounds 109 and 110 (Equation 54), is the most useful. The ring system containing the 1,6-anhydro linkage is sufficiently strain-free that reaction even in the presence of strong bases does not open the ring. Acid-catalyzed hydrolysis of a 1,6-anhydro compound restores the hydroxyl groups to C-1 and C-6; thus, 1,6-anhydro-3,4-di-*O*-benzyl-2-deoxy-2-fluoro-β-D-glucopyranoside [129] is hydrolyzed by heating in 50% methanesulfonic acid (Equation 59) (72).

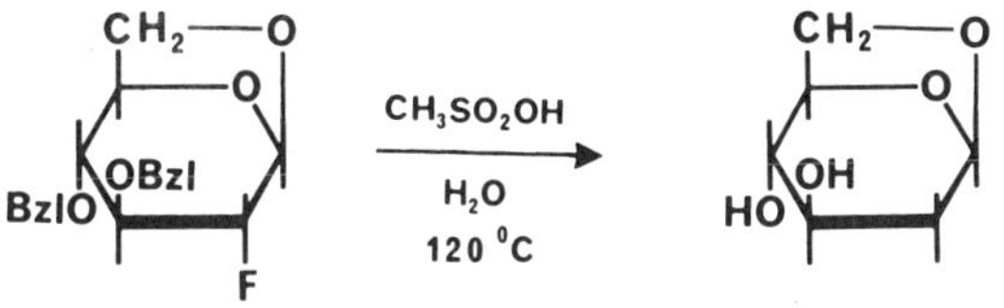

(59)

129

REFERENCES

1. T. W. Greene, *Protective Groups in Organic Synthesis*, John Wiley and Sons, New York, 1980.
2. O. T. Schmidt, *Methods Carbohydr. Chem.*, *1*, 191 (1962).
3. M. L. Wolfrom and A. Thompson, *Methods Carbohydr. Chem.*, *2*, 211 (1963).
4. H. G. Fletcher, Jr., *Methods Carbohydr. Chem.*, *2*, 228 (1963).
5. J. S. Brimacombe, *Methods Carbohydr. Chem.*, *6*, 376 (1972).
6. G. N. Bollenback, *Methods Carbohydr. Chem.*, *2*, 326 (1963).
7. N. K. Richtmyer, *Methods Carbohydr. Chem.*, *1*, 107 (1962).
8. G. O. Aspinall, D. Chatterjee, and L. Khondo, *Can. J. Chem.*, *62*, 2728 (1984).
9. I. I. Cubero, M. D. P. Olea, and D. G. Poza, *Carbohydr. Res.*, *134*, 327 (1984).
10. A. H. Haines, *Adv. Carbohydr. Chem. Biochem.*, *33*, 11 (1976).
11. A. N. de Belder, *Adv. Carbohydr. Chem. Biochem.*, *34*, 179 (1977).

12. D. M. Clode, *Chem. Rev.*, *79*, 491 (1979).
13. J. Gelas, *Adv. Carbohydr. Chem. Biochem.*, *39*, 71 (1981).
14. A. H. Haines, *Adv. Carbohydr. Chem. Biochem.*, *39*, 13 (1981).
15. R. W. Binkley and T. W. Flecthner in *Synthetic Photochemistry* (W. M. Horspool, ed.), Plenum, New York, 1984, p. 375.
16. H. G. Fletcher, Jr., *Methods Carbohydr. Chem.*, *2*, 386 (1963).
17. H. A. El-Shenawy and C. Schuerch, *Carbohydr. Res.*, *131*, 227 (1984).
18. H. A. El-Shenawy and C. Schuerch, *Carbohydr. Res.*, *131*, 239 (1984).
19. J. Thiem and P. Ossowski, *J. Carbohydr. Chem.*, *3*, 287 (1984).
20. J. Conchie and G. A. Levvy, *Methods Carbohydr. Chem.*, *2*, 335 (1963).
21. R. W. Binkley, *J. Carbohydr. Chem.*, *4*, 227 (1985).
22. A. Hasegawa, M. Ozaki, M. Kiso, and I. Azuma, *J. Carbohydr. Chem.*, *3*, 331 (1984).
23. J. Gelas and D. Horton, *Carbohydr. Res.*, *71*, 103 (1979).
24. M. L. Wolfrom, A. B. Dwadkar, J. Gelas, and D. Horton, *Carbohydr. Res.*, *35*, 87 (1974).
25. This procedure was developed by M. E. Evans, *Carbohydr. Res.*, *21*, 473 (1972).
26. P. J. Garegg and C. G. Swahn, *Methods Carbohydr. Chem.*, *8*, 317 (1980).
27. K. Takeo and K. Shinmitsu, *Carbohydr. Res.*, *133*, 13 (1984).
28. H. Sasai, K. Matsuno, and T. Suami, *J. Carbohydr. Chem.*, *4*, 99 (1985).
29. R. Johansson and B. Samuelsson, *J. Chem. Soc. Perkin Trans. I*, 2371 (1984).

30. S. S. Bhattacharjee and P. A. J. Gorin, *Can. J. Chem.*, *47*, 1195 (1969).
31. R. W. Binkley, G. S. Goewey, and J. C. Johnston, *J. Org. Chem.*, *49*, 992 (1984).
32. J. R. King and A. D. Allbutt, *Can. J. Chem.*, *47*, 1455, (1969) and *48*, 1754 (1970).
33. P. Deslongchamps, P. Atlani, D. Frehel, and A. Malaval, *Can. J. Chem.*, *50*, 2405 (1972).
34. D. H. R. Barton, D.-K. Zheng, and S. D. Gero, *J. Carbohydr. Chem.*, *1*, 105 (1982).
35. T. B. Grindley and C. Wickramage, *J. Carbohydr. Chem.*, *4*, 171 (1985).
36. M. L. Wolfrom, A. Beattie, S. S. Bhattacharjee, G. G. Parekh, *J. Org. Chem.*, *33*, 3990 (1968).
37. J. D. Wander and D. Horton, *Adv. Carbohydr. Chem. Biochem.*, *32*, 15 (1976).
38. D. Horton and O. Varela, *Carbohydr. Res.*, *134*, 205 (1984).
39. E. Tanahashi, M. Kiso, and A. Hasegawa, *J. Carbohydr. Chem.*, *2*, 129 (1983).
40. A. F. Cirelli, M. Sznaidman, L. Jeroncic, and R. M. de Lederkremer, *J. Carbohydr. Chem.*, *2*, 167 (1983).
41. R. D. Dawe and B. Fraser-Reid, *J. Carbohydr. Chem.*, *1*, 21 (1982).
42. P. Sinay, J.-C. Jacquinet, M. Petitou, P. Duchaussoy, I. Lederman, J. Choay, and G. Torri, *Carbohydr. Res.*, *132*, C5 (1984).
43. F. A. Carey and K. O. Hodgson, *Carbohydr. Res.*, *12*, 463 (1970).
44. S. Kim, H. Chang, and W. J. Kim, *J. Org. Chem.*, *50*, 1751 (1985).
45. K. Kinomura, M. Tamura, T. Oga, and H. Okai, *J. Carbohydr. Chem.*, *3*, 229 (1984).

46. C. P. J. Glaudemans and M. J. Bertolini, *Methods Carbohydr. Chem.*, *8*, 271 (1980).
47. N. V. Bovin, S. E. Zurabyan, and A. Y. Khorlin, *J. Carbohydr. Chem.*, *2*, 249 (1983).
48. Y. Ishido, N. Sakairi, M. Sekiya, and N. Nakazaki, *Carbohydr. Res.*, *97*, 51 (1981).
49. T. Yasumori, K. Sato, H. Hashimoto, J. Yoshimura, *Bull. Chem. Soc. Jpn.*, *57*, 2538 (1984).
50. R. Albert, K. Dax, A. E. Stutz, and H. Weidmann, *J. Carbohydr. Chem.*, *2*, 279 (1983).
51. H. A. El-Shenawy and C. Schuerch, *J. Carbohydr. Chem.*, *4*, 215 (1985).
52. S. Zen, S. Tashima, and S. Koto, *Bull. Chem. Soc. Jpn.*, *41*, 3025 (1968).
53. J. Honeyman and T. C. Stening, *J. Chem. Soc.*, 537 (1958).
54. J. Honeyman and J. W. W. Morgan, *Adv. Carbohydr. Chem.*, *12*, 117 (1957).
55. R. J. Ferrier, A. J. Hannaford, W. G. Overend, and B. C. Smith, *Carbohydr. Res.*, *1*, 38 (1965).
56. R. J. Ferrier, *Adv. Carbohydr. Chem. Biochem.*, *35*, 31 (1978).
57. P. J. Garegg and S. Oscarson, *Carbohydr. Res.*, *136*, 207 (1985).
58. R. Gigg, S. Payne, and R. Conant, *J. Carbohydr. Chem.*, *2*, 207 (1983).
59. T. Iversen and D. R. Bundle, *J. Chem. Soc. Chem. Commun.*, 1240 (1981).
60. J. M. Berry and L. D. Hall, *Carbohydr. Res.*, *47*, 307 (1976).
61. P. J. Garegg, T. Iversen, and S. Oscarson, *Carbohydr. Res.*, *50*, C12 (1976).
62. S. S. Rana, R. Vig, and K. L. Matta, *J. Carbohydr. Chem.*, *1*, 261 (1982).

63. P. Kovac and C. P. J. Glaudemans, *J. Carbohydr. Chem., 3*, 349 (1984).
64. Y. Oikawa, T. Tanada, K. Horita, T. Yoshioka, and O. Yonemitsu, *Tetrahedron Lett., 25*, 5393 (1984).
65. P. Kovac and C. P. J. Glaudemans, *J. Carbohydr. Chem., 2*, 313 (1983).
66. R. van Rijsbergen, M. J. O. Anteunis, and A. D. Bruyn, *J. Carbohydr. Chem., 2*, 395 (1983).
67. P. Weizel, H. P. Bullan, A. Maulahagen, D. Muller, and G. Snatzke, *Tetrahedron, 40*, 3657 (1984).
68. S. Hakomori, *J. Biochem. (Tokyo), 55*, 205 (1964).
69. M. Prashad and B. Fraser-Reid, *J. Org. Chem., 50*, 1564 (1985).
70. A. O. Stewart and R. M. Williams, *Carbohydr. Res., 135*, 167 (1984).
71. J. J. K. Wright and P. Lee, *J. Carbohydr. Chem., 1*, 157 (1982).
72. T. Haradahira, M. Maeda, Y. Kai, and M. Kojima, *J. Chem. Soc., Chem. Commun.*, 364 (1985).

9 Nucleophilic Substitution Reactions

I. THE IMPORTANCE OF NUCLEOPHILIC SUBSTITUTION REACTIONS

If it were necessary to select two or three types of reaction as the most important for modifying the structure of monosaccharides, one of these would be nucleophilic substitution. When one pictures the basic structure of carbohydrates, it is not surprising that substitution should be such an important process because every hydroxyl group in these polyhydroxy compounds is a potential leaving group. Since a hydroxyl group can be displaced from a molecule only with great difficulty, a prerequisite for its displacement is its conversion into a more reactive leaving group. A natural starting point for discussion of displacement reactions is, therefore, to consider how hydroxyl groups are activated. As a preface to this discussion, some general observations about nucleophilic substitution will be made.

II. GENERAL FACTORS INFLUENCING NUCLEOPHILIC SUBSTITUTION

A. Type of Carbon Atom Bearing the Leaving Group

From the point of view of reactivity, a typical monosaccharide contains three types of hydroxyl groups: anomeric, primary, and secondary. When primary and secondary hydroxyl groups are converted into effective leaving groups and are displaced, they nearly always experience S_N2 substitution (Figure 1); in contrast, when anomeric hydroxyl groups are transformed into leaving groups, they usually are involved in S_N1 reactions (Figure 2). Even though a leaving group attached to an anomeric center can experience S_N2 displacement, ionization to a resonance stabilized cation [1] is so facile that S_N1 substitution generally prevails. The mechanistic and reactivity differences between S_N1 and S_N2 substitution dictate that, with a few exceptions, the leaving groups most useful for reaction at the anomeric center are not the ones best suited for reaction at primary and secondary carbon atoms. As particular leaving groups are discussed later in this chapter, one of the items which will be considered is the ability of a group to depart from various positions in a molecule.

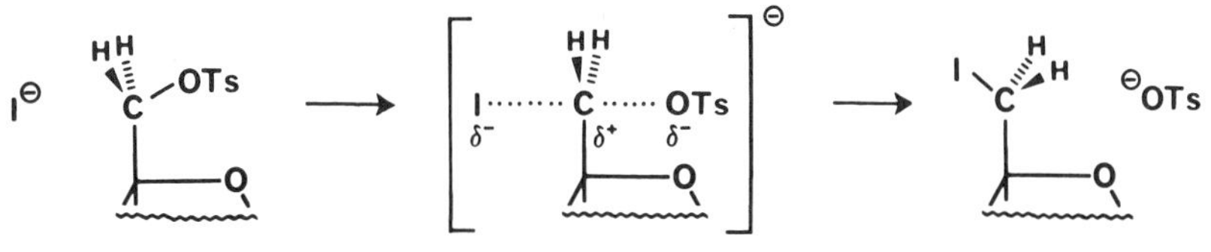

FIG. 1. Example of an S_N2 substitution in which tosylate is displaced by iodide.

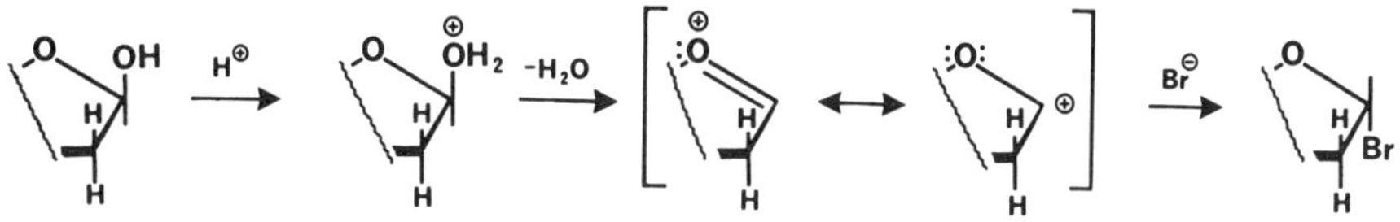

1

FIG. 2. Example of S_N1 substitution in which the protonated hydroxyl group is the leaving group and bromide is the nucleophile.

B. Stereoelectronic Effects

Displacement reactions at primary and secondary carbon atoms are significantly more difficult in carbohydrates than in less highly functionalized molecules. The reluctance of carbohydrates to participate in displacement reactions arises from a combination of steric and electronic effects. The most easily identified of these effects is one associated with the high degree of functionalization in a typical sugar or sugar derivative. With the large number of functional groups present in a carbohydrate, an approaching nucleophile is likely to encounter steric repulsion from at least one group. It is not only the size and position of a functional group that are important to an incoming nucleophile but also the atoms present in the substituent are influential. An electron-rich oxygen atom repels a nucleophile if the path that the nucleophile must take forces it into the vicinity of the oxygen atom. Consider the possible displacement of the tosyloxy group from the epimeric compounds 1,2:5,6-di-*O*-isopropylidene-3-*O*-*p*-tolylsulfonyl-α-D-glucofuranose [2] and 1,2:5-6-di-*O*-isopropylidene-3-*O*-*p*-tolylsulfonyl-α-D-allofuranose [3] (Figure 3). A nucleophile can approach C-3 in the allo compound [3] more easily than it can C-3 in the gluco epimer [2]. The reason for this difference is that

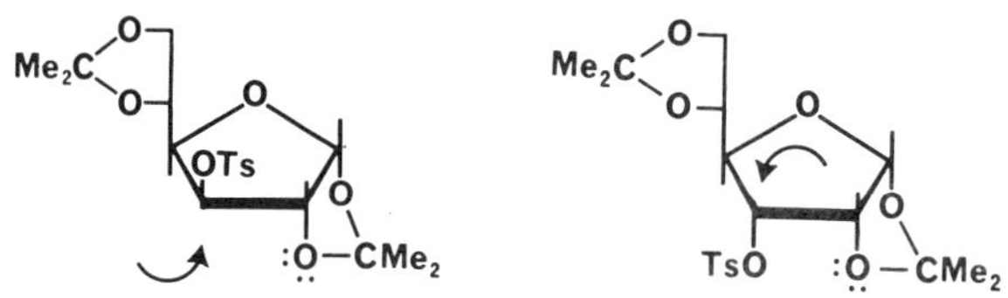

2 3

FIG. 3. Approach of a nucleophile to C-3 in 1,2:5,6-di-*O*-isopropylidene-3-*O*-*p*-tolylsulfonyl-α-D-glucofuranose [2] and 1,2:5,6-di-*O*-isopropylidene-3-*O*-*p*-tolylsulfonyl-α-D-allofuranose [3]. Arrows indicate pathways of nucleophiles.

in reacting with the allo tosylate [3], it is not necessary for the electron-rich nucleophile to pass near (and be repelled by) the unshared electrons on the oxygen atom attached to C-2. Since there are a large number of oxygen atoms present in a typical carbohydrate, electrostatic repulsion by unshared electrons on these atoms is a common phenomenon, one which contributes to the difficulty in leaving group displacement which carbohydrates experience.

The inherent electron demand of electronegative atoms is another factor which tends to make S_N2 substitution difficult in carbohydrates. In an S_N2 reaction (as in essentially all chemical reactions) there is more bond breaking early in the reaction than bond formation. This situation requires that at the transition state in an S_N2 process, such as the one shown in Figure 1, the bond between the reacting carbon atom and the leaving group is broken to a greater extent than the carbon-to-nucleophile bond is formed. (This net decrease in bonding at the transition state is the primary reason that the transition state structure is energetically less stable than that for either the reactants or products.) Unequal bond breaking and bond formation at the transition state mean that electron density has been taken from the reacting carbon atom by the leaving group to a greater extent than it has been supplied by the nucleophile; therefore, the carbon atom experiencing substitution develops a partial positive charge (Figure 1). This charge development places an inductive, electron-withdrawing demand on the rest of the molecule. It is difficult for a carbohydrate to meet this demand due to the large number of electronegative oxygen atoms which are part of the carbohydrate structure. This situation makes the development of a partial positive charge on the reacting carbon atom difficult. Such difficulty acts in concert with the steric and electronic repulsion created by the remainder of the molecule to cause S_N2 reactions to have higher energies of activation in carbohydrates than in most other molecules.

C. Reactivity at Various Carbon Atoms in Pyranoid and Furanoid Ring Systems

The observations which have been made about stereoelectronic effects permit some conclusions to be drawn about S_N2 substitution at various positions in a pyranoid ring. Consider methyl α-D-glucopyranoside [4] as a representative compound containing a pyranoid ring (Figure 4), and assume that the displacement

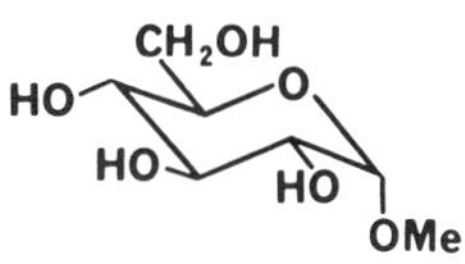

4

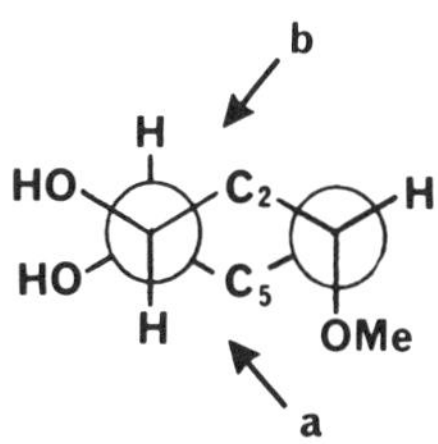

4

FIG. 4. (Left) Methyl α-D-glucopyranoside [4]. (Right) Approach of a nucleophile to C-3 (path a) and C-4 (path b) in the methyl α-D-glucopyranosyl system.

reactions under discussion take place after conversion of one of its four hydroxyl groups into an effective leaving group. (A leaving group attached to C-6 will be included in this discussion even though C-6 is not a ring carbon atom.) Displacement of a group from C-6 will occur most easily. Steric hindrance at this primary carbon atom (C-6) will be less than at other carbon atoms in the molecule and unfavorable electronic interaction between the nucleophile, and electronegative atoms also will be less important because these atoms are more remote from C-6 than from the other possible reaction sites.

In general, displacement at C-2 will be the most difficult. Since C-2 is bonded to the acetal carbon atom (C-1), it has more nearby oxygen atoms than C-3, C-4, or C-6; consequently, electron withdrawal from C-2 will be greater than from these other three atoms. In an S_N2 displacement reaction, which requires a build-up of positive charge on the reacting carbon atom as the transition state is reached (Figure 1), C-2 will be the carbon atom least able to accept such a charge; thus, the transition state for displacement at C-2 will be the most difficult to reach.

Displacement of leaving groups from C-3 and C-4 would appear to be approximately equal in difficulty if the α-methoxy group were not present at C-1. The α-methoxy group hinders the approach of a nucleophile to C-3 (path a, Figure 4) but not C-4 (path b, Figure 4); thus, in the methyl α-D-glucopyranosyl system displacement of a group at C-4 should be second in ease only to displacement of one at C-6.

An important aspect of the generalizations made in the previous paragraph concerns the extent to which they hold for other compounds. Do these generalizations apply to other molecules with pyranoid rings and to compounds with furanoid rings? The answer to this question is that two of these observations can be transferred to other molecules, including those with furanoid rings.

S_N2 reaction at a primary position in a molecule is nearly always the easiest and displacement from a position next to an anomeric carbon atom is usually the most difficult. S_N2 substitution at other positions is too dependent upon the particular substituents present and the stereochemistry at various ring carbon atoms to allow generalization. Finally, it should be noted that these comments about relative reactivity only pertain to bimolecular reactions. Internal displacements usually occur much more rapidly than other substitution reactions. Some examples showing the differences in ease of intra- and intermolecular displacement are given later in this chapter (Section III.A.1.).

D. Neighboring Group Participation (Anchimeric Assistance)

Neighboring group participation (anchimeric assistance) is not a particularly important process in synthetic transformations involving monosaccharides. It is, however, quite significant to oligosaccharide synthesis; consequently, discussion of this topic will be delayed until Chapter 14 (Oligosaccharide Syntheses).

III. LEAVING GROUPS

A. Stable Compounds Which Contain Effective Leaving Groups

1. Sulfonic Acid Esters

a. *p*-Toluenesulfonates (Tosylates) and Methanesulfonates (Mesylates). The *p*-tolylsulfonyloxy and methylsulfonyloxy groups are among the most widely used leaving groups in carbohydrate chemistry. They are introduced easily and quantitatively by reaction of a partially protected sugar either with *p*-toluenesulfonyl chloride or methanesulfonyl chloride at or below room temperature. Since primary hydroxyl groups are more rapidly esterified than secondary ones, regioselective reaction is possible

(Equation 1) (1). Pyridine commonly serves both as a solvent for reaction and as a base to neutralize the hydrogen chloride formed.

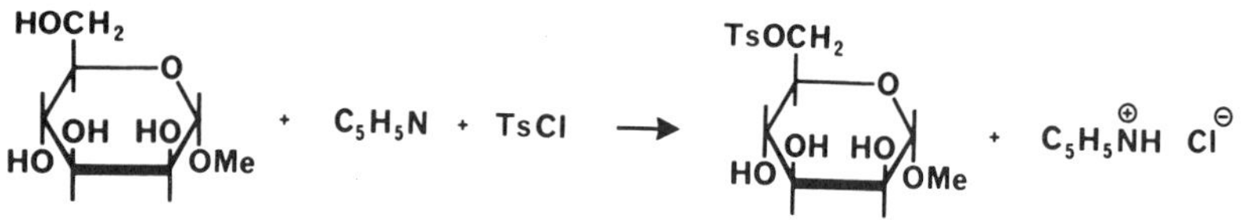

4 5 (1)

Pyridine also acts to establish an equilibrium in the solution in which a low concentration of the much more reactive alkoxide ion of the sugar is present (Equation 2).

$$C_5H_5N + ROH \rightleftharpoons C_5H_5\overset{\oplus}{N}H + \overset{\ominus}{O}R \qquad (2)$$

For compounds in which sulfonylation is particularly difficult, it can become necessary either to raise the temperature of the reaction mixture or to increase the alkoxide ion concentration in order for esterification to occur at an appreciable rate. Adding 4-(*N*,*N*-dimethylamino)pyridine [6] (DMAP) to a reaction mixture increases the rate of esterification by increasing the concentration of alkoxide ion present. DMAP is a stronger base than pyridine because it forms a resonance-stabilized pyridinium ion [7] (Equation 3). The formation of this more stable ion shifts the

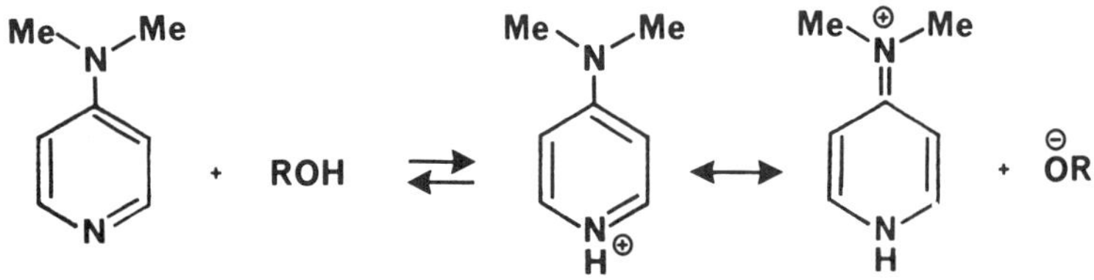

6 7 (3)

equilibrium shown in Equation 3 to the right (when compared to the equilibrium shown in Equation 2), thus increasing the alkoxide ion concentration. An example of the value of DMAP in sulfonylation is found in the reaction of methyl 3-*O*-benzoyl-4,6-*O*-benzylidene-2-*C*-methyl-α-D-glucopyranoside [8] with mesyl chloride, an esterification which requires the addition of DMAP [6] for formation of the tertiary mesylate (Equation 4) (2).

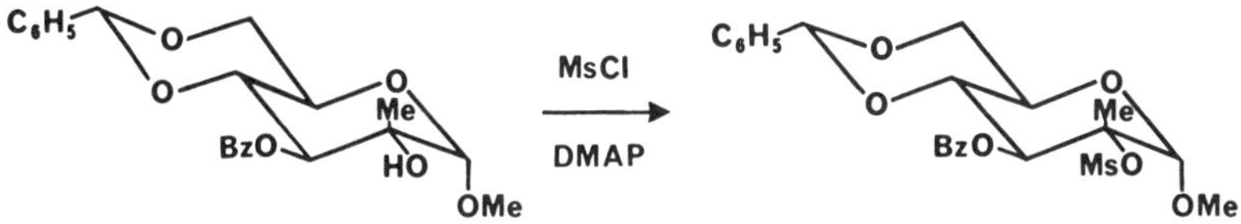

(4)

8 **9**

Another method for increasing the concentration of alkoxide ion present in a reaction mixture is through phase transfer catalysis using tetrabutylammonium hydroxide. Reaction of methyl 2-*O*-benzoyl-4,6-*O*-benzylidene-3-*C*-methyl-α-D-glucopyranoside [10] with tosyl chloride in dichloromethane-water in the presence of tetrabutylammonium hydroxide results in the formation of the tertiary tosylate 11 (Equation 5) (2). The

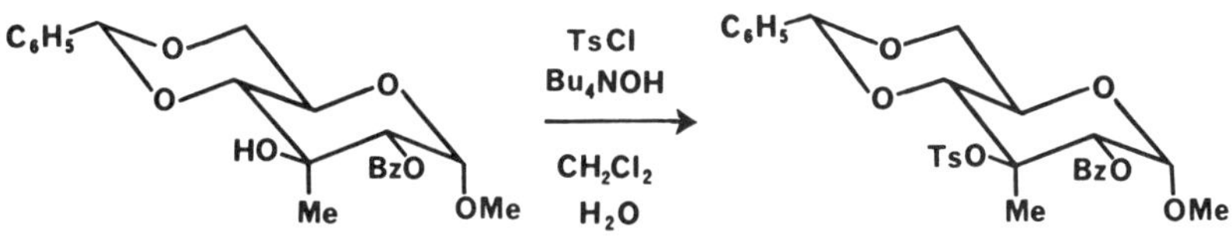

(5)

10 **11**

tetrabutylammonium salt 12 is soluble in dichloromethane where it is easily esterified by tosyl chloride.

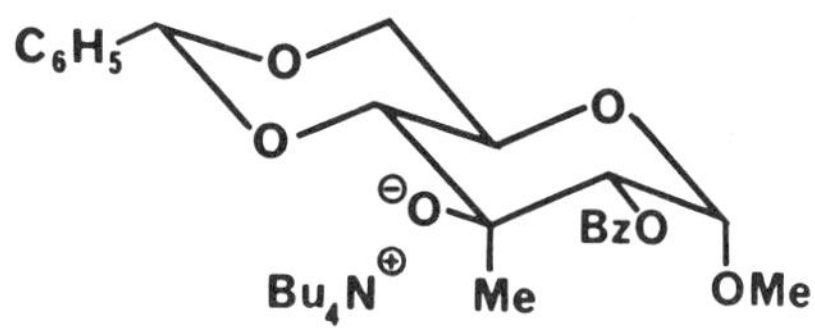

12

There is not a large difference in ease of displacement of tosyloxy and mesyloxy groups. They are both sufficiently reactive to allow substitution relatively easily when they are attached to primary carbon atoms. A typical example is the reaction of 3-*O*-acetyl-1,2-*O*-isopropylidene-5-*O*-*p*-tolylsulfonyl-α-D-ribofuranose [13] with sodium iodide in butanone at 80° C, a reaction which requires a single hour for completion (Equation 6) (3).

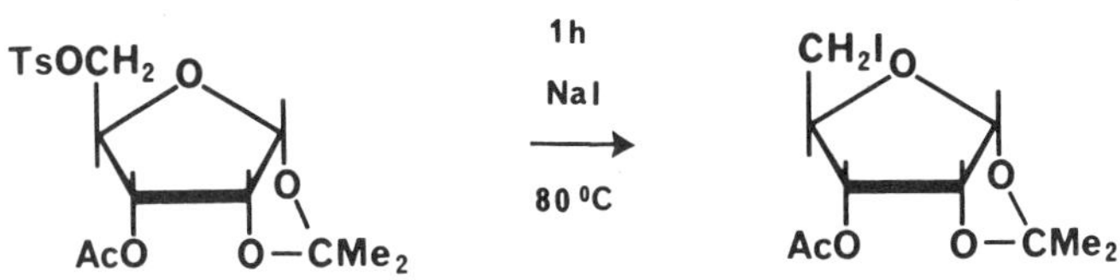

13 14

(6)

Exceptions to facile reaction of primary tosylates (and mesylates) do exist. Heating 1,2:3,4-di-*O*-isopropylidene-6-*O*-*p*-tolylsulfonyl-α-D-galactopyranose [15] with sodium iodide in acetone in a sealed tube requires 36 hours at 110° C for complete reaction (Equation 7) (4). Difficulty in displacement at C-6 in the galacto

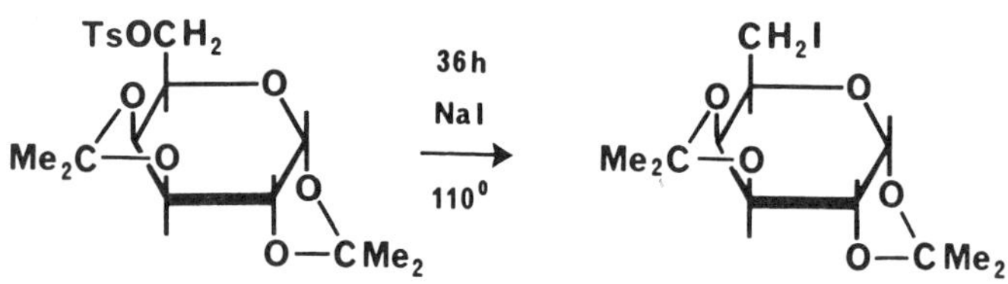

15 16

(7)

FIG. 5. Newman projection along the C-5, C-6 bond in the two significantly populated conformations of 1,2:3,4-di-*O*-isopropylidene-6-*O*-*p*-tolylsulfonyl-α-D-galactopyranose [15].

system is understandable in terms of the stereoelectronic effects discussed in Section II.B. of this chapter. Consider the two most heavily populated conformers (Figure 5) arising from rotation about the C-5, C-6 bond in the tosylate 15. In the conformation shown on the left, the nucleophile is repelled by the oxygen atom attached to C-4, while in the conformation on the right, the nucleophile must follow a path which causes it to interact with the ring oxygen atom. Stereoelectronic interactions in each case discourage substitution.

Secondary tosylates and mesylates vary considerably in their reactivity. Intramolecular reaction of the secondary tosylate 1,6-anhydro-4-*O*-(4,6-*O*-isopropylidene-β-D-glucopyranosyl)-2-*O*-*p*-tolylsulfonyl-β-D-galactopyranose [17] leading to the epoxide 18 occurs with ease, requiring only treatment with sodium methoxide for one hour at room temperature (Equation 8) (5). Most intermolecular reactions of secondary tosylates and mesylates require forcing conditions involving some combination of high temperatures, long reaction times, and dipolar aprotic solvents. For example, displacement of the mesyloxy group from methyl 2-acetamido-4,6-*O*-benzylidene-2-deoxy-3-*O*-methylsulfonyl-α-D-allopyranoside [19] by azide ion requires heating at reflux for eight days in dimethoxyethane (bp 160° C) (Equation 9) (2). Some

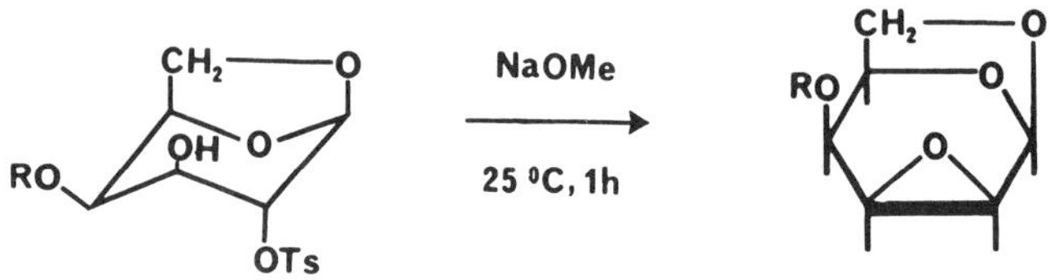

(8)

17 18

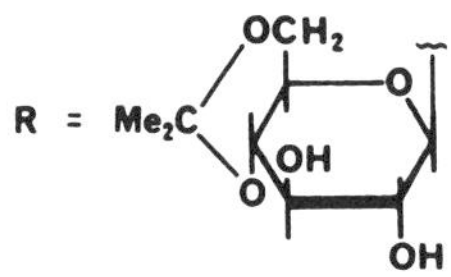

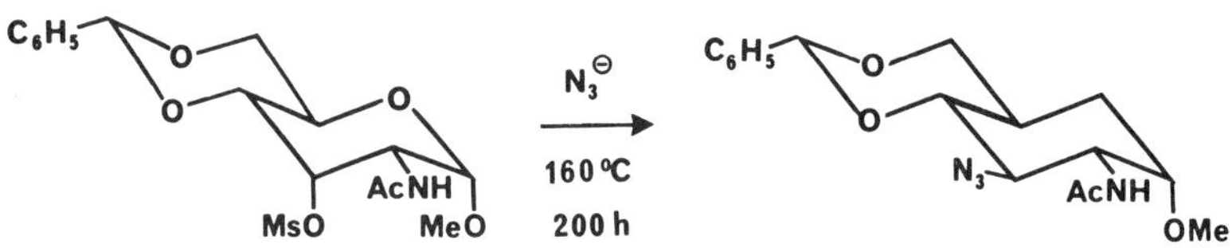

(9)

19 20

tosylates (e.g., compound 2) and mesylates require such drastic conditions for reaction that decomposition occurs before displacement with most nucleophiles. As a result of the relatively low reactivity of secondary tosyloxy and mesyloxy groups in intermolecular substitution reactions, more reactive leaving groups generally are used when displacement at secondary positions is required.

Sulfonates are rarely used as leaving groups at the anomeric position unless a glycosyl halide is desired (Equation 10) (6). Although anomeric tosylates and mesylates have been prepared, their normal method of preparation (from tosyl or mesyl chloride) generates chloride ion which can displace the sulfonyloxy group even under mild conditions (Equation 10) (6).

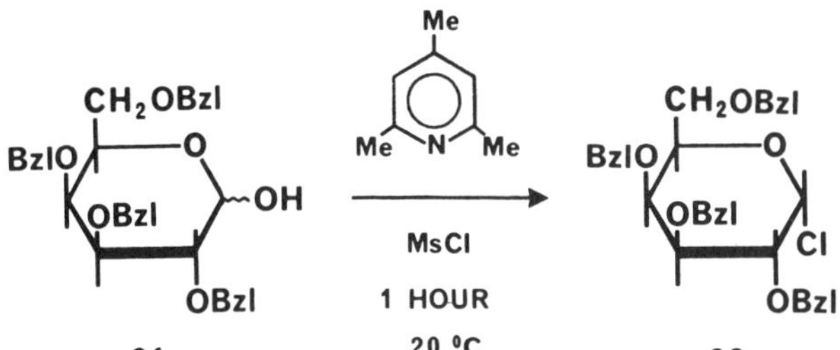

21 22 (10)

b. **Trifluoromethanesulfonates (Triflates).** Trifluoromethanesulfonates are sufficiently reactive to allow displacement at both primary and secondary positions. These esters are easily formed by reaction of the partially protected sugar with trifluoromethanesulfonic (triflic) anhydride. Since triflates are thousands of times more reactive than tosylates or mesylates (7), displacement at secondary positions does not require drastic reaction conditions; in fact, triflate displacement often occurs at room temperature or slightly above. 1,2,3,4-Tetra-*O*-acetyl-6-*O*-trifluoromethylsulfonyl-β-D-glucopyranose [23] actually experiences substitution below 0°C (Equation 11) (8). The

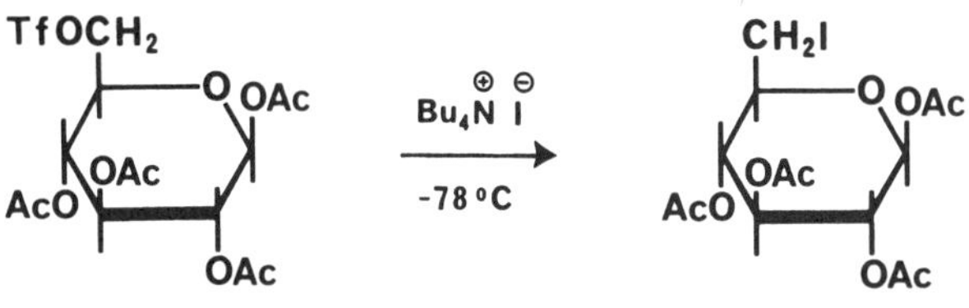

23 24 (11)

substitution reaction shown in Equation 12, which cannot be achieved for the corresponding tosylate 2 under any conditions, takes place for the triflate 25 in boiling benzene (9), while the displacement of the secondary triflyloxy group from 1,6-anhydro-4-*O*-allyl-3-*O*-benzoyl-2-*O*-trifluoromethylsulfonyl-α-D-mannose

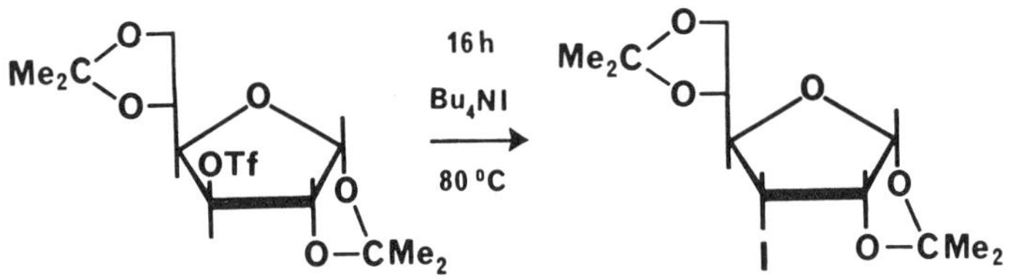

(12)

25 26

[27] with azide ion (Equation 13) occurs at room temperature (10). A number of additional examples of the displacement of

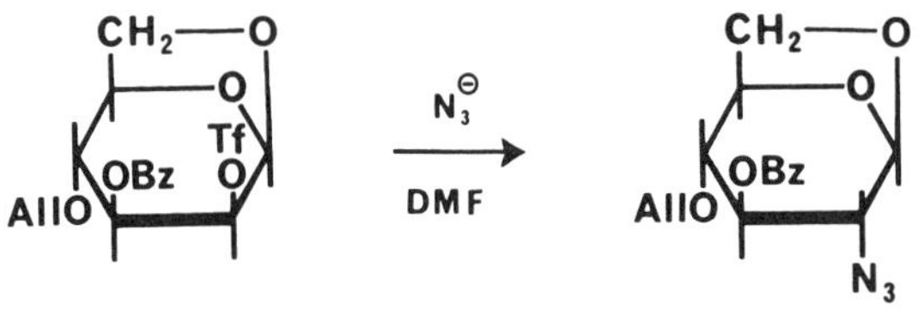

(13)

27 28

the triflyloxy group can be found in a recent review of the chemistry of carbohydrate triflates (11).

The unusually facile displacement of the triflyloxy group is directly related to the stability of the departing anion. Triflate is one of the most stable of all anions because its negative charge is dispersed by the considerable electron-withdrawing ability of the trifluoromethyl group. One measure of the stability of the triflate anion is that triflic acid is one of the strongest acids known. Another measure is that the desire of the triflyloxy group to become a triflate ion is so great that anomeric triflates have never been isolated.

c. **Other Sulfonates.** A number of other sulfonic acid esters have been used, although rather infrequently, in displacement reactions involving carbohydrates. These esters include the *p*-bromobenzenesulfonate [29], the *p*-nitrobenzenesulfonate [30], the 2,2,2-trifluoroethanesulfonate [31], and the imidazolesulfonate [32]. The first three of these have similar or

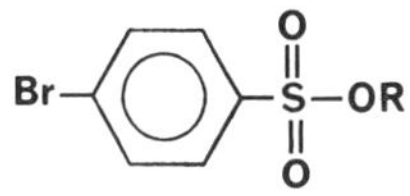

29

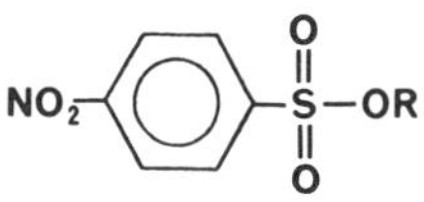

30

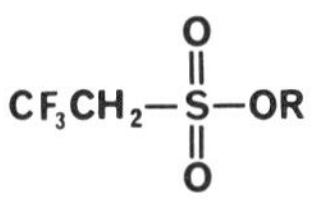

31

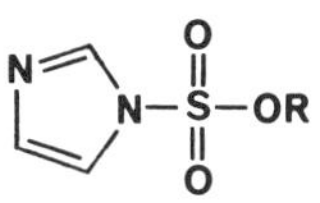

32

slightly greater reactivity than mesylates and tosylates. The imidazolesulfonate approaches the triflate in reactivity (12).

2. Halogens

a. Deoxyhalogeno Sugars. Iodide and bromide ions can be displaced from primary positions if powerful nucleophiles are used. Since an iodine atom normally is introduced at a primary position by displacement, iodide functions as a leaving group only in special situations. For instance, methyl 2-acetamido-4-*O*-benzoyl-6-bromo-2,3,6-trideoxy-α-D-*ribo*-hexopyranoside [33] is produced by a method [*N*-bromosuccinimide bromination, described later in this chapter (Section IV.B.)], which cannot be used to generate the corresponding iodide [34]. Since, however, the iodide 34 is a better starting material for the next reaction in the sequence (an elimination process), displacement of bromide is accomplished by heating 33 with sodium iodide in boiling butanone (Equation 14) (13).

33 34 (14)

b. Glycosyl Halides. Although halogens are of limited value as leaving groups from nonanomeric carbon atoms, they are extremely important ones from the anomeric position. Their importance derives from their widespread use in the synthesis of oligosaccharides; consequently, detailed discussion of the reactivity of glycosyl halides will be delayed to Chapter 14 (Oligosaccharide Syntheses), although a few general comments will be made in the following paragraph. Due to their importance in glycoside synthesis, it is not surprising that a wide variety of methods exist for glycosyl halide formation; in fact, most of the reagents which activate groups at the anomeric position cause the formation of glycosyl halides. These reagents will be discussed later in this chapter (Sections III.B.4. and III.B.5.).

Anomeric iodides are too unstable to be of significant synthetic value; however, chlorides and bromides both form reasonably stable yet reactive compounds. Ionization of a glycosyl bromide or chloride in a reaction mixture is usually catalyzed by the addition of a silver or mercury salt. These metal ions complex with the halide to form an intermediate [35] which rapidly ionizes (Figure 6). Even glycosyl fluorides, which are much more stable than the corresponding chlorides and bromides, are effective leaving groups in the presence of activating agents (14).

3. Epoxides

Epoxides experience ring opening upon reaction with nucleophiles. The relief of ring strain is sufficiently great that reaction

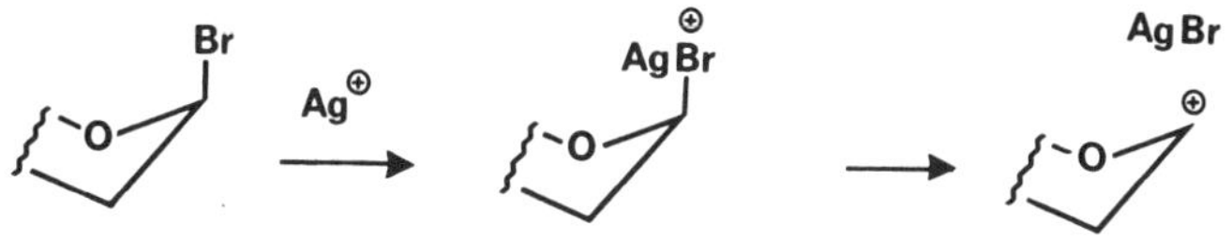

35

FIG. 6. Proposed mechanism for reaction between a glycosyl bromide and silver ion.

with effective nucleophiles takes place under relatively mild conditions; thus, when 1,6:2,3-dianhydro-4-*O*-(4,6-*O*-isopropylidene-β-D-glucopyranosyl)-β-D-mannose [<u>18</u>] is heated with sodium azide in 60% ethanol under reflux for 38 hours, the corresponding azide [<u>36</u>] is formed (Equation 15) (5). The

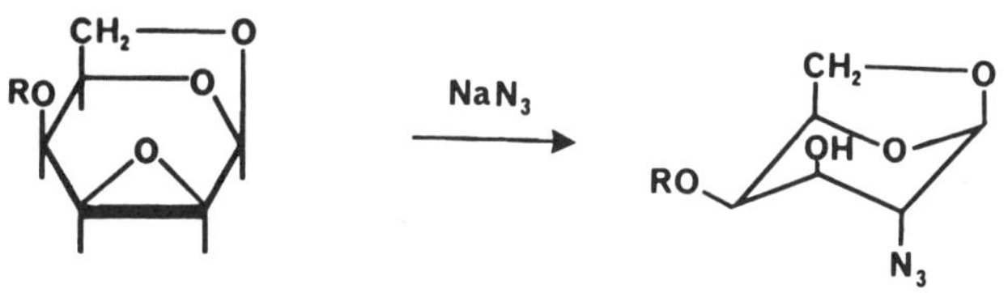

(15)

18 36

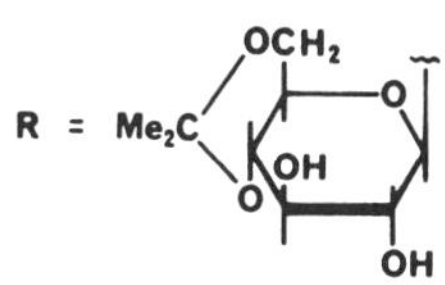

combination of epoxide formation by internal S_N2 displacement (Equation 8) followed by ring opening by a nucleophile (Equation 15), a second S_N2 process, is a useful reaction sequence for achieving a substitution with retention of configuration. Configuration is retained in this reaction sequence because there are definite stereochemical requirements placed on substitution reactions which proceed by epoxide formation and subsequent ring opening.

When an epoxide is formed by internal S_N2 displacement, the leaving group and the nucleophile must be: (a) on adjacent carbon atoms, (b) on opposite sides of the molecule, and (c) in the same plane or, at least, nearly coplanar. If the nucleophile and the leaving group are attached to a pyranoid ring, then the ring needs to be able to assume a conformation in which both groups are axial. For a compound with a 1,6-anhydro linkage, such as that found in the tosylate 17 (Equation 8), the ring is held in the proper conformation for reaction. These stereochemical requirements demand that for the conformationally mobile methyl 2,6-dideoxy-3-*O*-*p*-tolylsulfonyl-α-D-*arabino*-hexopyranoside [37] (15), reaction must take place when 37 or its corresponding alkoxide anion is in a conformation which is not heavily populated (Figure 7).

Ring opening of an epoxide also has definite steric requirements. The approaching nucleophile attacks the ring in such a

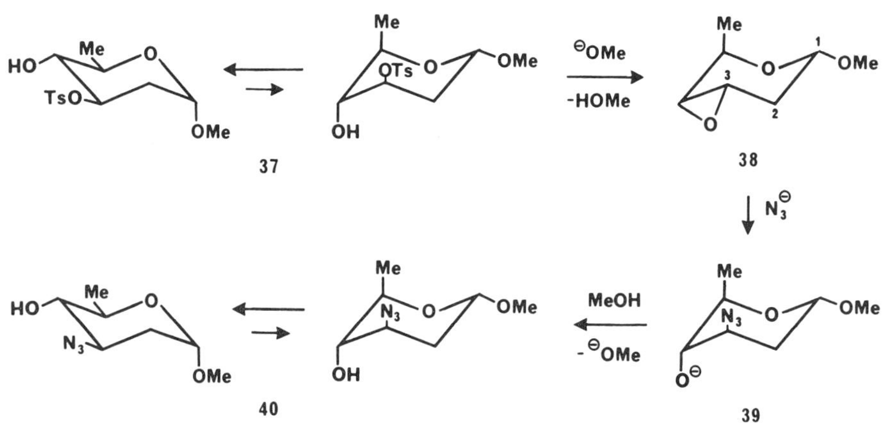

FIG. 7. Formation of an epoxide by internal S_N2 substitution. Ring opening of an epoxide by S_N2 substitution with azide ion.

fashion that the two substituents produced by ring opening have a *trans*, coplanar relationship; that is, ring opening follows a pathway which is essentially the reverse of that for ring formation (Figure 7). In a conformationally rigid molecule such as 18, reaction only at C-2 can open the ring to give two axial substituents; thus, the attack of the nucleophile is highly regioselective. In conformationally mobile epoxides, such as methyl 3,4-anhydro-2,6-dideoxy-α-D-*ribo*-hexopyranoside [38], either one of the C-0 bonds of the epoxide can be attacked. Stereoelectronic factors not only determine the point of attack but dictate that ring opening occur in such a manner as to give (at least initially) two axial substituents. Opening of an epoxide to give two axial substituents is described as a ring opening which obeys the Furst-Plattner rule.

B. Reagents Which Generate Transitory Leaving Groups

Some of the most effective leaving groups are ones which have only a transitory existence. Certain combinations of reagents generate such groups which then are displaced by a nucleophile also present in the reaction mixture. It is difficult in some cases to determine the precise structure for the leaving group and the structures for the intermediates involved in the reaction but in most cases reasonable proposals have been made.

1. Phosphorus-containing Reagents

When partially protected carbohydrates are treated with any of several combinations of phosphorus-containing reagents, a nucleophilic substitution reaction takes place (16, 17). The phosphorus-containing component of these combinations usually is triphenylphosphine [41]. Some examples of these combinations are: (a) 41 and diethyl azodicarboxylate [42] (DEAD), (b) 41 and carbon tetrahalides, (c) 41, imidazole [43], and iodine or bromine, (d) 41 and tribromo-[44] or triiodoimidazole [45], (e) methyl triphenoxyphosphonium halides [46], and (f) halotriphenoxyphosphonium halides [47]. The halides 46 and 47 are

$(C_6H_5)_3P$

41

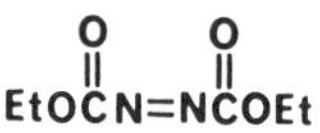

42

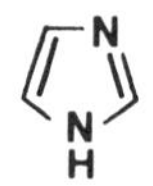

43

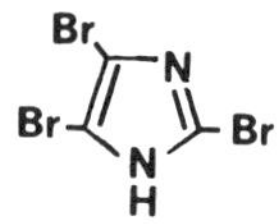

44

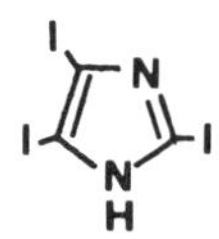

45

$Me\overset{\oplus}{P}(OC_6H_5)_3$ $X^{\ominus}$

46

X = I

X = Br

$X\overset{\oplus}{P}(OC_6H_5)_3$ $X^{\ominus}$

47

X = I

X = Br

sometimes referred to as the Rydon reagents. The common factor in reactions using these reagents is that a leaving group containing a phosphorus atom, usually a phosphonium ion, is believed to be formed. The actual method of formation of the leaving group depends upon the particular reagents used. For example, a reaction involving triphenylphosphine [41] and DEAD [42], a type of reaction known as a Mitsunobu reaction (17), is believed to generate the phosphonium salt 48 according to the mechanism outlined in Figure 8. One reason for the effectiveness of triphenylphosphine [41] in these reactions is the formation of triphenylphosphine oxide [49], a product so stable that its generation represents a significant "driving force" for the reaction. An impressive array of compounds can be prepared by the Mitsunobu reaction (17). These include aminodeoxy sugars, deoxyhalogeno sugars, and acylated carbohydrates. Examples of these reactions

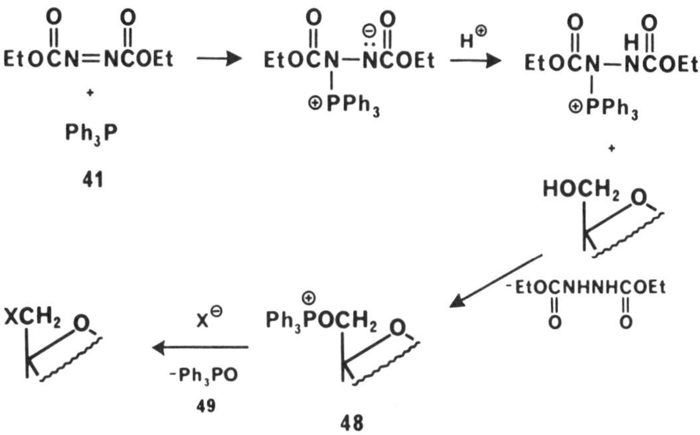

FIG. 8. Proposed mechanism for the Mitsunobu reaction.

will be considered later in this chapter when various nucleophiles are discussed. The other reagents using phosphorus-containing compounds usually lead to deoxyhalogeno sugars.

2. *N*-Bromosuccinimide (NBS) and Benzylidene Acetals

The acetal linkage in a benzylidene acetal can be converted into an excellent leaving group by reaction with molecular bromine. This process, a free-radical chain reaction in which a bromine atom replaces the hydrogen atom attached to the acetal carbon atom (Figure 9), is the first step in a chain of events leading to the formation of a bromodeoxy sugar [51]. Halogenation of carbohydrates in this manner is known as the Hanessian-Hullar reaction (18, 19). Bromination is followed by ionization of the carbon-bromine bond to generate a resonance-stabilized carbocation [50]. Bromide ion then displaces the oxygen atom attached to C-6, a reaction which results in simultaneous formation of a benzoyloxy group at C-4. The bromine for this reaction comes from *N*-bromosuccinimide (NBS) and is continuously generated by reaction of

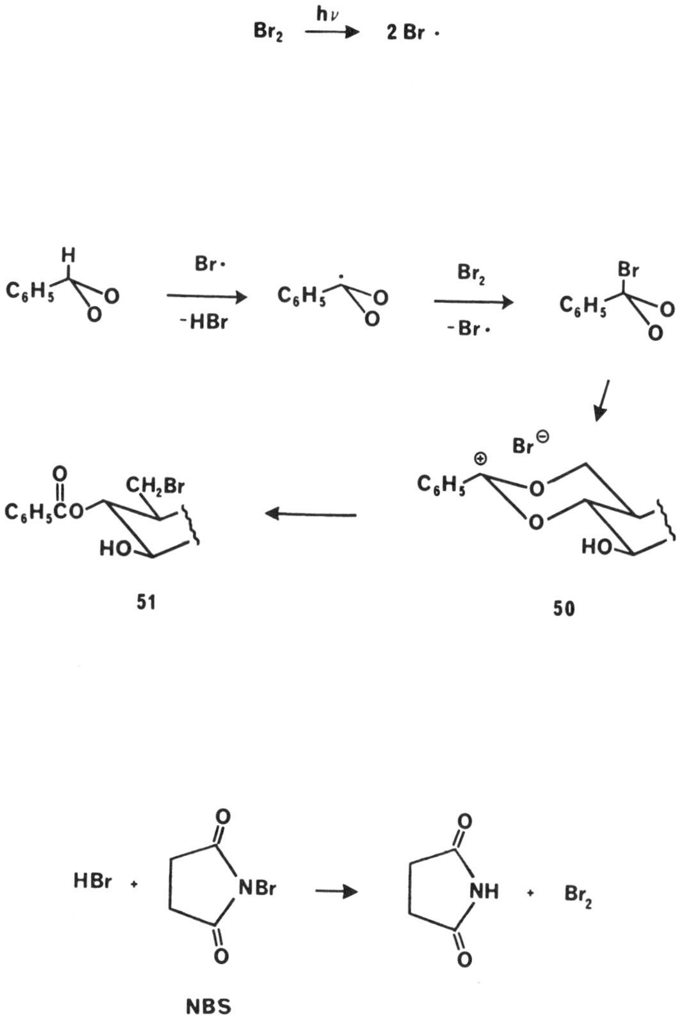

FIG. 9. Proposed mechanism for the Hanessian-Hullar reaction.

NBS with hydrogen bromide, one of the products from the free-radical bromination (Figure 9). Since bromide ion is always present in the Hanessian-Hullar reaction, it is always the nucleophile involved in the displacement process.

Since, as with other reactions, substitution at a primary carbon atom is greatly preferred over substitution at a secondary one, the Hanessian-Hullar reaction is an excellent method for synthesis of 6-bromo-6-deoxy sugars. If, however, the benzylidene acetal protects two secondary hydroxyl groups, as is the case with methyl 3,4-*O*-benzylidene-6-deoxy-α-D-galactopyranoside [52], reaction with NBS gives a mixture of bromodeoxy compounds (Equation 16) (20).

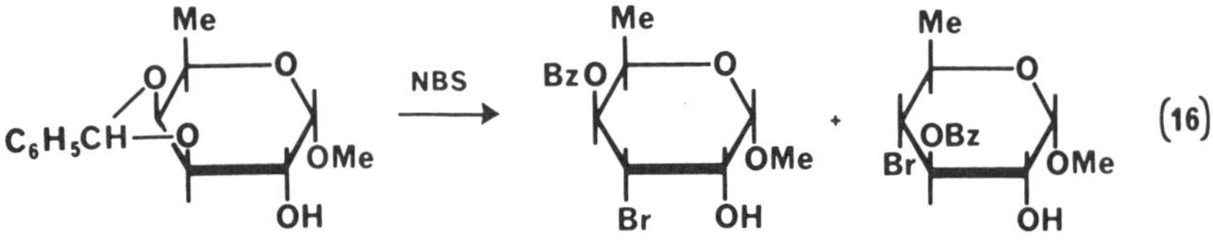

52 53 54 (16)

3. Diethylaminosulfur Trifluoride (DAST)

Diethylaminosulfur trifluoride [55] (DAST) is a reagent which reacts with a hydroxyl group to generate an effective leaving group and at the same time to produce fluoride ion (Figure 10); consequently, DAST is an excellent reagent for the formation of deoxyfluoro sugars. Methyl 6-deoxy-6-fluoro-α-D-glucopyranoside [56], for example, is formed when methyl α-D-glucopyranoside [4] reacts with DAST (Equation 17) (21). Prolonged treatment of 4

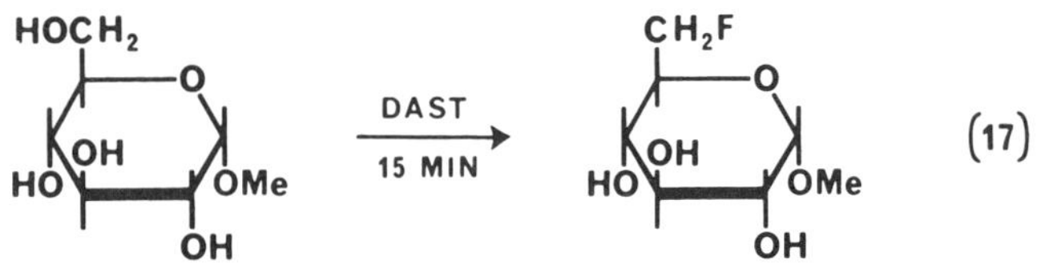

4 56 (17)

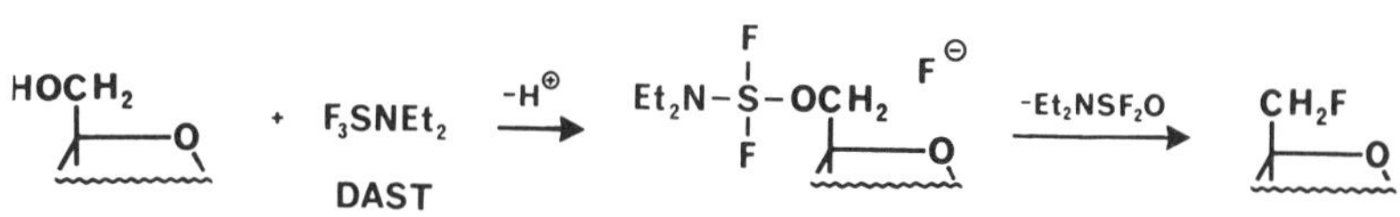

55

FIG. 10. Proposed mechanism for the reaction of DAST [55] with a partially protected sugar.

with DAST introduces a second fluorine atom into the molecule at C-4 (Equation 18) (22). DAST is, therefore, quite regioselective in

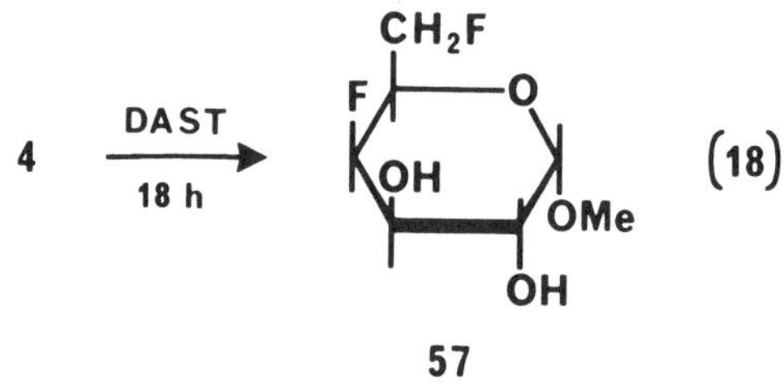

its reaction. (Regioselective reaction at C-4 rather than C-2 or C-3 is understandable in terms of the factors, discussed earlier in this chapter [Section II.C.], important to reactivity of leaving groups attached to a pyranoid ring.) If methyl β-D-glucopyranoside [58] is used as the starting material in this reaction, the steric hindrance imposed by the methoxy group on the approach of fluoride ion to C-3 (Figure 4) is eliminated; also, any repulsion between electrons on the methoxy atom and the approaching fluoride ion is greatly reduced. When methyl β-D-glucopyranoside [58] reacts with DAST, the major product from prolonged reaction is methyl 3,6-dideoxy-3,6-difluoro-β-D-allopyranoside [59] (Equation 19)

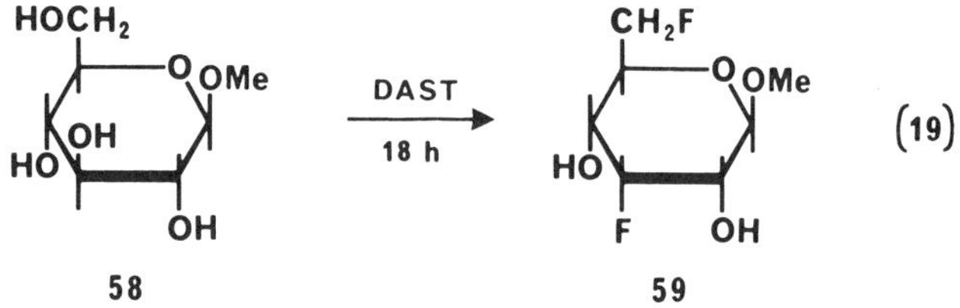

(23). DAST is also an effective reagent for synthesis of glycosyl fluorides (Equation 20) (24). The reactions of DAST with carbohydrates have been recently reviewed (25).

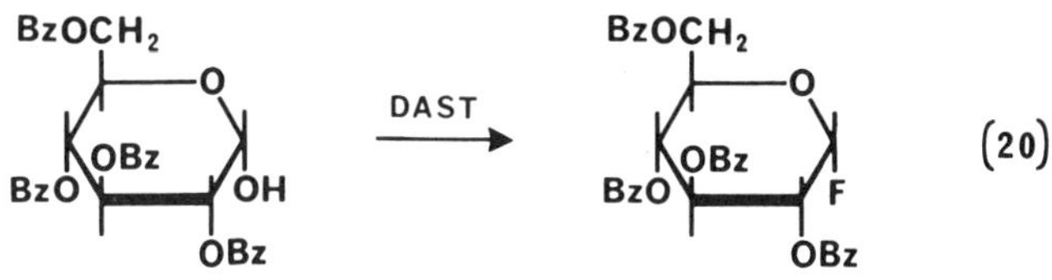

60 61

4. Protonic (Bronsted) Acids

The most common method for forming a leaving group at the anomeric center is by protonation. Protonation converts an anomeric hydroxyl, alkoxy, or acyloxy group into a leaving group which ionizes to form a carbocation. Carbocation formation in this fashion is involved in acid-catalyzed glycoside formation from unprotected sugars and alcohols (Chapter 7, Section III.A.) and in acid-catalyzed hydrolysis of a methyl glycoside (Chapter 8, Section III.A.) Acetoxy groups are often converted into leaving groups during the synthesis of glycosyl halides; thus, reaction of 1,2,4-tri-*O*-acetyl-3,6-di-*O*-methyl-α-D-glucopyranoside [62] with hydrogen chloride produces the corresponding glycosyl chloride 64 (Figure 11) (26). Although all of the oxygen atoms in compound 62 undoubtedly are protonated at some time, only protonation of the C-1 acetoxy group is productive because only this protonation leads to formation of a resonance-stabilized

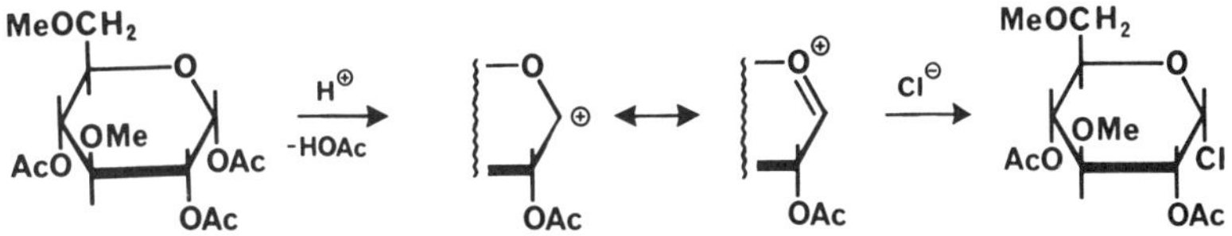

62 63 64

FIG. 11. Proposed mechanism for glycosyl halide formation from reaction of 1,2,4-tri-*O*-acetyl-3,6-di-*O*-methyl-α-D-glucopyranose [62] with hydrogen chloride.

cation [63]. The anomeric effect dictates preferential formation of a compound with the electronegative substituent in the axial orientation; thus, only the α chloride 64 is formed.

Simple protonation of nonanomeric hydroxyl groups to convert them into leaving groups is normally not a useful method for group activation because it leads to dehydration and molecular rearrangement (Chapter 7, Section IV). Protonation is the first step in some deprotection processes, such as debenzylidenation and detritylation (Chapter 8, Sections III.B.4. and V.C., respectively), but in these reactions protonation leading to deprotection takes place on oxygen atoms in the protecting groups.

5. Nonprotonic (Lewis) Acids

In addition to acids which protonate an anomeric hydroxyl group (or other oxygen-containing substituent at the anomeric carbon atom) to generate a leaving group, there are a number of other reagents which can activate a group at the anomeric carbon atom. Some of these are: α,α-dihalomethyl methyl ethers [65 and 66], titanium tetrabromide and tetrachloride, trimethylsilyl trifluoromethanesulfonate [67] (TMS-triflate), and (chloromethylene)dimethylimminium chloride [68]. These reagents

X_2CHOMe	$Me_3SiOSO_2CF_3$	$Me_2\overset{\oplus}{N}{=}CHCl\ \ Cl^{\ominus}$
65 X=Cl	67	68
66 X=Br		

generate leaving groups at the anomeric center in a variety of ways. Even a single reagent may have more than one pathway for reaction; consequently, a thorough discussion of the possibilities would be lengthy and would involve considerable speculation; however an indication of the probable, essential intermediates in the most important of these reactions is worthwhile along with

$$Br_2CH{-}O{-}Me \longrightarrow BrCH{=}\overset{\oplus}{O}{-}Me\ \ Br^{\ominus}$$

66 69

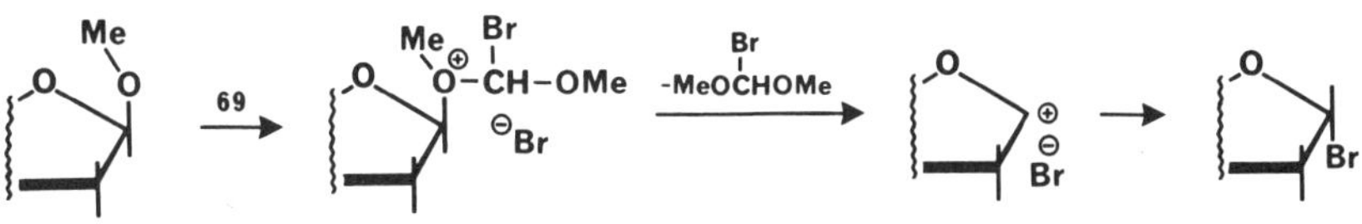

FIG. 12. Proposed mechanism for formation of a glycosyl bromide using α,α-dibromomethyl methyl ether [66].

some illustrative examples. α,α-Dibromomethyl methyl ether [66] eliminates bromide ion to form an oxonium salt [69] which reacts with an oxygen atom attached to the anomeric carbon atom to begin a reaction sequence leading to formation of a glycosyl bromide (Figure 12). Reaction of 66 with methyl 3,4-di-*O*-benzoyl-2,6-dideoxy-α-D-*ribo*-hexopyranoside [70] converts it into the corresponding glycosyl bromide [71] (Equation 21) (27). A similar type of reaction is thought to occur

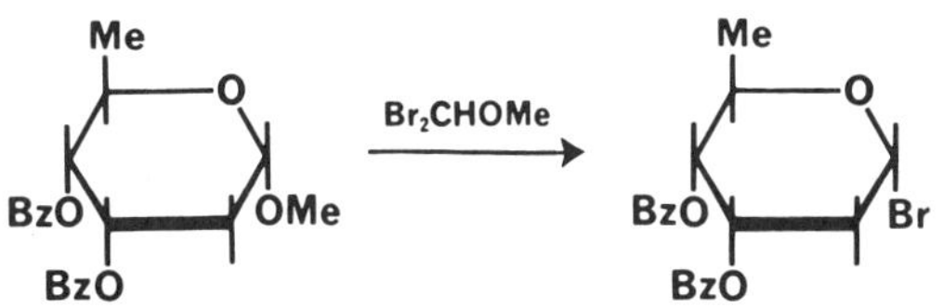

70 71 (21)

when (chloromethylene)dimethylimminium chloride [68] reacts with an anomeric hydroxyl group (Figure 13) (28). TMS-triflate [67] transfers a trimethylsilyl group to an anomeric hydroxyl group to generate a protonated silyl ether as a leaving group (Figure 14). TMS-triflate [67], therefore, differs from the other

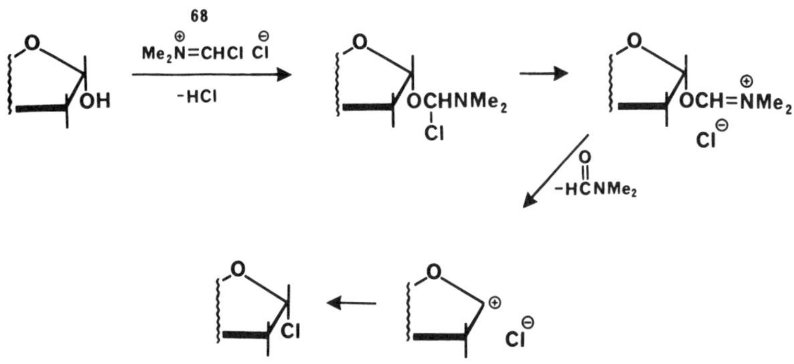

FIG. 13. Proposed mechanism for glycosyl chloride formation using (chloromethylene) dimethylimminium chloride [68].

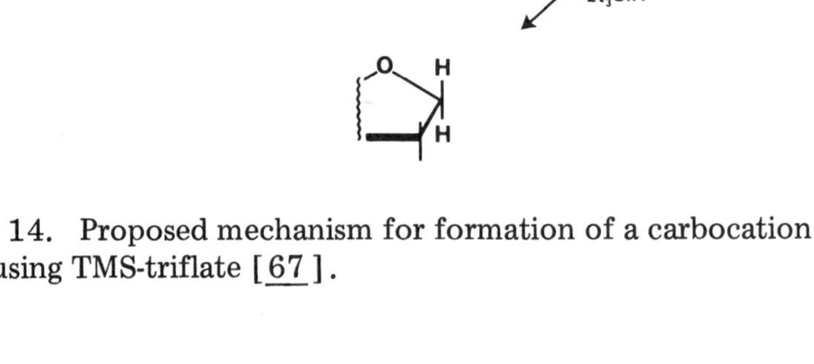

FIG. 14. Proposed mechanism for formation of a carbocation at C-1 using TMS-triflate [67].

reagents mentioned in this section in that the result of reaction is not necessarily formation of a glycosyl halide (Equation 22) (29).

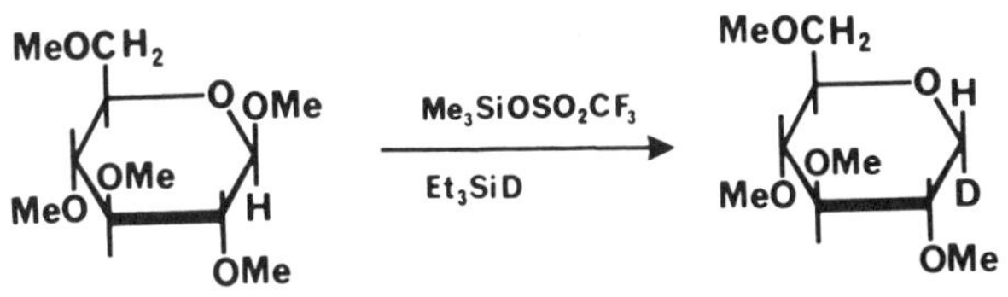

72 73

(22)

IV. NUCLEOPHILES

A. Characteristics of Nucleophiles

Nucleophilicity is the ability of an ion or molecule to share electrons during a substitution reaction with a carbocation (an S_N1 reaction) or with a carbon atom from which a group is being displaced (an S_N2 reaction). The ions and molecules most effective at this task are the best nucleophiles. Some generalizations about the reactions of nucleophiles in solution are listed below.

(a) When comparing nucleophiles within the same row of the periodic table, nucleophilicity roughly parallels basicity; thus, $R_3C^- > R_2N^- > RO^- > F^-$ is the order of nucleophilicity for these four ions.

(b) An ion with a negative charge is a better nucleophile than its conjugate acid; thus, RO^- is a better nucleophile than ROH.

(c) Nucleophilicity increases when proceeding down in the periodic table. RS^- is a better nucleophile than RO^-.

(d) The more strongly an anion is solvated or the more closely it is associated with a counter ion, the less nucleophilic it is.

(e) Nucleophilicity of an anion or molecule is decreased by substituents in the vicinity of the nucleophilic center if they sterically hinder the reaction of the nucleophile. CH_3O^- is a better nucleophile than $(CH_3)_3CO^-$.

(f) An atom will become a better nucleophile when substituted by a substituent bearing an unshared electron pair. (This phenomenon is known as the α-effect.) NH_2NH_2 is a better nucleophile than NH_3.

The rationale behind these six generalizations will be presented in the remainder of this chapter as specific nucleophiles and situations are discussed.

B. Factors Influencing Nucleophilicity

1. Solvation

Although all anions are solvated in solution, polar solvent molecules have a much greater attraction for smaller ions because smaller ions present a concentrated and, hence, more attractive negative charge. A strong attraction between an anion and solvent molecules makes participation by the anion in displacement reactions more difficult because the solvent molecules must be separated from the ion as a part of the substitution process. Due to the energy which must be expended to separate an anion from solvent molecules, solvation has a dramatic effect on the reactivity of nucleophiles. This is particularly true for ions such as fluoride where nucleophilicity is greatly reduced by solvating molecules. Solvation undoubtedly plays an important role in the difference in ease of displacement of the triflyloxy group from 1,2,3,4-tetra-*O*-acetyl-6-*O*-trifluoromethylsulfonyl-β-D-glucopyranose [<u>23</u>] by halide ions (8). Reaction of <u>23</u> with tetrabutylammonium iodide takes place at -78°C (Equation 11), while substitution with tetrabutylammonium fluoride trihydrate requires heating at reflux in benzene (Equation 23) (8). This difference in reactivity reflects,

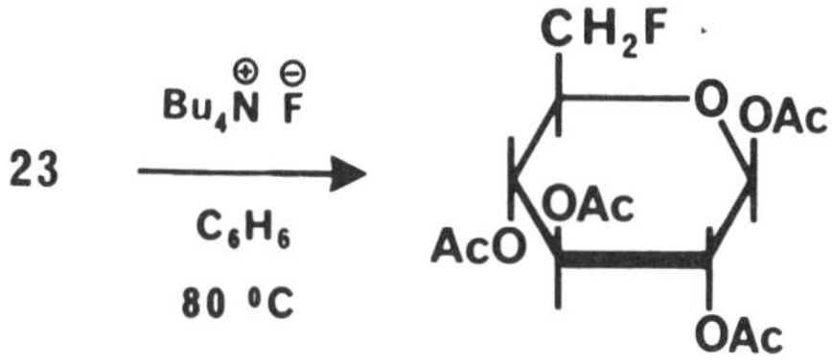

74

(23)

at least in part, the strength with which the water molecules are held in association with fluoride ion.

2. Polarizability

Another factor which is credited with imparting greater reactivity to nucleophiles in which the reactive center is on a large atom is the polarizability of the valence electrons in this atom. The valence electron distribution associated with a large atom (e.g., iodine or bromine) can be distorted more easily than that associated with a small one (e.g., fluorine). For the smaller atom these electrons are much closer to the nucleus and experience greater nuclear attraction. Since the valence electrons in a larger atom experience less nuclear attraction, their electron clouds can be distorted (polarized) more easily. Polarization of valence electrons allows bonding between the nucleophile and the carbon atom bearing the leaving group to begin at a greater distance between these two reactants than would be possible without polarization. The sooner the stabilizing effect of bond formation becomes important in a reaction, the lower the energy of the transition state for the reaction will be.

3. Separation of the Nucleophile and the Counter Ion

The nucleophilicity of an anion can be increased by selecting a different counter ion for the nucleophile or by altering the reaction medium in such a manner as to "free" the nucleophile for reaction. Methods for increasing nucleophilicity by taking advantage of one of these changes include use of tetraalkylammonium salts, dipolar aprotic solvents, and crown ethers.

a. Tetrabutylammonium Salts. Tetrabutylammonium halides are often used in substitution reactions (Equations 11, 12, and 23) because the tetrabutylammonium counter ion has advantages over ions such as sodium and potassium. Homogeneous reaction mixtures are easily obtained with tetrabutylammonium salts because they are soluble in most organic solvents. The positively charged nitrogen atom in the cation is necessarily considerably separated from the anion by the presence of the butyl groups. This renders the anion more "free" and, consequently, more reactive; that is, the energy necessary to overcome the attraction between the charged nucleophile and its counter ion is less for tetrabutylammonium salts than for salts with smaller cations. Using as starting materials salts with anions and cations already well separated has the effect of reducing the energy required in a reaction to reach the transition state. Tetrabutylammonium salts are not the only ones which achieve the desired anion-cation separation but they are the ones which are most often used for this purpose.

b. Dipolar Aprotic Solvents. Another method for activating an anion by separating it from its counter ion involves the use of a dipolar aprotic solvent. Dipolar aprotic solvents, such as methyl sulfoxide (DMSO, [75]) and *N,N*-dimethylformamide (DMF, [76]), tend to solvate cations well but anions poorly. This

DMSO

75

DMF

76

difference in solvating ability is directly related to the structure of the solvent molecule. Both DMF and DMSO have an oxygen atom bearing considerable negative charge on the exterior of the molecule where it can easily solvate a cation. The nitrogen atom in the DMF and the sulfur atom in DMSO both bear a partial positive charge, but this charge is on a highly substituted atom, a situation which prevents close approach by an anion and makes solvation of an anion relatively ineffective. Using dipolar aprotic solvents is then another method for increasing nucleophilicity by generating anions which are poorly solvated and well separated from their counter ions. DMF is often the solvent of choice in substitution reactions because it is more stable than DMSO at elevated temperatures and in the presence of certain reagents. (DMSO functions as an oxidizing agent in the presence of a number of compounds such as acetic anhydride and dicyclohexylcarbodiimide. These and other oxidation reactions will be discussed in Chapter 10.)

a. **Crown Ethers.** Crown ethers, such as 12-crown-4 [77], are large-ring polyethers which solvate cations by complexation with

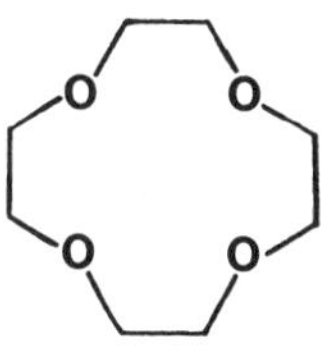

77

oxygen atoms. If the size of the "cavity" in the interior of the crown ether corresponds to the ionic diameter of an ion, complexation is exceedingly strong. The lithium ion is the proper size to be complexed well by 12-crown-4 [77]. Complexation of a cation by a crown ether increases the solubility of an inorganic salt in organic solvents. Surrounding the cation by the crown ether separates it from the anion, which becomes poorly solvated and more nucleophilic.

C. The Principle of Hard and Soft Acids and Bases

The greater reactivity of larger nucleophiles in S_N2 substitution is frequently explained in terms of the principle of hard and soft acids and bases (30), a concept which takes polarizability, electronegativity, and other factors into account in a convenient, qualitative manner. According to this principle, reaction between soft bases and soft acids (also between hard bases and hard acids) is favored. Nucleophiles of low electronegativity and high polarizability (e.g., iodide and bromide) are soft bases, while the carbon atom bearing the leaving group in an S_N2 substitution is considered to be a soft acid; thus, nucleophiles such as iodide and bromide should be especially effective in S_N2 reactions. Fluoride (very electronegative and not easily polarized) is classified as a hard base and, therefore, according to the principle of hard and soft acids and bases, it should be a poor nucleophile in S_N2 reactions.

D. Nitrogen-containing Nucleophiles

1. Nitrite Ion

An essential aspect of the ability of a molecule or ion to function as a nucleophile in an S_N2 substitution is the electronegativity of the atom bearing the electrons to be shared. In the same row in the periodic table, less electronegative atoms are better nucleophilic centers than more electronegative ones. One way to appreciate this phenomenon is to consider the reactivity of an ambident nucleophile, that is, a nucleophile which has more than one potential center for reaction. The nitrite ion [78] is an example of such a

78

nucleophile. Even though nitrite can react at either oxygen or nitrogen atoms in S_N2 substitution, reaction at the nitrogen atom often prevails; thus, when iodide is displaced from methyl 5-deoxy-5-iodo-2,3-*O*-isopropylidene-β-D-ribofuranoside [79] by nitrite, a nitro compound is formed (Equation 24) (31). This

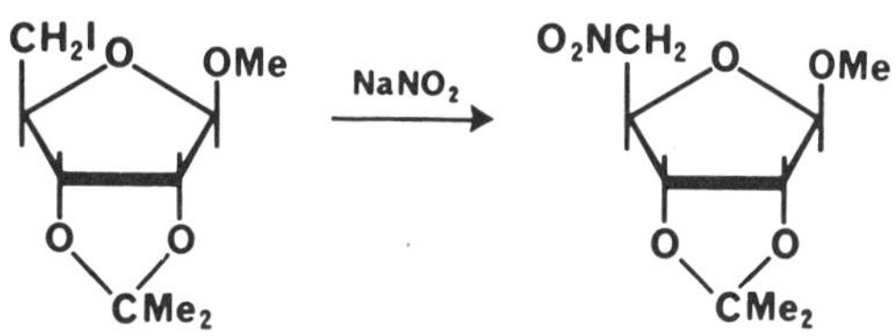

(24)

79 **80**

situation provides a rather striking example of the influence of electronegativity on nucleophilicity since reaction occurs at the nitrogen atom even though the negative charge in the nitrite ion [78] is shared primarily between the oxygen atoms. Qualitatively, this phenomenon can be attributed to the more electronegative atom exerting a greater "pull" on unshared electron pairs, thus holding them more tightly and making them less available for bonding.

Reaction at an oxygen atom should become more important when nitrite interacts with a carbon atom which has a significant amount of positive charge. Under these conditions the negative charge on the oxygen atom and the positive charge on the carbon atom would draw these atoms together electrostatically. Such a situation exists in a S_N1 reaction or even in a S_N2 process in which there is a highly polarized bond between the reacting carbon atom and the leaving group. An example of displacement by a nitrite oxygen atom is found in the reaction of 1,2-*O*-isopropylidene-5-*O*-trifluoromethylsulfonyl-α-D-glucofuranourono-6, 3-lactone [81] with tetrabutylammonium nitrite (Figure 15) (32). Such a reaction suggests that there is considerable accumulation of positive charge at C-5 in 81, a reasonable possibility since the bond between C-5 and the very electronegative triflyloxy group should be quite polar.

Another factor contributes to the difference in reactivity exhibited by nitrite in the reactions shown in Equation 24 and Figure 15. The steric requirements for substitution using an oxygen atom in the nitrite ion as a nucleophile are less demanding than those for substitution using the nitrogen atom. When nitrite is displacing a group from a secondary carbon atom, the probability

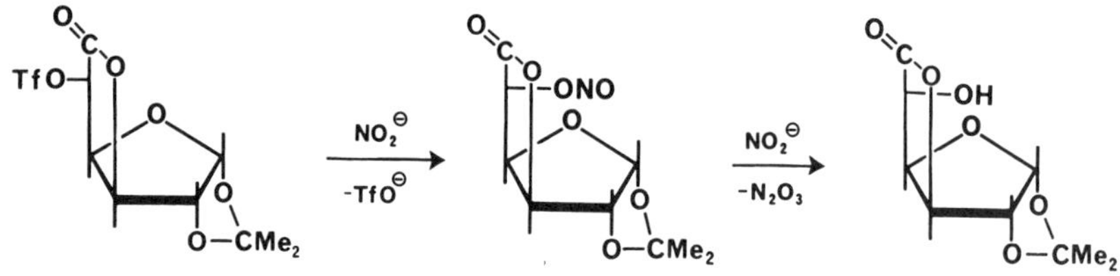

81 82

FIG. 15. Displacement of the trifluoromethylsulfonyloxy group from 1,2-*O*-isopropylidene-5-*O*-trifluoromethylsulfonyl-α-D-glucofuranourono-6,3-lactone [81] by nitrite ion.

of an oxygen atom becoming the reactive nucleophilic center increases due to the greater steric requirements for substitution at a secondary carbon atom. From consideration of these two reactions (Equation 24 and Figure 15), it is clear that the nucleophilicity of nitrite (and other nucleophiles) is determined by a combination of steric and electronic effects; further, although it is usually possible to identify important factors influencing nucleophilicity, it is often difficult to determine which factor is of greatest importance in a particular reaction.

2. Azide Ion

Azide ion [83] has two characteristics which make it a synthetically valuable nucleophile; that is, it is easily introduced and its

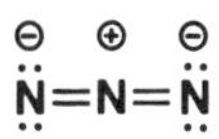

83

subsequent reduction to an amino group constitutes one of the best reaction sequences for synthesis of aminodeoxy sugars. Azide ion displaces leaving groups from primary positions without difficulty. If the leaving group is an excellent one, displacement from secondary positions also occurs with ease (Equation 13); however, forcing conditions are required for less reactive leaving groups attached to secondary carbon atoms (Equation 9). Azide ion is also capable of opening epoxides under relatively mild conditions; thus, heating compound 38 with sodium azide in methanol at reflux for 24 hours results in complete ring opening (Figure 7).

A combination of factors is responsible for the pronounced nucleophilicity of the azide ion [83]. One of these is that its linear structure minimizes steric hindrance. Another is that each of the terminal nitrogen atoms in the azide ion [83] is electron rich because each has a formal negative charge. Finally, the nucleophilic centers in this ion have the reactivity advantage associated with atoms of relatively low electronegativity.

3. Ammonia, Amines, Amides, Imides, and Their Corresponding Anions

a. Neutral Compounds. Ammonia and amines are relatively weak nucleophiles which are used rarely in displacement reactions. Due to their low nucleophilicity, it is necessary to heat ammonia and low molecular weight amines well above their boiling points in order for reaction to occur in most instances. This situation requires that substitution reactions be conducted under pressure. For example, reaction of 1,6:3,4-dianhydro-2-*O*-(1-ethoxyethyl)-β-D-talopyranose [84] with methyl amine takes two hours at 140°C in a Parr bomb to form the ring-opened, amino alcohol 85 (Equation 25) (33). The nitrogen atom in an amide or an imide is

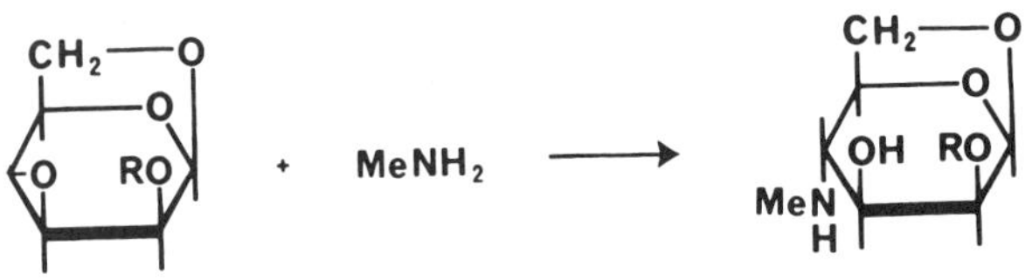

(25)

84 R = $CH_2CH_2OCH_2CH_3$ 85

considerably less nucleophilic than the nitrogen atom in an amine; consequently, amides and imides do not function as nucleophiles in S_N2 substitution reactions. This reduced nucleophilicity is due to withdrawal of electron density from the nitrogen atom both by inductive and resonance effects.

b. Ionic Compounds. Deprotonation of ammonia increases considerably its nucleophilicity but also creates such a powerful base that elimination becomes a significant competing process

(Chapter 13). Imides and acylated amines can be deprotonated to form resonance-stabilized anions [86 and 87], which are effec-

86

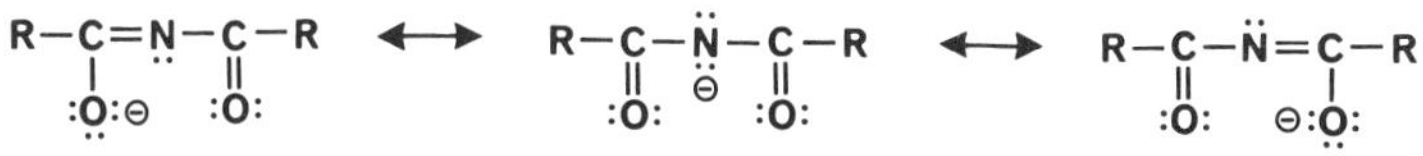

87

tive nucleophiles. Imides, such as phthalimide [88], ionize so readily that they provide both the nucleophile and the catalyst for the Mitsunobu reaction (Equation 26) (34), a process which

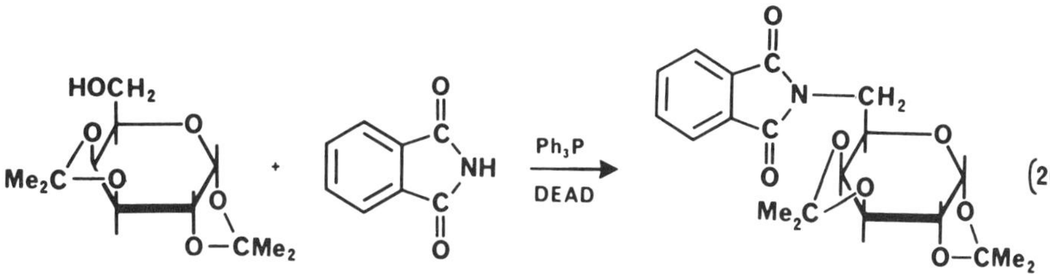

(26)

89 88 90

requires catalysis by a weak acid (Figure 8). Substitution to give phthaloylaminocarbohydrates is an important synthetic process because removal of the phthaloyl group gives the corresponding aminodeoxy sugar (Equation 27) (34). Since the nitrogen atom in phthalmide [88] has two carbonyl substituents, it is not surprising

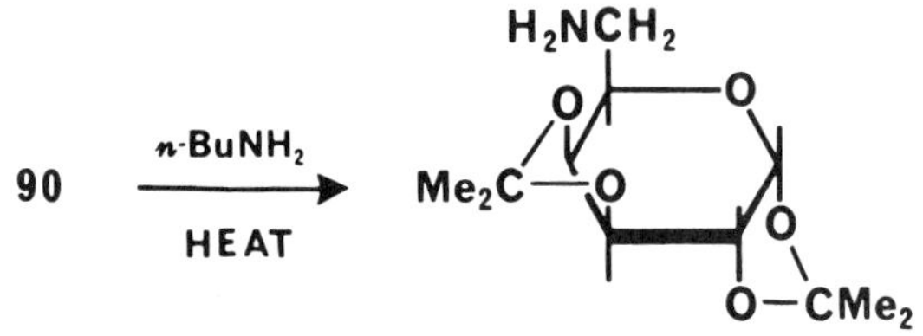

(27)

91

that the anion derived from 88 is sterically hindered. This steric hindrance is sufficient to prevent formation of a phthalimido derivative when the Mitsunobu reaction is attempted with methyl 4,6-*O*-benzylidene-α-D-glucopyranoside [92] (34).

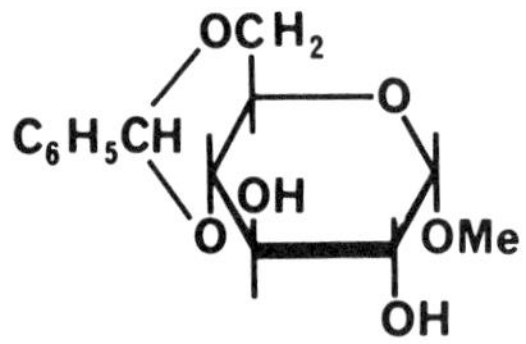

92

4. Hydrazine

In contrast to ammonia and low molecular weight amines, if hydrazine is used as the nucleophile in an S_N2 substitution, displacement can take place at atmospheric pressure. For example, heating methyl 6-deoxy-2,3-*O*-isopropylidene-4-*O*-methylsulfonyl-α-L-mannopyranoside [93] in hydrazine under reflux results in complete displacement after 52 hours (Equation 28) (35). Not only is the boiling point of hydrazine (113°C) a significant factor in permitting displacement without resorting to a sealed reaction vessel, but the nucleophilicity of hydrazine is also greater than that of ammonia or amines. The increased nucleophilicity of hydrazine is attributed to a well-documented but incompletely

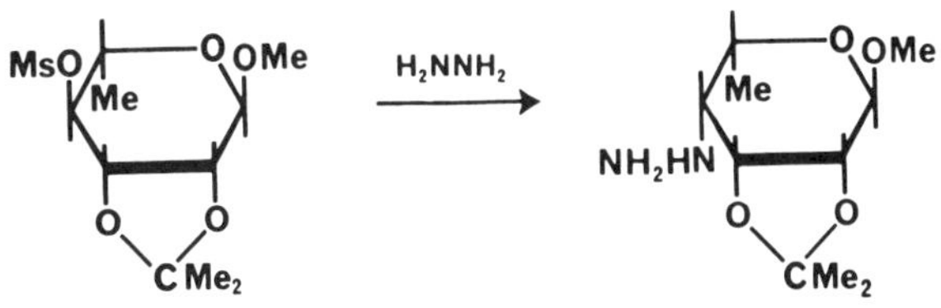

(28)

93 94

understood phenomenon known as the α-effect, a term used to describe the enhanced reactivity of an atom which bears a substituent with an unshared pair of electrons; thus, the nucleophilicity of each nitrogen atom in hydrazine is increased by the α-effect. As with other nitrogen-containing nucleophiles, the ultimate goal of substitution reactions using hydrazine usually is the synthesis of aminodeoxy sugars. This goal can be realized by hydrogenolysis of the hydrazino derivative produced by substitution (Equation 29) (35).

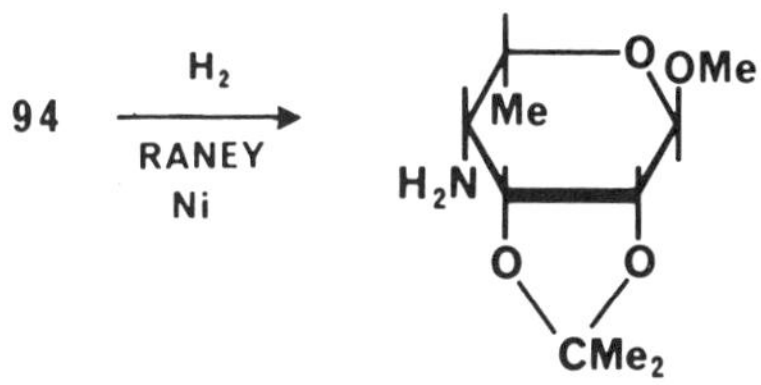

(29)

95

E. Oxygen Nucleophiles

1. Alcohols and Alkoxides

Formation of an epoxide (Equation 8 and Figure 7) by internal displacement, a reaction discussed earlier in this chapter (Section III.A.3.), is the most important of the reactions using oxygen nucleophiles. A tosyloxy or mesyloxy group is the usual leaving group and typical reaction conditions call for the presence of a strong base. These conditions indicate that the nucleophile is an alkoxide; however, if the departing anion is triflate, the hydroxyl group itself may be the nucleophile since a strong base is not required for reaction (Equation 30) (36).

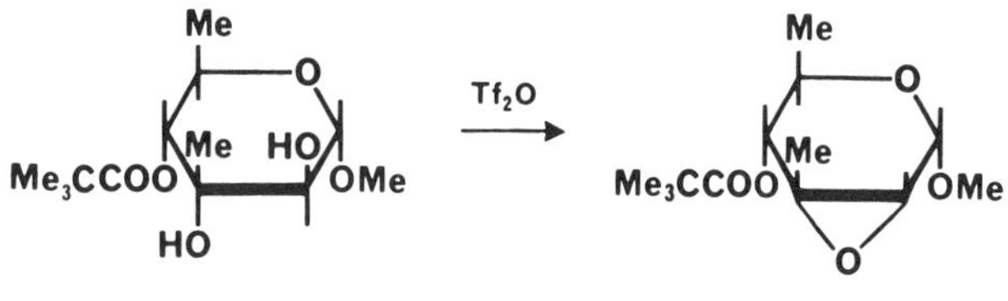

96 97 (30)

Acid-catalyzed S_N1 substitution at the anomeric carbon atom with an alcohol as the nucleophile (the Fischer method) is the primary process for the synthesis of glycosides from unprotected sugars and low molecular weight alcohols. This reaction was discussed in Chapter 7 (Section III.A.).

2. Carboxylates

Displacement of a leaving group at a primary carbon atom by a carboxylate anion is a method for regioselective protection; thus, the Mitsunobu reaction between methyl 2,3-di-*O*-acetyl-α-D-galactopyranoside [98] and benzoic acid gives the corresponding 6-*O*-benzoyl derivative [99] (Equation 31) (37). If the leaving

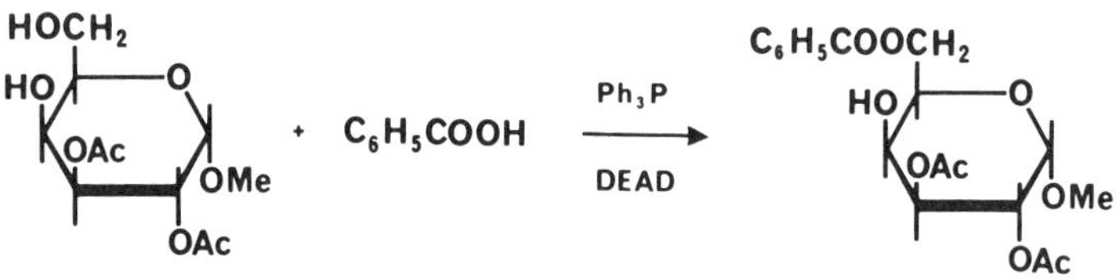

98 99 (31)

group is attached to a chiral center, displacement constitutes a method for inversion of configuration and simultaneous protection (Equation 32) (38).

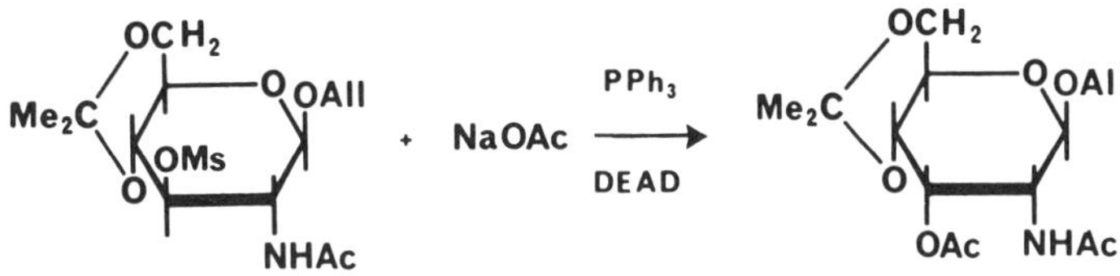

(32)

100 101

3. Nitrate

Substitution using the nitrate ion also provides a means for group protection as well as a method for inversion of configuration at a chiral center. For example, treatment of 1,2:5,6-di-*O*-isopropylidene-3-*O*-triflyl-α-D-allofuranose [<u>102</u>] with tetrabutylammonium nitrate results in formation of the corresponding gluco nitrate [<u>103</u>] (Equation 33) (39). As described in Chapter 8

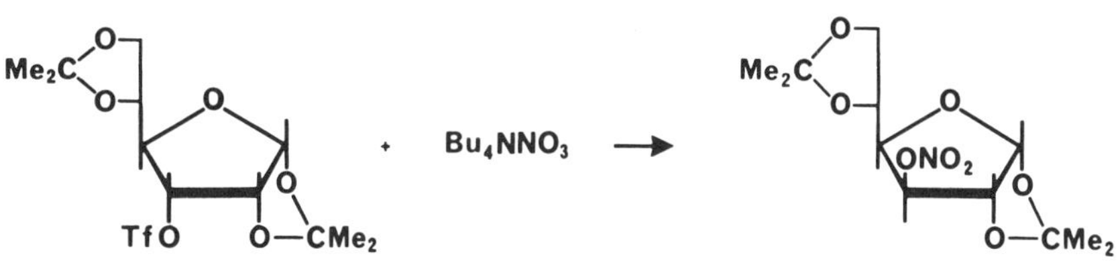

(33)

102 103

(Section IV.C.), nitrate protection can be removed by reaction with hydrazine.

F. Halides

The usual order of reactivity of halide ions in S_N2 substitution reactions in solution is iodide > bromide > chloride > fluoride.

This order is followed, for instance, when 1,2,3,4-tetra-*O*-acetyl-6-*O*-trifluoromethylsulfonyl-β-D-glucopyranose [23] reacts with tetrabutylammonium halides (8). This difference in reactivity of halide ions is caused by a combination of factors, among which solvation and polarizability are the most important. In discussing solvation effects and polarizability earlier in this chapter (Section IV.B.), fluoride ion was used as an example (Equation 23) of a nonpolarizable anion for which nucleophilicity was greatly reduced by strong solvation while iodide was cited as an equally good example (Equation 11) of a powerfully nucleophilic, highly polarizable ion which was not strongly solvated.

Iodide is such an effective nucleophile that it is often not necessary to use tetrabutylammonium salts or dipolar aprotic solvents to increase reactivity; in addition, since sodium iodide is quite soluble in low molecular weight ketones, displacement of primary leaving groups by iodide often is conducted by reaction in acetone (Equation 7) or butanone (Equation 6). Use of tetrabutylammonium halides, which are readily available, becomes increasingly important to nucleophilicity as the size of the halide ion decreases (8). In a like manner, using a dipolar aprotic solvent such as DMF is valuable in increasing solubility and nucleophilicity when sodium halides other than sodium iodide are used (40).

The examples of halide substitution reactions considered thus far have involved displacement of sulfonyloxy groups. This method for formation of deoxyhalogeno sugars is the most prevalent one; nevertheless, these compounds are also frequently formed using phosphorus-containing reagents (Section III.B.1.). These include the Rydon reagents 46 and 47, triphenylphosphine [41] in combination with carbon tetrahalides, and triphenylphosphine [41] in conjunction with a trihaloimidazole [44 and 45]. An example of the use of a Rydon reagent to introduce a bromine atom into a carbohydrate is shown in Equation 34 (41). In recent years the reagent of choice increasingly has been the combination

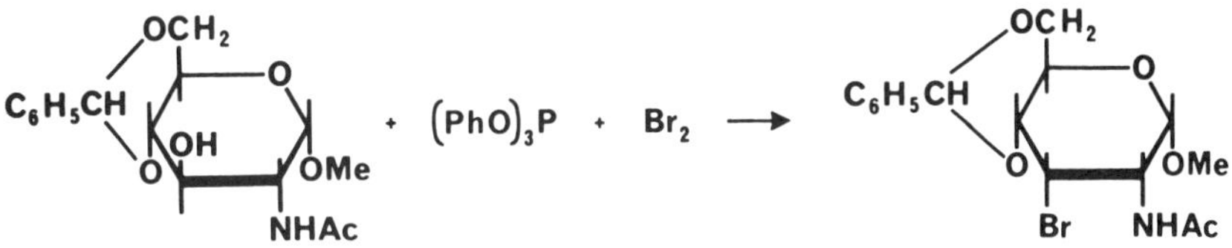

(34)

104 105

of triphenylphosphine [41] with 44 or 45 (42). Alternatively, a modification of this reagent is sometimes used which replaces 44 and 45 with imidazole [43] and iodine or bromine, respectively (43). A proposed mechanism for halogenation using 41, 43, and iodine is shown in Figure 16, and an example of halogenation involving 41 and 44 is provided by the reaction shown in Equation 35 (44).

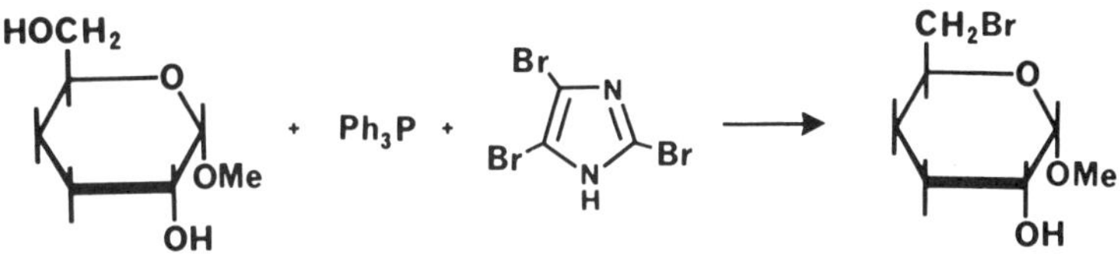

(35)

106 107

Several of the most important methods for synthesis of deoxyhalogeno sugars by nucleophilic substitution reaction were discussed earlier in this chapter. The sections in which this information is found are: *N*-Bromosuccinimide (NBS) and Benzylidene Acetals (III.B.2.), Diethylaminosulfur Trifluoride (DAST) (III.B.3.), Protonic (Bronsted) Acids (III.B.4.), and Nonprotonic (Lewis) Acids (III.B.5.).

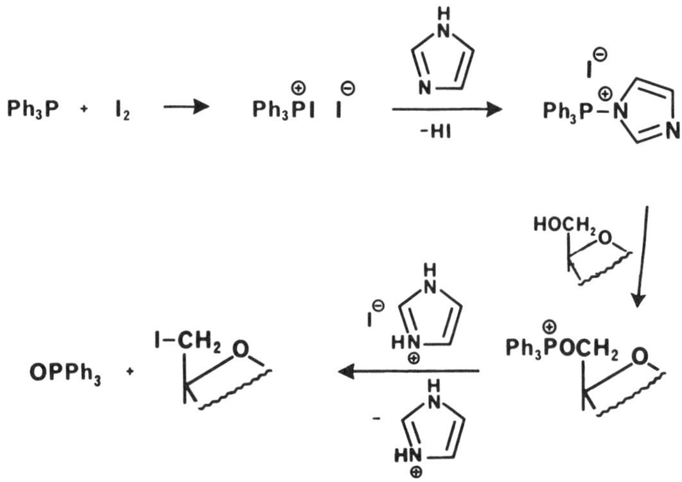

FIG. 16. Proposed mechanism for halogenation using triphenylphosphine [41], imidazole [43], and iodine.

G. Hydride

Metal hydrides such as sodium borohydride, lithium aluminum hydride (LAH), and lithium triethylborohydride (LTBH) react with carbohydrates containing sulfonyloxy groups or halogens to give the corresponding deoxy compounds. These reactions appear to be analogous to other nucleophilic substitution reactions and will be treated here as displacement processes in which hydride is the nucleophile. Although sodium borohydride is not a very reactive hydride transfer agent, it is capable of displacing some primary tosyloxy groups in DMSO [75] at 80°C (Equation 36)

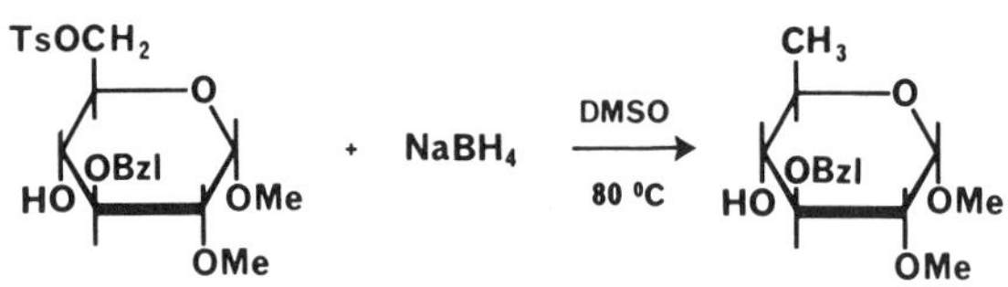

108 109 (36)

(2). LAH and LTBH are more reactive hydride reagents which displace primary sulfonyloxy groups with ease. For example, methyl 2,3,6-tri-*O*-*p*-tolylsulfonyl-α-D-galactopyranoside [110] reacts with LTBH at 25°C to give the corresponding 6-deoxy compound 111 (Equation 37) (45). LTBH is a particularly

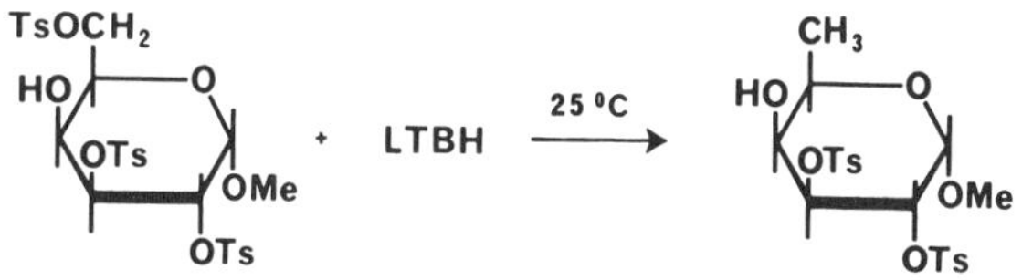

(37)

110 111

attractive hydride source because it is quite nucleophilic and readily soluble in organic solvents.

A significant limiting factor in the use of metal hydrides in displacement reactions at other than a primary position is that molecular rearrangement often takes place; thus, treatment of methyl 6-deoxy-2-*O*-*p*-tolylsulfonyl-α-D-galactopyranoside [112] with LTBH causes ring contraction (Equation 38) (46). Reaction

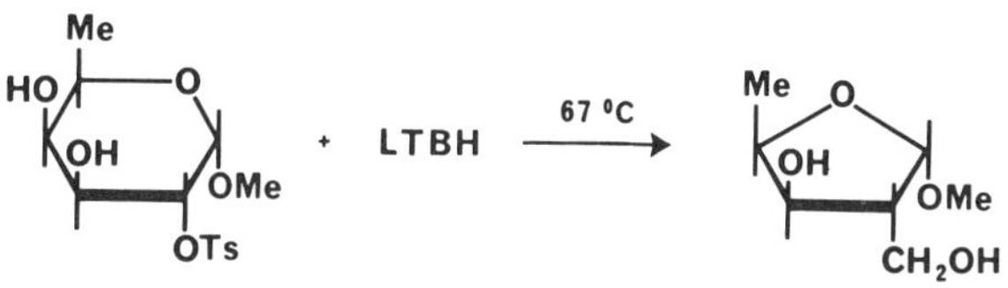

(38)

112 113

of sulfonates with LTBH also has resulted in hydride (45) and methyl (36) migrations. Internal displacement by alkoxide, another possible competing reaction, accompanies substitution by hydride in the reaction of 1,5-anhydro-6-*O*-*p*-tolylsulfonyl-D-*arabino*-hex-1-enitol [114] with LAH (Equation 39) (47).

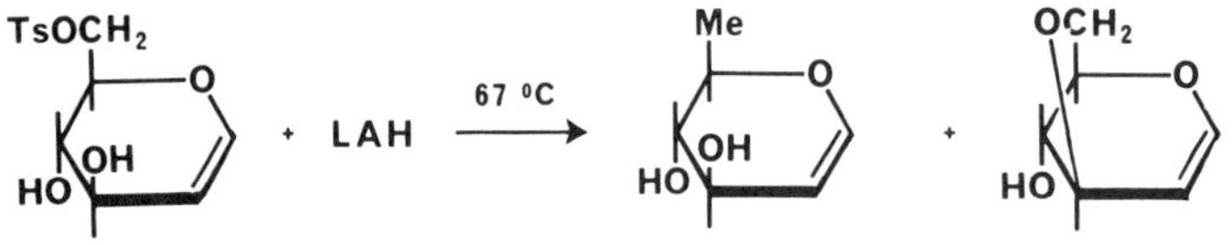

114 115 116

(39)

Hydride reagents which are less nucleophilic than aluminum- and boron-containing hydrides can be used as hydride transfer agents in S_N1 processes (Figure 14). This type of reaction occurs when methyl 2,3,4,6-tetra-*O*-methyl-β-D-glucopyranoside [<u>72</u>] reacts with TMS-triflate [<u>67</u>] and deuterotriethylsilane (Equation 22) (29).

Ring-opening reactions involving hydride reagents represent a synthetically valuable group of displacement processes. The opening of benzylidene acetals using metal hydrides was discussed in Chapter 8 (Section III.B.4.b.). Reaction of epoxides with metal hydrides is an excellent method for synthesis of deoxy sugar derivatives. Ring opening, which obeys the Furst-Plattner rule discussed earlier in this chapter (Section III.A.3.), is often highly regioselective (Equation 40) (48).

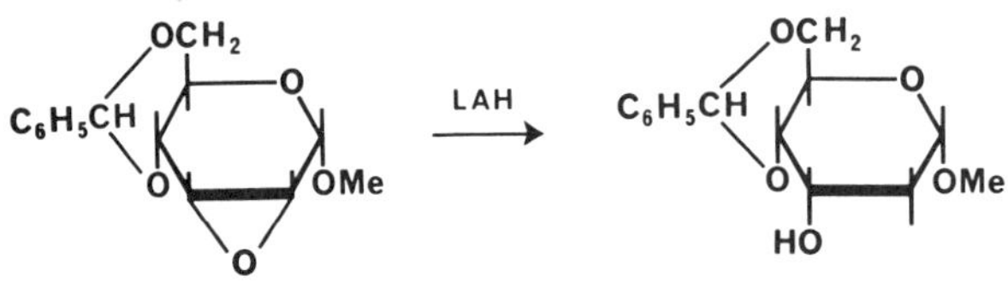

117 118

(40)

H. Carbon Nucleophiles

Although carbon nucleophiles are used extensively in addition reactions to carbonyl groups (Chapter 12), their use in S_N2 displacements is much more limited. Lithium dimethylcuprate

[119] is useful as an agent for simultaneous addition of a methyl group and ring opening of an epoxide. Reaction of methyl 2,3-anhydro-5-deoxy-α-D-ribofuranoside [120] with 119 opens the three-membered ring to give a mixture of branched-chain compounds (Equation 41) (49). (An extensive study of the

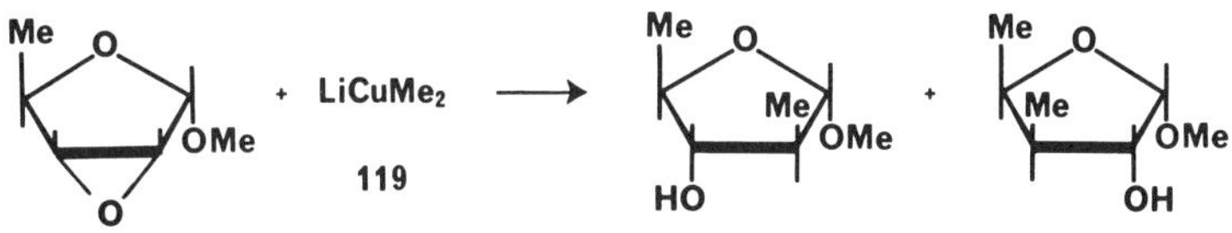

120 121 122 (41)

addition of organocopper reagents to epoxides is found in reference 2.) Chain extension is also possible with carbon nucleophiles; thus, cyanide ion displaces triflate from compound 123 to extend the carbon atom chain (Equation 42) (50).

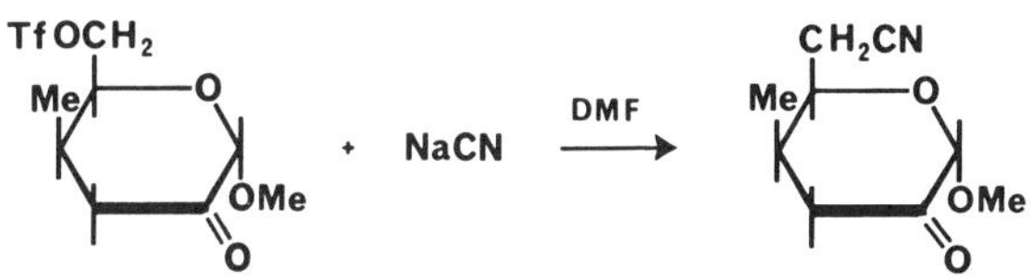

123 124 (42)

I. Phosphorus-containing Nucleophiles

Although phosphorus-containing compounds often are involved in nucleophilic substitution reactions, using a phosphorus-containing nucleophile to displace a leaving group in a carbohydrate is not a common occurrence. A phosphorus atom is, nevertheless, an excellent nucleophilic center which reacts with even moderately active leaving groups; thus, heating the iodide 125 with triethylphosphite for four hours under reflux gives the corresponding phosphonate [126] (Equation 43) (the Arbuzov reaction) (51).

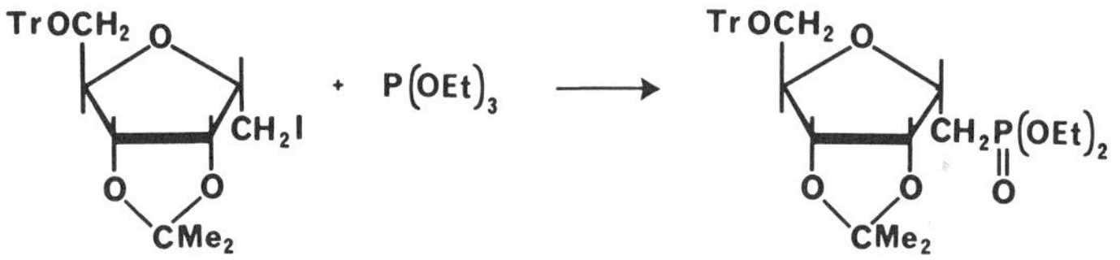

(43)

125 126

J. Sulfur-containing Nucleophiles

Sulfur-containing nucleophiles, often thioacetate, can be used to introduce sulfur into a carbohydrate by a substitution reaction. For example, thioacetate displaces the secondary mesyloxy group in allyl 2-acetamido-2-deoxy-4,6-*O*-isopropylidene-3-*O*-mesyl-β-D-allopyranoside [127] under relatively mild conditions (Equation 44) (38). Even though thioacetate is an ambident nucleophile,

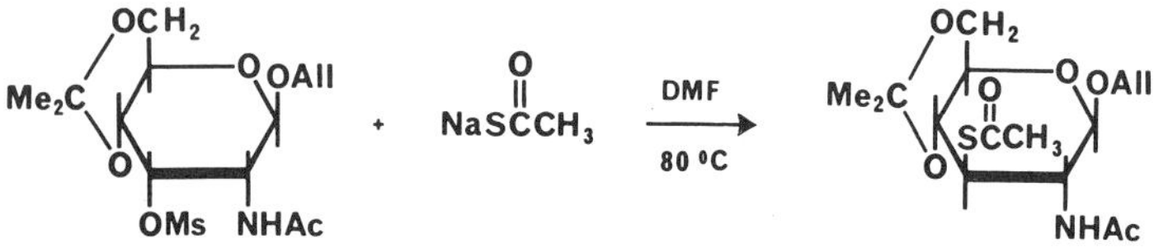

(44)

127 128

only reaction at the sulfur atom, the softer of the two nucleophilic centers, is observed.

REFERENCES

1. M. Haque and A. Freestone, *J. Chem. Soc. Perkin Trans. II*, 1509 (1977).

2. J. Yoshimura, N. Kawauchi, T. Yasumori, K. I. Sato, and H. Hashimoto, *Carbohydr. Res.*, *133*, 255 (1984).
3. H. Sasai, K. Matsuno, and T. Suami, *J. Carbohydr. Chem.*, *4*, 99 (1985).
4. A. L. Raymond and E. F. Schroeder, *J. Am. Chem. Soc.*, *48*, 2785 (1948).
5. T. K. M. Shing and A. S. Perlin, *Carbohydr. Res.*, *130*, 65 (1984).
6. B. Doboszewski and A. Zamojski, *Carbohydr. Res.*, *132*, 29 (1984).
7. P. J. Stang, M. Hanack, and L. R. Subramanian, *Synthesis*, 85 (1982).
8. M. G. Ambrose and R. W. Binkley, *J. Org. Chem.*, *48*, 674 (1983).
9. R. W. Binkley, M. G. Ambrose, and D. G. Hehemann, *J. Org. Chem.*, *45*, 4387 (1980).
10. M. Kloosterman, M. P. de Nijs, J. H. van Boom, *Recl. Trav. Chim. Pays-Bas.*, *103*, 243 (1984).
11. R. W. Binkley and M. G. Ambrose, *J. Carbohydr. Chem.*, *3*, 1 (1984).
12. S. Hanessian and J.-M. Vatele, *Tetrahedron Lett.*, *22*, 3579 (1981).
13. M. M. Abuaan, J. S. Brimacombe, and L. C. N. Tucker, *Carbohydr. Res.*, *132*, 51 (1984).
14. K. C. Nicolaou, R. E. Dolle, A. Chucholowski, and J. L. Randall, *J. Chem. Soc., Chem. Commun.*, 1153 (1984).
15. A. O. Stewart and R. M. Williams, *Carbohydr. Res.*, *135*, 167 (1984).
16. W. A. Szarek, *Adv. Carbohydr. Chem. Biochem.*, *28*, 225 (1973).
17. O. Mitsunobu, *Synthesis*, 1 (1981).
18. S. Hanessian and N. R. Pleassas, *J. Org. Chem.*, *34*, 1035, 1045, and 1053 (1969).

19. T. L. Hullar and S. B. Siskin, *J. Org. Chem.*, *35*, 225 (1970).
20. B. M. Pinto and D. R. Bundle, *Carbohydr. Res.*, *133*, 333 (1984).
21. P. J. Card, *J. Org. Chem.*, *48*, 393 (1983).
22. C. W. Somawardhana and E. G. Brunngraber, *Carbohydr. Res.*, *94*, C14 (1981).
23. P. J. Card and G. S. Reddy, *J. Org. Chem.*, *48*, 4734 (1983).
24. G. H. Posner and S. R. Haines, *Tetrahedron Lett.*, *26*, 5 (1985).
25. P. J. Card, *J. Carbohydr. Chem.*, *4*, 453 (1985).
26. R. Gigg, S. Payne, R. Conant, *J. Carbohydr. Chem.*, *2*, 207 (1983).
27. J. Thiem and S. Kopper, *J. Carbohydr. Chem.*, *2*, 75 (1983).
28. V. Dourtoglou and B. Gross, *J. Carbohydr. Chem.*, *2*, 57 (1983).
29. D. Rolf, J. A. Bennek, and G. R. Gray, *J. Carbohydr. Chem.*, *2*, 373 (1983).
30. For a description of the principle of hard and soft acids and bases and references to further reading see: J. A. March, *Advanced Organic Chemistry*, John Wiley and Sons, New York, 1985, 3rd Ed., p. 228.
31. T. Suami, Y. Fukuda, J. Yamamoto, Y. Saito, M. Ito, and S. Ohba, *J. Carbohydr. Chem.*, *1*, 9 (1982).
32. R. Alber, K. Dax, R. W. Link, and A. E. Stutz, *Carbohydr. Res.*, *118*, C5 (1983).
33. M. Georges, D. MacKay, and B. Fraser-Reid, *Carbohydr. Res.*, *130*, 115 (1984).
34. A. Zamojski, W. A. Szarek, and J. K. N. Jones, *Carbohydr. Res.*, *23*, 460 (1972).
35. J. Jary and A. Zobacova, *Methods Carbohydr. Chem.*, *6*, 229 (1972).
36. R. W. Binkley, *J. Carbohydr. Chem.*, *4*, 227 (1985).

37. G. Alfredsson and P. J. Garegg, *Acta. Chem. Scand.*, *27*, 724 (1973).
38. A. Hasegawa, M. Ozaki, M. Kiso, and I. Azuma, *J. Carbohydr. Chem.*, *3*, 331 (1984).
39. R. W. Binkley and D. J. Koholic, *J. Carbohydr. Chem.*, *3*, 85 (1984).
40. K. Takeo and K. Shinmitsu, *Carbohydr. Res.*, *133*, 135 (1984).
41. A. Calvo-Mateo, M. J. Camarasa, and F. G. De las Heras, *J. Carbohydr. Chem.*, *3*, 461 (1984).
42. P. J. Garegg and B. Samuelsson, *Synthesis*, 813 (1979).
43. P. J. Garegg and B. Samuelsson, *Synthesis*, 469 (1979).
44. K. Bock and M. Meldal, *Acta Chem. Scand.*, *B 38*, 255 (1984).
45. R. W. Binkley, *J. Org. Chem.*, *50*, 5646 (1985).
46. H. H. Baer, D. J. Astles, H.-C. Chin, and L. Siemsen, *Can. J. Chem.*, *63*, 432 (1985).
47. B. Fraser-Reid, D. R. Kelly, D. B. Tulshian, and P. S. Ravi, *J. Carbohydr. Chem.*, *2*, 105 (1983).
48. H. H. Baer and H. R. Hanna, *Carbohydr. Res.*, *110*, 19 (1982).
49. H. Yamamoto, H. Sasaki, and S. Inokawa, *Carbohydr. Res.*, *132*, 287 (1984).
50. Y. Nakahara, A. Fujita, and T. Ogawa, *J. Carbohydr. Chem.*, *3*, 487 (1984).
51. F. Nicotra, L. Panza, F. Ronchetti, and L. Toma, *Tetrahedron Lett.*, *25*, 5937 (1984).

10 Oxidation Reactions

I. INTRODUCTION

Although oxidation reactions are the topic of this chapter, it should be noted that a number of these reactions already have been encountered. In Chapter 7 (Section III.D.) the oxidation of unprotected sugars with several reagents, including nitric acid, halogens, and hypohalites to give aldonic and aldaric acids, was described. Oxidative deprotection of benzylidene acetals and *p*-methoxybenzyl ethers was discussed in Chapter 8 (Sections

III.B.4. and V.B., respectively). In the present chapter attention will be focused on hydroxyl-to-carbonyl oxidation in partially protected carbohydrates. The carbonyl compounds produced by such oxidation reactions provide some of the most useful intermediates in carbohydrate synthesis.

II. THE STABILITY OF CARBONYL COMPOUNDS

Although oxidation is one of the most important and extensively studied reactions of carbohydrates, the success of a particular reagent in a given situation is sometimes difficult to predict. The carbohydrate literature contains numerous examples in which one oxidizing agent was unsuccessful in a particular situation but another proved satisfactory. Unfortunately, the reason for the success of one reagent and not another often is not clear; however, much of the difficulty with oxidation reactions certainly must be associated with the instability of the carbonyl compounds produced. Many carbohydrates which contain carbonyl groups experience facile elimination reactions. For example, the oxidation of 1,2,3,4-tetra-*O*-acetyl-β-D-glucopyranose [1] with a DMSO-based reagent gives 1,2,3-tri-*O*-acetyl-4-deoxy-6-aldehydo-α-L-*threo*-hex-4-enodialdo-1,5-pyranose [2] as the oxidation product (Figure 1) (1). There is little doubt that the initial product from this reaction is 1,2,3,4-tetra-*O*-acetyl-6-aldehydo-β-D-glucopyranose [3] and that 3 experiences loss of the elements of acetic acid by an E2 elimination under the reaction conditions; in fact, when a sample of 3, prepared by a photochemical method (2), is treated with triethylamine, it rapidly forms the unsaturated compound 2 (Figure 1).

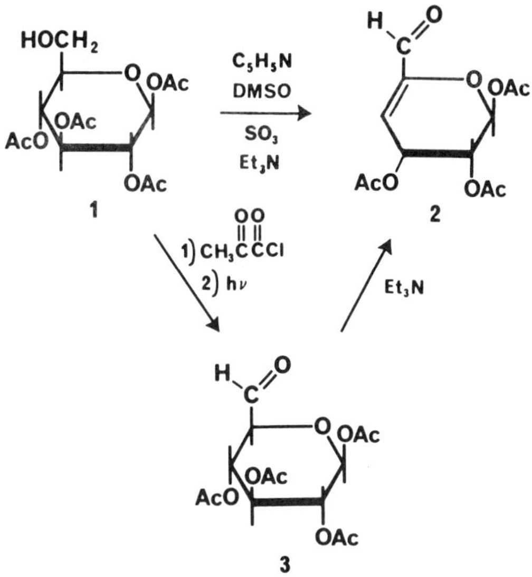

FIG. 1. Oxidation of 1,2,3,4-tetra-*O*-acetyl-β-D-glucopyranose [1] and subsequent elimination reaction.

III. OXIDIZING AGENTS AND OXIDATION REACTIONS

Oxidation of compounds which contain a single unprotected hydroxyl group usually is accomplished either with a combination of reagents which includes DMSO (methyl sulfoxide) or with a chromium-containing compound. Both types of oxidizing agent convert primary and secondary alcohols into aldehydes and ketones, respectively.

A. DMSO-based Reagents

Reagent combinations which include DMSO also typically contain either an acid chloride (e.g., oxalyl chloride), an acid anhydride (e.g., acetic anhydride, phosphorus pentoxide, or sulfur trioxide), or dicyclohexylcarbodiimide (DCC). The purpose of these "additional" compounds is to interact with DMSO to form a reactive intermediate; for example, the oxidation process using DCC is initiated by acid-catalyzed reaction of DMSO with DCC to generate the transient intermediate 4 (Figure 2). This intermediate provides a means for coupling the partially protected sugar (the compound to be oxidized) with DMSO (the oxidizing agent) to generate a new charged species [5]. The positively charged sulfur atom in 5 increases the acidity of the hydrogen atoms on the methyl groups to the point that deprotonation occurs with ease to form the zwitterion 6, a compound which experiences a concerted bond reorganization to give the oxidized carbohydrate and dimethyl sulfide.

Oxidations using DMSO are sometimes accompanied by side reactions such as the formation of (methylthio)methyl ethers (3).

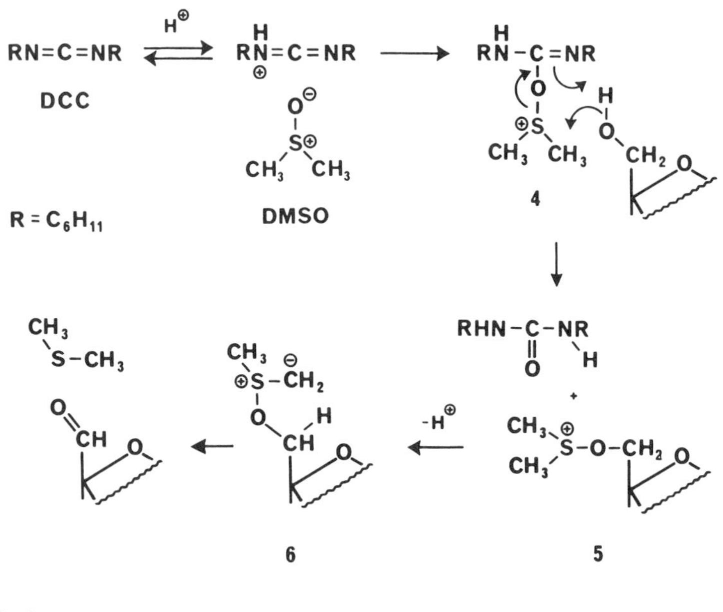

FIG. 2. Proposed mechanism for oxidation of a partially protected carbohydrate using DMSO and DCC.

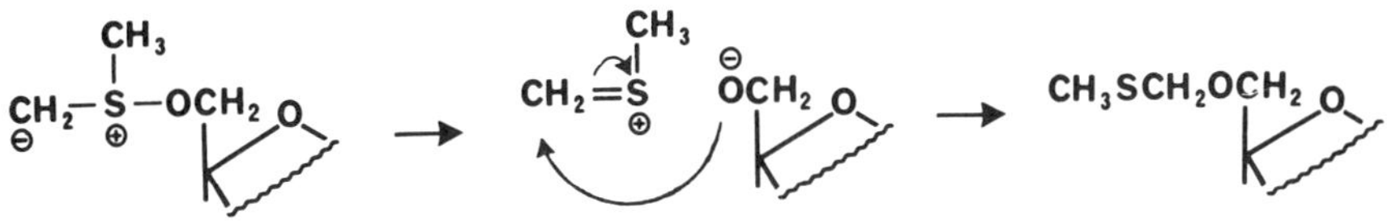

6

FIG. 3. Proposed mechanism for (methylthio)methyl ether formation.

(Methylthio)methyl ethers probably arise from the zwitterion 6 by an alternate mode of decomposition such as that shown in Figure 3. DMSO-based oxidations also generate compounds with unpleasant odors which may require chromatography for removal.

The DMSO-based oxidation which has enjoyed the greatest popularity over the longest period of time is the Pfitzner-Moffatt reaction (4). Pfitzner-Moffatt oxidation is accomplished using DMSO, dicyclohexylcarbodiimide (DCC), and a proton source (the reagents selected to illustrate the mechanism for DMSO-based oxidation shown in Figure 2). An example of this type of oxidation is given in Equation 1 (5). Reaction which involves DMSO in

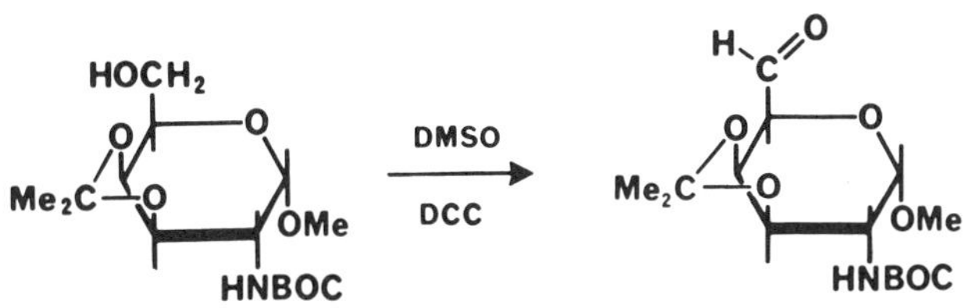

7 8 (1)

$$\mathrm{BOC} = -\overset{\overset{\displaystyle \mathrm{O}}{\|}}{\mathrm{C}}\mathrm{OCH_2C_6H_5}$$

conjunction with oxalyl chloride (Equation 2) (6) recently has become increasingly important in the oxidation of carbohydrates. This reaction, known as the Swern oxidation (7), is conducted at

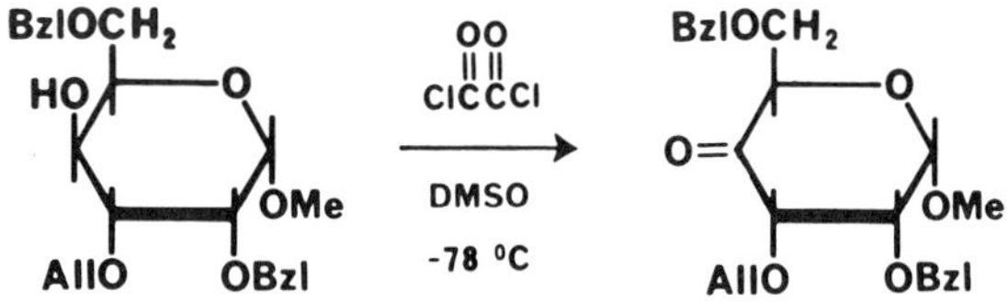

9 10 (2)

at -78°C, a temperature at which side reactions are minimized; consequently, product yields from Swern oxidations usually are high. A review of DMSO based oxidations is found in reference 8.

B. Chromium-containing Compounds

Chromium-containing compounds are equal to DMSO-based reagents in importance in the oxidation of carbohydrates. Chromium trioxide in pyridine and pyridinium chlorochromate (PCC) are the two members of this group most often selected for synthetic reactions. An example of a chromium trioxide oxidation is provided by the reaction shown in Equation 3 (9) and one using

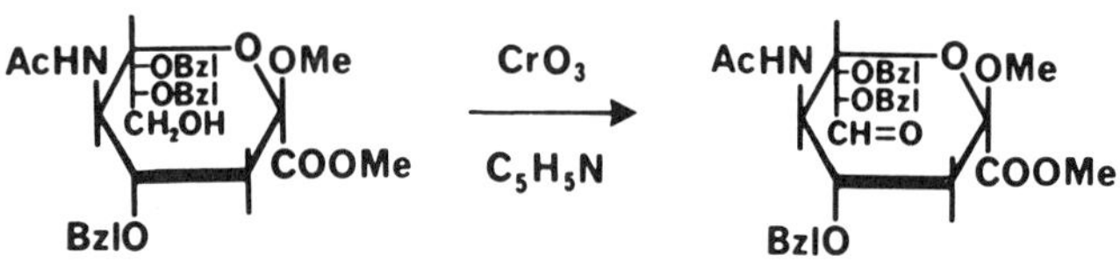

11 12 (3)

PCC is given in Equation 4 (10). A proposed mechanism for a typical reaction involving a chromium-containing compound (chromium trioxide) is pictured in Figure 4. This process begins with the formation of a chromate ester [15] which then loses the elements of H_2CrO_3 in an E2 elimination reaction. Reactions using chromium-containing compounds are conducted under

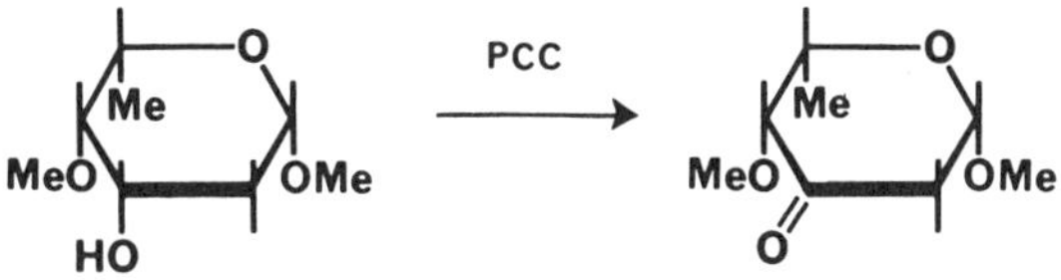

13 14 (4)

$$PCC = C_5H_5\overset{\oplus}{N}H \ \ CrClO_3^{\ominus}$$

anhydrous conditions to prevent further oxidation of the type shown in Figure 5.

Anhydrous conditions often are maintained during oxidation with pyridinium chlorochromate (PCC) by adding molecular sieves to the reaction mixture. Reaction mixtures in which PCC is used are slightly acidic; therefore, if the starting material or the product is acid sensitive, a compound such as sodium acetate is added to the reaction mixture to minimize acid-catalyzed reaction. Oxidation with chromium-containing compounds produces tarry chromium salts which sometimes can be removed by filtration, but other times require chromatography.

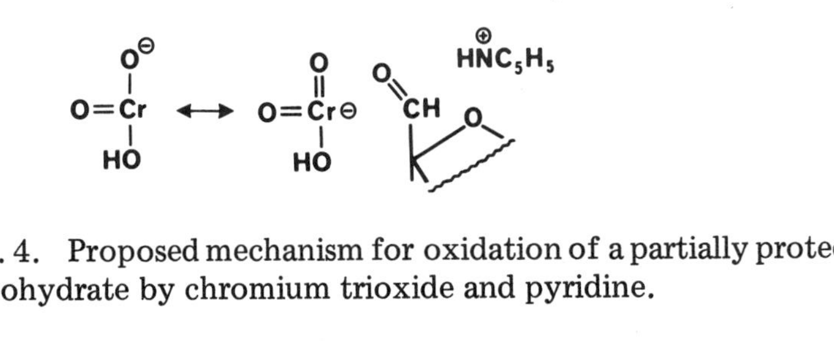

FIG. 4. Proposed mechanism for oxidation of a partially protected carbohydrate by chromium trioxide and pyridine.

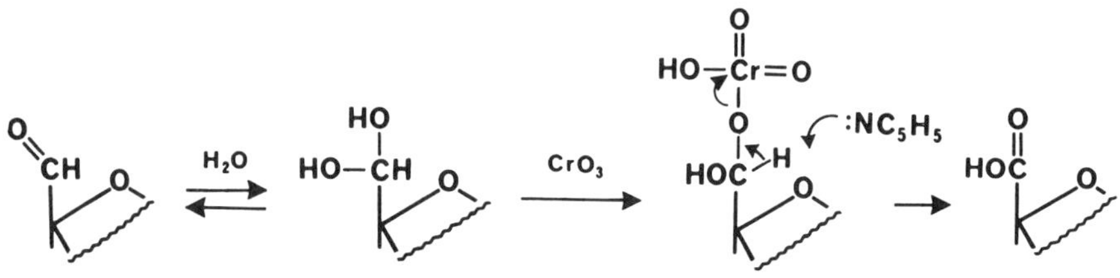

FIG. 5. Hydration of an aldehyde and subsequent oxidation to a carboxylic acid.

C. Ruthenium Tetraoxide

Another compound which often has been used to oxidize partially protected carbohydrates is ruthenium tetraoxide (ruthenium (VIII) oxide). This reagent converts 6-deoxy-1,2-*O*-isopropylidene-3-*O*-methyl-6-nitro-α-D-glucofuranose [16] into the corresponding ketose [17] (Equation 5) (11). Oxidation also can be

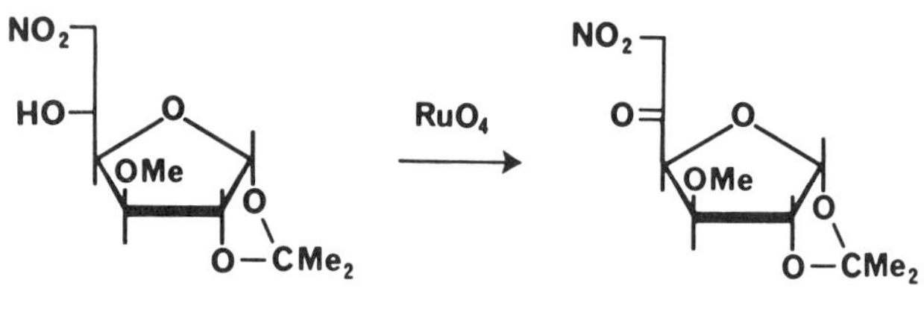

16 17 (5)

conducted using a catalytic amount of ruthenium tetraoxide if sodium or potassium periodate is present to continuously regenerate the ruthenium tetraoxide. Ruthenium tetraoxide is such a powerful oxidizing agent that it is sometimes difficult to prevent further oxidation of the initially formed products. Oxidations with this reagent are two-phase reactions which are quite dependent upon reaction conditions including such factors as the efficiency of stirring of the reaction mixture; thus, reaction times can vary considerably. Because of the possibility of overoxidation, it is important to monitor the progress of the reaction and remove the oxidizing agent as soon as reaction is complete. Since either a DMSO-based reagent or a chromium-containing oxidant usually can accomplish the same oxidation with fewer difficulties, ruthenium tetraoxide is used less frequently now than in the past, although it remains an important oxidizing agent.

D. Photochemical Oxidation

A photochemical process exists for oxidation of carbohydrates which requires synthesis of a pyruvate ester and subsequent photolysis of this ester to give a carbonyl compound (2). An example of this photochemical oxidation is shown in Figure 1. The "key" steps in the photochemical reaction are an internal transfer of a hydrogen atom and fragmentation of the resulting diradical [18] (Figure 6). The primary advantage of photochemical oxidation is that it is an extremely mild process; however, the sequence of reactions necessary for the oxidation is sufficiently involved that its use is important only in situations where other oxidizing agents are unsuccessful.

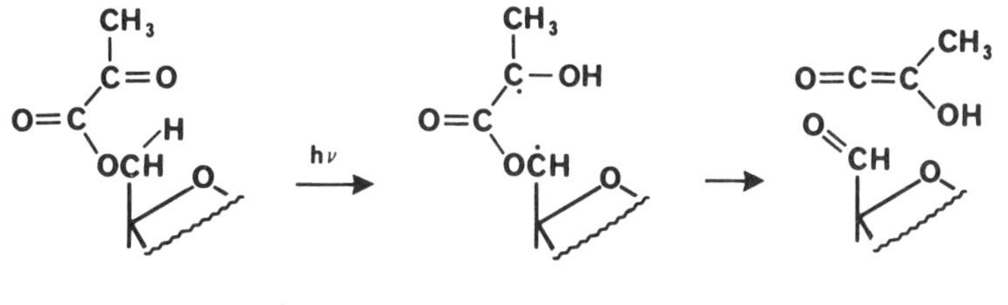

18

FIG. 6. Proposed mechanism for photochemical oxidation.

IV. OXIDATION OF DIOLS

Oxidation of a diol with an appropriate oxidizing agent, such as sodium periodate, is a process which cleaves the carbon-carbon bond in the diol to generate two carbonyl groups. This type of reaction was used extensively for structural determination before the advent of modern instrumentation. Although its use is now much less frequent as a structural tool, oxidation of a terminal diol by periodate represents an excellent method for reducing the length of a carbon atom chain by one atom and leaving an aldehydo group at the end of the chain. An example of this oxidative chain reduction is given in Equation 6 (12). A widely

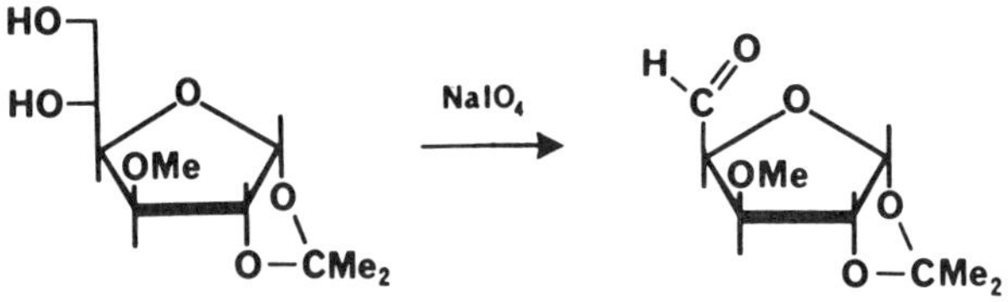

19 20 (6)

accepted mechanism for periodate oxidation involves the formation and concerted decomposition of a cyclic periodate ester [21] (Figure 7).

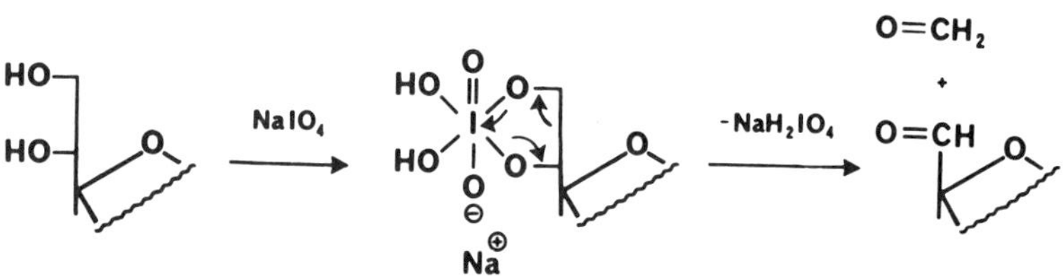

21

FIG. 7. Proposed mechanism for periodate oxidation.

REFERENCES

1. D. M. Mackie and A. S. Perlin, *Carbohydr. Res.*, *24*, 67 (1972).
2. R. W. Binkley, *J. Org. Chem.*, *42*, 1216 (1977).
3. J. L. Godman and D. Horton, *Carbohydr. Res.*, *6*, 229 (1968).
4. K. E. Pfitzner and J. G. Moffatt, *J. Am. Chem. Soc.*, *87*, 5661 and 5670 (1965).
5. T. Suami, Y. Fukuda, J. Yamamoto, Y. Saito, M. Ito, and S. Ohba, *J. Carbohydr. Chem.*, *1*, 9 (1982).
6. P. Weitzel, H. P. Bullan, A. Maulahagen, D. Muller, and G. Snatzke, *Tetrahedron*, *40*, 3657 (1984).
7. K. Omura and D. Swern, *Tetrahedron*, *34*, 1651 (1978).
8. A. J. Mancusco and D. Swern, *Synthesis*, 165 (1981).
9. M. Sharma and W. Korytnyk, *J. Carbohydr. Chem.*, *1*, 311 (1982).
10. J. Yoshimura, T. Yasumori, T. Kondo, and K. Sato, *Bull. Chem. Soc. Jpn.*, *37*, 2535 (1984).
11. J. M. J. Tronchet, K. D. Pallie, and F. Barbalat-Rey, *J. Carbohydr. Chem.*, *4*, 29 (1985).
12. H. Sasai, K. Matsuno, and T. Suami, *J. Carbohydr. Chem.*, *4*, 99 (1985).

11 Hydrogenation and Hydrogenolysis

I. INTRODUCTION

Hydrogenation and hydrogenolysis are two types of reaction which introduce hydrogen atoms into a molecule. Hydrogenation is the addition of hydrogen atoms to a double or triple bond. The bond being hydrogenated may be between two carbon atoms or it may be a bond involving one or more heteroatoms. Hydrogenolysis refers to the cleavage of a single bond and the subsequent (or simultaneous) addition of hydrogen atoms to the fragments produced by bond homolysis. An example of a hydrogenation reaction is given in Equation 1 (1) while a typical hydrogenolysis reaction is shown in Equation 2 (2).

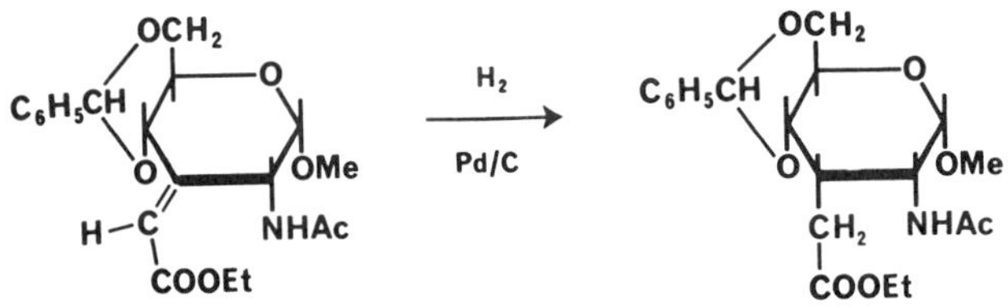

1 2 (1)

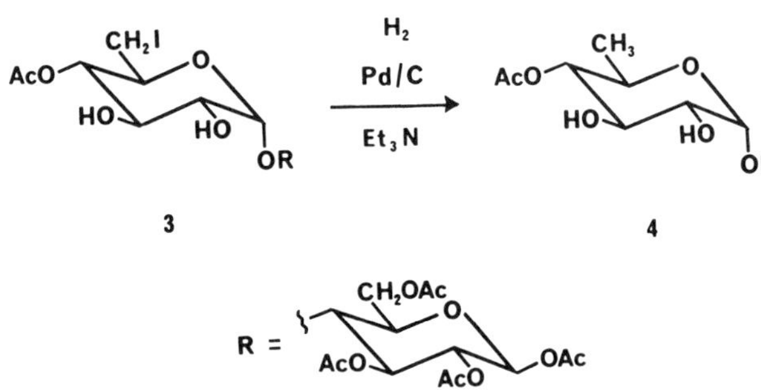

3 4 (2)

II. HYDROGENATION

A. Catalytic Hydrogenation

1. Heterogeneous Reactions

a. Addition to Unsaturated Carbon-Carbon Bonds. Catalytic hydrogenations of unsaturated carbohydrates using catalysts such as palladium, Raney nickel, platinum, or platinum oxide are heterogeneous reactions which take place on the surface of the catalyst. Although such reactions are of considerable theoretical and practical importance and have been extensively studied, their heterogeneous nature makes mechanistic investigation difficult. Even though this difficulty exists, considerable progress has been made in understanding heterogeneous reactions and, as a result, the mechanism for hydrogenation shown in Figure 1 has received wide acceptance (3). According to this mechanistic proposal, the unsaturated carbohydrate is adsorbed on two adjacent sites on the surface of the metal catalyst. Adsorption from the least hindered side of the molecule occurs preferentially. The nature of the bonding between the unsaturated compound and the catalyst is not fully understood; however, adsorption onto the metal surface definitely decreases the bonding between the two doubly bonded carbon atoms. Adsorption of a hydrogen molecule breaks the hydrogen-hydrogen bond to give two adsorbed hydrogen atoms. Since the unsaturated compound has its least hindered side facing the catalyst surface and since hydrogen is also on the surface, the hydrogen addition takes place on the least hindered side of the molecule (Figure 1). Addition of one hydrogen atom to the unsaturated compound creates a radical [5] which is still adsorbed on the catalyst. When the second hydrogen atom adds to this radical, the hydrogenated molecule [6] is released from the surface into solution. A comprehensive discussion of hydrogenation reactions is found in reference 3.

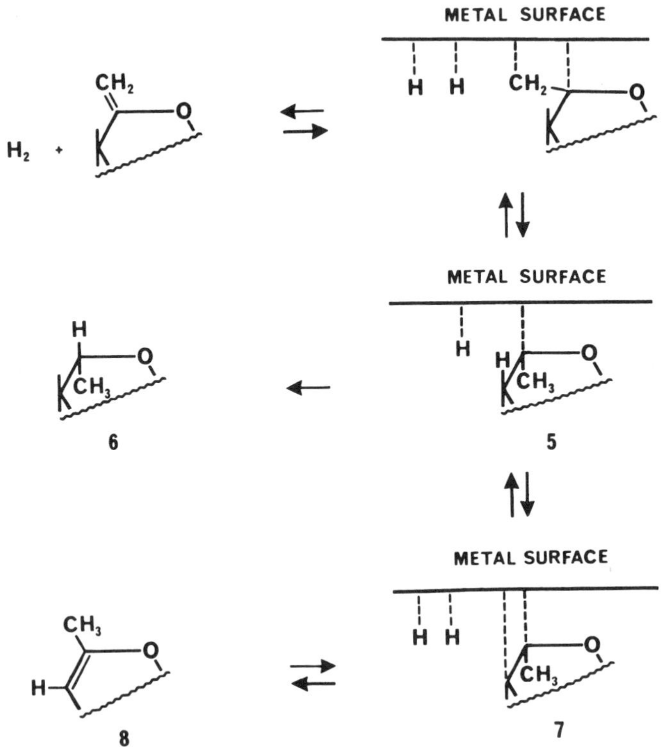

FIG. 1. Proposed mechanism for hydrogenation of a carbon-carbon double bond using a heterogeneous catalyst.

All of the steps in the mechanism for hydrogenation shown in Figure 1 up to and including the formation of the adsorbed radical 5 are reversible. This reversibility permits double-bond migration. The intermediate 5 can return a hydrogen atom to the metal surface to produce a new diadsorbed species 7. Deadsorption of 7 gives a new unsaturated compound [8].

Catalytic hydrogenation is an important step in several common synthetic transformations. One of these is the inversion of configuration at C-5 in a pyranoid ring system. Usually this type of stereochemical change is needed to convert a compound in the D series into one in the L series; thus, when methyl 4-*O*-benzoyl-6-bromo-2,3,6-trideoxy-2-fluoro-3-trifluoroacetamido-α-D-altropyranoside [9] is treated with silver fluoride in pyridine and the resulting unsaturated compound [10] is catalytically hydrogenated, methyl 4-*O*-benzoyl-2,3,6-trideoxy-2-fluoro-3-trifluoroacetamido-β-L-galactopyranoside [11] is produced (Figure 2) (4). Since a number of the substituents in compound 10 project onto its α-face, particularly the substituents in the region of the double bond, hydrogenation occurs as expected from the less hindered β-face. The resulting compound [11] is a member of the L family.

Another important synthetic application of catalytic hydrogenation is in the synthesis of branched-chain sugars. One example of this application has been encountered (Equation 1). Another example is the reaction of methyl 6-*O*-acetyl-2,3-dideoxy-2-*C*-methyl-4-*C*-methylene-α-D-*threo*-hexopyranoside [13] with hydrogen in the presence of palladium. The methylene group in 13, which was introduced into the molecule by a Wittig reaction, is converted into a second methyl branch by hydrogenation (Figure 3) (5). (The Wittig reaction will be discussed in Chapter 12.) The formation of 14 and 15 in a ratio of 1.5/1.0 suggests

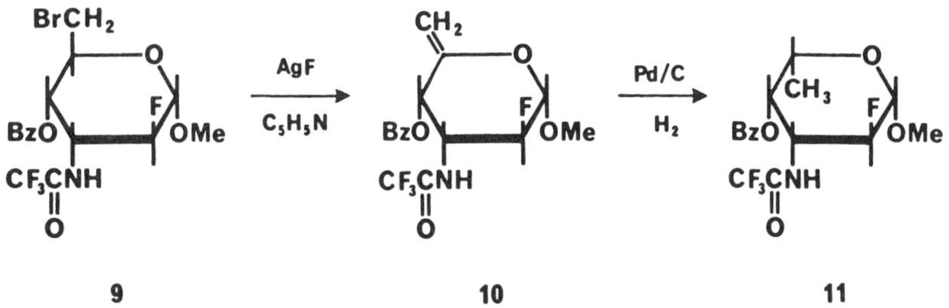

FIG. 2. Reaction sequence for inversion of configuration at C-5 in methyl 4-*O*-benzoyl-2,3,6-trideoxy-2-fluoro-3-trifluoroacetamido-α-D-altropyranoside [9].

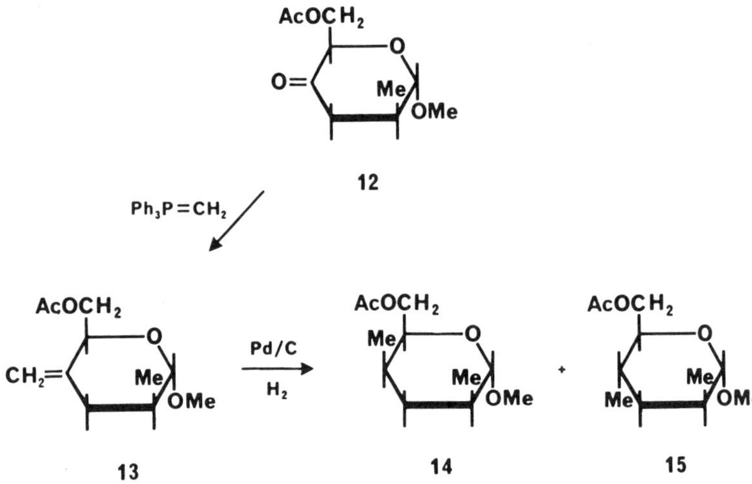

FIG. 3. Reaction sequence for introduction of a methyl branch at C-4 in methyl 6-*O*-acetyl-2,3-dideoxy-2-*C*-methyl-4-methylene-α-D-*threo*-hexopyranoside [13].

that there is little preference for the orientation of 13 during its adsorption onto the catalyst surface.

Although palladium on carbon is the most widely used of the hydrogenation catalysts, it has a tendency to isomerize double bonds (Figure 1). Isomerization of double bonds is controlled by the adsorption-deadsorption process. Some catalysts (e.g., platinum oxide) adsorb molecules so strongly that deadsorption usually does not take place until hydrogenation is complete. Such catalysts permit little isomerization. Palladium does not adsorb molecules strongly and, consequently, the possibility is increased for deadsorption of an isomerized compound. For example, double-bond isomerization is a significant side reaction when methyl 3-benzamido-4-*O*-benzoyl-2,3,6-trideoxy-2-fluoro-α-D-*arabino*-hex-5-enopyranoside [16] reacts with hydrogen in the presence of palladium (Equation 3) (6). Compounds 17 and 18 are formed in nearly equal amounts

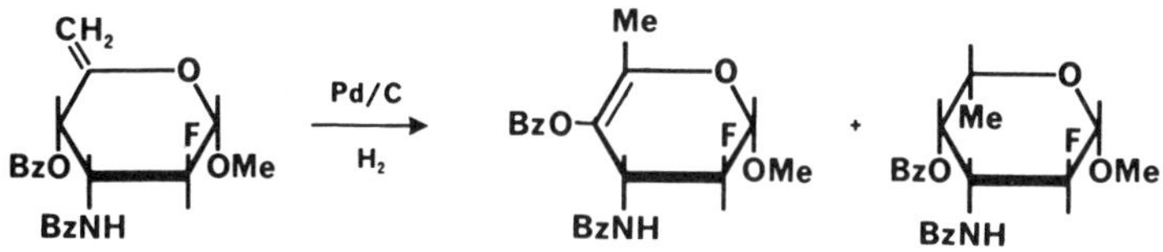

16 17 18 (3)

b. Addition to Unsaturated Nitrogen-Nitrogen Bonds. Azides can be reduced catalytically to the corresponding aminodeoxy compounds. When displacement by azide (Chapter 9, Section IV.D.2.) is coupled with catalytic hydrogenation, this sequence of reactions represents the most widely used method for synthesis of aminodeoxy sugars. Primary (7) and secondary (8) azides react with ease (Equations 4 and 5); in fact, the conversion of

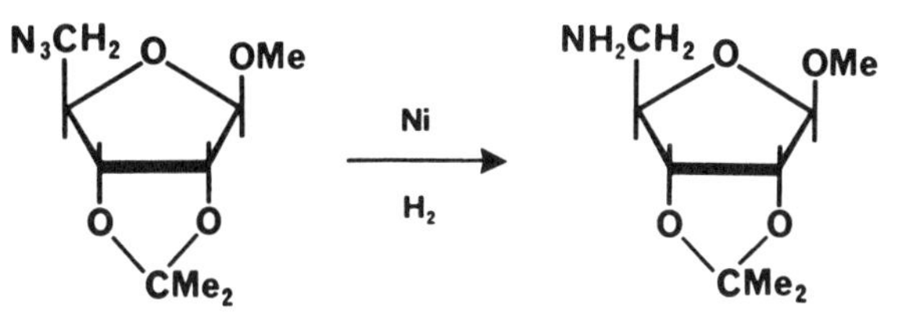

19 20 (4)

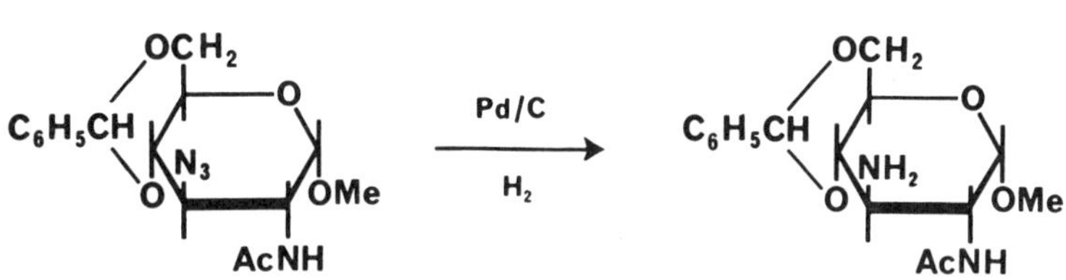

21 22 (5)

azides into amines is so facile that the azido function can be hydrogenated in the presence of other easily reduced groups. The initial steps in the hydrogenation process resemble those for hydrogenation of compounds containing carbon-carbon unsaturation (Figure 1); however, the latter stages of reaction must involve cleavage of nitrogen-nitrogen bonds and addition of hydrogen atoms (hydrogenolysis).

2. Homogeneous Reactions

Although catalytic hydrogenation usually is a heterogeneous process, it also can take place in a single phase if a catalyst is used which is soluble in the reaction solvent. One such catalyst, which is soluble in organic solvents, is tris(triphenylphosphine)rhodium chloride [$(Ph_3P)_3RhCl$] [23], known as Wilkinson's catalyst.

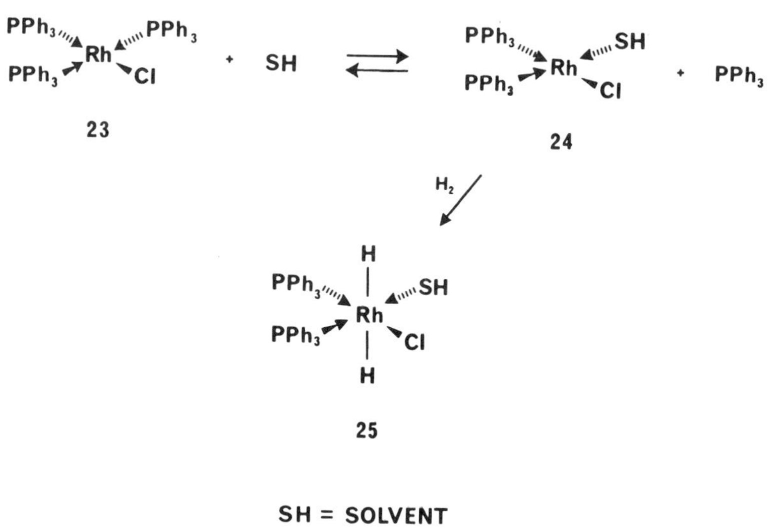

FIG. 4. The first part of the proposed mechanism for hydrogenation of a carbon-carbon double bond using a homogeneous catalyst.

When compound 13 reacts with hydrogen in the presence of Wilkinson's catalyst [23], the same products [14 and 15] that were produced by the palladium-catalyzed reaction (Figure 3) are formed. The only difference in product formation is a greater stereoselectivity when 23 is the catalyst (14/15 = 6/1 (5). In general, homogeneous reactions are more selective than heterogeneous ones. This selectivity includes greater discrimination in the preferential hydrogenation of a less hindered double bond in the presence of a more hindered one.

The mechanism for homogeneous hydrogenation using Wilkinson's catalyst can be viewed as the series of reactions shown in Figures 4 and 5. In the first phase of this process $(PPh_3)_3RHCl$ [23] exchanges a PPh_3 ligand for a solvent molecule (Figure 4).

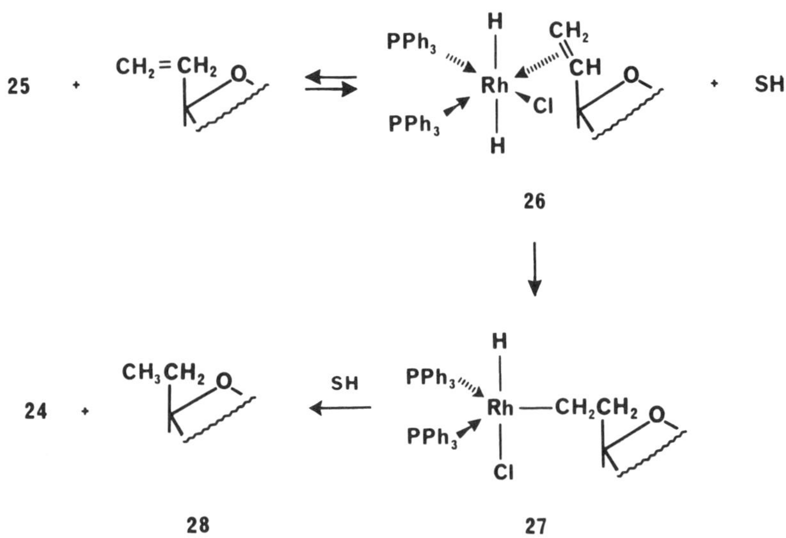

FIG. 5. The second part of the proposed mechanism for hydrogenation of a carbon-carbon double bond using a homogeneous catalyst.

(The arrows in the rhodium-containing compounds indicate dative bonds; that is, bonds in which both electrons are considered to "belong" to the ligand for purposes of determining formal charge, but these electrons do contribute to establishing the rare gas configuration about rhodium.) Compound 24 reacts with molecular hydrogen to form a new complex [25] in which two hydrogen atoms are bonded to rhodium. The compound to be hydrogenated next replaces the solvent molecule in 25 to give an intermediate [26] in which the unsaturated ligand is ready to undergo hydrogenation (Figure 5). Two hydrogen atoms then are transferred in a stepwise fashion to give the saturated carbohydrate [28] and regenerate the catalyst [24].

B. Reactions with Hydride Reagents

The other important type of hydrogenation reaction involves the addition of hydride reagents to polar double bonds (e.g., C=0 and C=N). This topic will be considered in Chapter 12 along with other nucleophilic addition reactions to unsaturated systems.

III. HYDROGENOLYSIS

Hydrogenolysis is a type of reaction which has been encountered several times in previous chapters. In Chapter 8 hydrogenolysis was seen to be the principal method for removing the benzyl group from benzyl glycosides (Section III.A.1.), benzyl ethers (Section V.A.3.), and benzyloxycarbonyl (BOC)-protected compounds (Section VI). It also was the reaction used to convert hydrazino compounds, produced by nucleophilic substitution, into the corresponding amines (Chapter 9, Section IV.D.3.). Finally, as mentioned earlier in this chapter (Section II.A.1.b.), hydrogenolysis is involved in the catalytic conversion of azides into amines.

A. Catalytic Hydrogenolysis

Heterogeneous catalytic hydrogenolysis usually takes place in compounds which have a carbon to heteroatom single bond. Most frequently the compound reacting is a deoxyhalogeno sugar and the catalyst is Raney nickel or palladium on carbon. If necessary, a base is added to the reaction mixture to neutralize any acids (e.g., HI or HBr) produced. This reaction occurs most readily for deoxyiodo sugars (Equation 2) (2) but it also takes place easily with bromodeoxy compounds (Equation 6) (9). The normal order

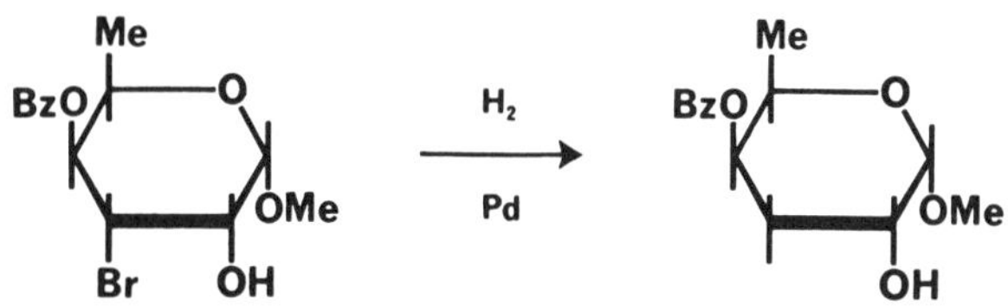

29 30 (6)

of reactivity for halogenated carbohydrates in catalytic hydrogenolysis is I > Br > Cl >> F. When hydrogenolysis is combined with the synthesis of deoxyhalogeno sugars by nucleophilic substitution, this pair of reactions represents an excellent sequence for deoxy sugar synthesis.

B. Free-Radical Additions of Hydrogen Atoms

1. Reaction of Tri-*n*-butylstannane (Tri-*n*-butyltin Hydride) with Halogenated Carbohydrates

Tri-*n*-butylstannane reacts with deoxyiodo or bromodeoxy sugars by a free-radical, chain process in which the carbon-halogen bond in the starting material is replaced by a carbon-hydrogen bond in the product. Reaction is initiated by normal free-radical

$$n\text{-}Bu_3Sn\cdot + R{-}I \longrightarrow n\text{-}Bu_3SnI + R\cdot$$

$$R\cdot + n\text{-}Bu_3SnH \longrightarrow R{-}H + n\text{-}Bu_3Sn\cdot$$

FIG. 6. Proposed mechanism for hydrogenolysis using tri-*n*-butyltin hydride.

initiators such as azobisisobutyronitrile or benzoyl peroxide. The halogen atom replacement by a hydrogen atom is accomplished by the two-step sequence of reactions shown in Figure 6. (The reactions shown in Figure 6 are the propagation steps in the chain process.) This type of hydrogenolysis reaction occurs readily with both secondary (Equation 7) (10) and primary (Equation 8) (11)

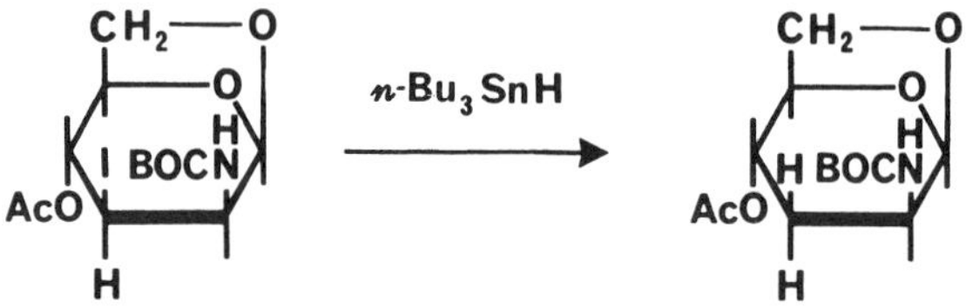

31 32 (7)

$BOC = C_6H_5CH_2OC(=O)$

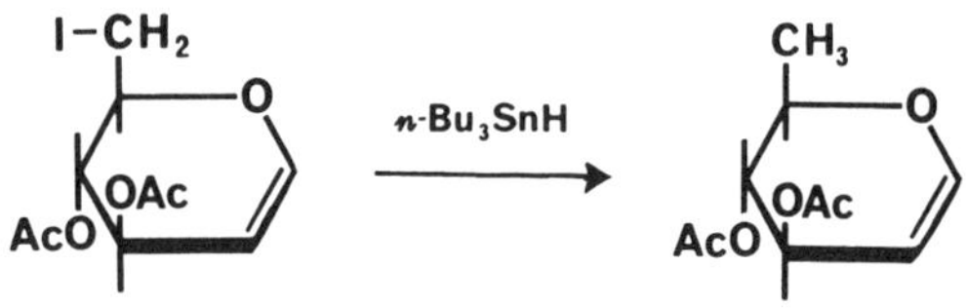

33 34 (8)

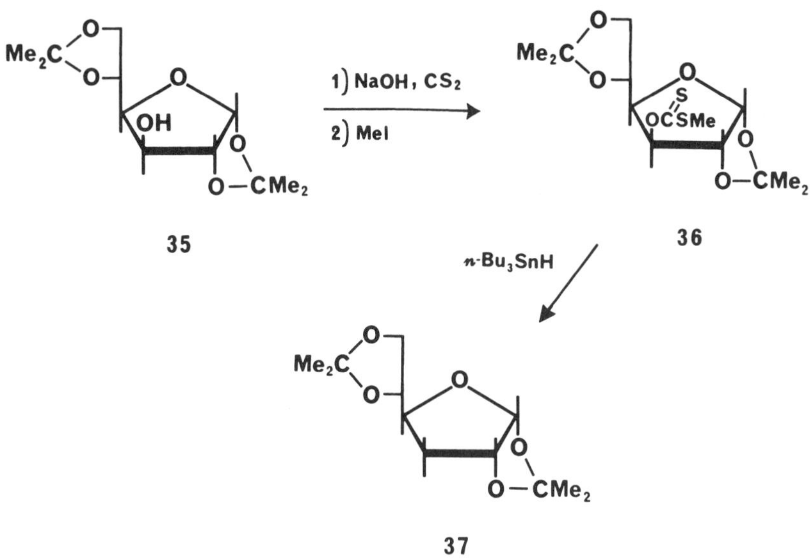

FIG. 7. Reaction sequence for hydroxyl group removal by hydrogenolysis of a dithiocarbonate.

bromides and iodides and represents an alternative to catalytic hydrogenolysis.

A related process is the reaction of dithiocarbonates with tri-*n*-butylstannane to give deoxy compounds (Figure 7) (12). The dithiocarbonate can be prepared directly from a partially protected sugar without displacement of a group attached to a carbon atom; as a result, this reaction is attractive when a deoxyhalogeno compound is difficult to prepare.

2. Photochemical Reaction of Deoxyhalogeno Sugars

Ultraviolet irradiation of deoxyiodo and bromodeoxy sugars cleaves the carbon-halogen bond homolytically to give a carbon radical [38] and a halogen atom (Figure 8). If reaction is conducted in a solvent which is a good hydrogen atom donor, such as

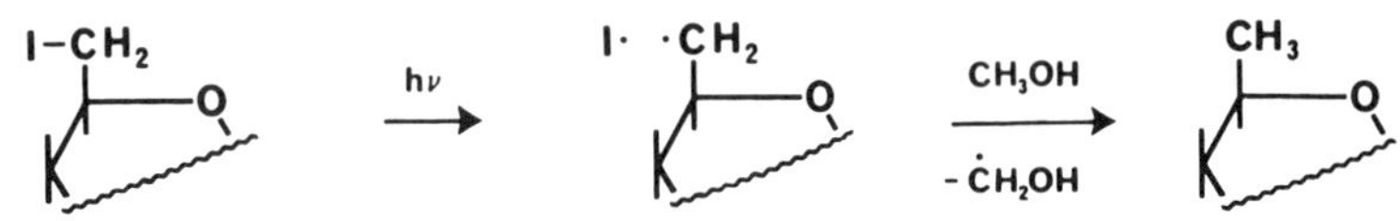

38 39

$$2\,\cdot CH_2OH \longrightarrow HOCH_2CH_2OH$$

$$2\,I\cdot \longrightarrow I_2$$

$$\cdot CH_2OH + I\cdot \longrightarrow CH_2O + HI$$

FIG. 8. Proposed mechanism for hydrogenolysis by photochemical reaction of deoxyhalogeno sugars.

methanol or 2-propanol, the carbon radical [38] abstracts a hydrogen atom from the solvent to give the corresponding deoxy sugar [39]. Halogen acids are produced by the reaction and must be neutralized. Additional mechanistic information on the photolysis of deoxyiodo sugars is found in reference 13.

Although reaction occurs for both deoxyiodo (Equation 9 (14) and bromodeoxy (Equation 10) (15) compounds, the iodo compounds are preferred because they absorb light at longer wavelengths and, consequently, photochemical reaction can be

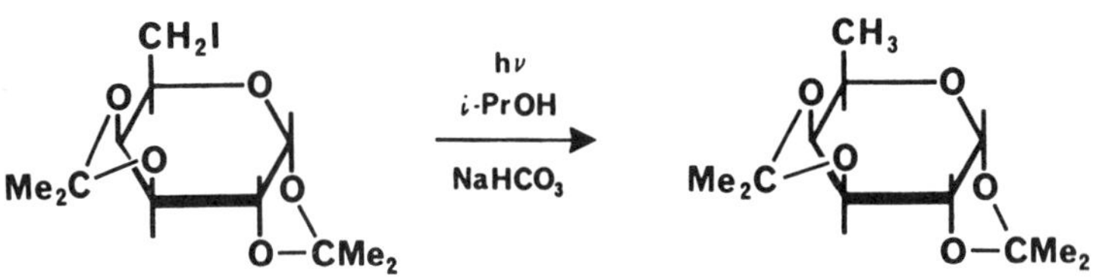

(9)

40 41

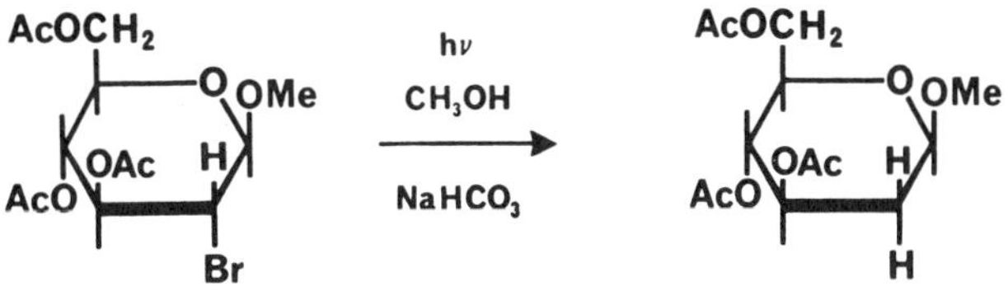

(10)

42 43

produced by less energetic radiation. In general, higher energy radiation causes a reduced yield of the desired product by allowing side reactions to occur. The primary difficulty with photolysis of halides as a synthetic technique is that the reactions are slowed by generation of light-absorbing materials during photolysis; thus, conversion of synthetically useful amounts of material can require long reaction times.

C. Nucleophilic Substitution with Hydride Reagents

Nucleophilic substitution reactions with hydride as the nucleophile constitute another method for hydrogenolysis. These reactions have already been discussed (Chapter 9, Section IV.G.) along with related displacement processes.

REFERENCES

1. A. Calvo-Mateo, M. J. Camarasa, and F. G. de las Heras, *J. Carbohydr. Chem.*, *3*, 475 (1984).
2. K. Takeo and K. Shinmitsu, *Carbohydr. Res.*, *133*, 135 (1984).
3. P. N. Rylander, *Catalytic Hydrogenation in Organic Synthesis*, Academic Press, Inc., New York, 1979, pp. 31-33.
4. D. Picq and D. Anker, *J. Carbohydr. Chem.*, *4*, 113 (1985).
5. Y. Chapleur, F. Germain, A. Aubry, and D. Bayeul, *J. Carbohydr. Chem.*, *3*, 433 (1984).

6. M. K. Gurjar, V. J. Patil, J. S. Yadav, and A. V. Rama Rao, *Carbohydr. Res.*, *135*, 174 (1984).
7. T. Suami, Y. Fukuda, J. Yamamoto, Y. Saito, M. Ito, and S. Ohba, *J. Carbohydr. Chem.*, *1*, 9 (1982).
8. A. Calvo-Mateo, M. J. Camarasa, and F. G. de las Heras, *J. Carbohydr. Chem.*, *3*, 461 (1984).
9. B. M. Pinto and D. R. Bundle, *Carbohydr. Res.*, *133*, 333 (1984).
10. I. Cerny, M. Budesensky, T. Trnka, M. Cerny, *Carbohydr. Res.*, *130*, 103 (1984).
11. B. Fraser-Reid, D. R. Kelly, D. B. Tulshian, and P. S. Ravi, *J. Carbohydr. Chem.*, *2*, 105 (1983).
12. B. L. Roy and P. Deslongchamps, *Can. J. Chem.*, *63*, 651 (1985).
13. R. C. Roth and R. W. Binkley, *J. Org. Chem.*, *50*, 690 (1985).
14. R. W. Binkley and W. W. Binkley, *Carbohydr. Res.*, *8*, 370 (1968).
15. R. W. Binkley and D. Bankaitis, *J. Carbohydr. Chem.*, *1*, 1 (1982).

12 Addition Reactions

I. NUCLEOPHILIC ADDITION REACTIONS
   A. Reversible and Irreversible Reactions
   B. Normal (1,2) and Conjugate (1,4) Addition
   C. Carbon Nucleophiles
      1. Organolithium Compounds and Grignard Reagents
      2. Lithium Dialkylcuprates
      3. Anions Derived from Nitro Compounds
         a. Nitromethane
         b. Carbohydrates Containing Nitro Groups
      4. Wittig Reagents
      5. Cyanide Ion
      6. The Anion Derived from Dithiane
   D. Hydride Reagents
      1. Lithium Aluminum Hydride and Sodium Borohydride
      2. Hindered Hydride Reagents
      3. Sodium Cyanoborohydride
   E. Nitrogen-containing Nucleophiles
   F. Oxygen-containing Nucleophiles

II. ELECTROPHILIC ADDITION REACTIONS
  A. Halogenation
  B. Epoxidation
  C. Mercury-catalyzed Ring Opening (Formation of Substituted Cyclohexanones)

III. CYCLOADDITION REACTIONS
  A. Diels-Alder Reaction
  B. Formation of Diols by Reaction with Potassium Permanganate and Osmium Tetraoxide

I. NUCLEOPHILIC ADDITION REACTIONS

A. Reversible and Irreversible Reactions

When a sufficiently reactive combination of a nucleophile and polar double bond interact with each other, addition of the nucleophile to the double bond can take place. With certain nucleophiles and double bonds this addition is reversible; for example, the reaction of water with 1,2:3,4-di-*O*-isopropylidene-α-D-*galacto*-hexodialdo-1,5-pyranose [1] is an equilibrium process which favors the hydrated aldehyde 2 when water is the reaction solvent but in chloroform saturated with water, the aldehyde 1 is the major component of the mixture (Equation 1) (1). Reversible

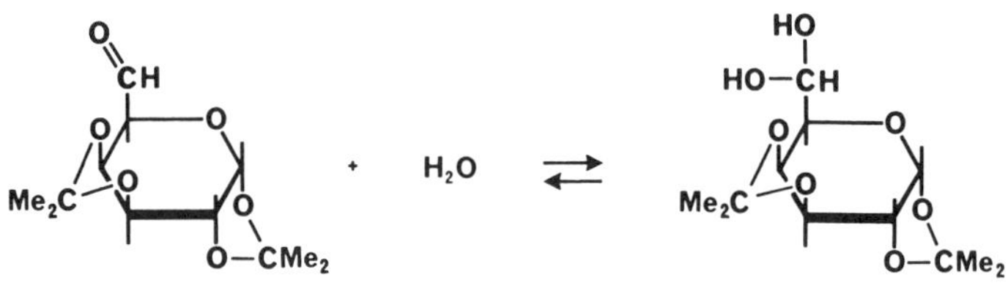

(1)

1 2

addition reactions of this type normally have limited synthetic application. They stand in contrast to irreversible addition such as that between a Grignard reagent and a ketone [e.g., the reaction shown in Equation 2 (2)]. Irreversible additions of nucleophiles to

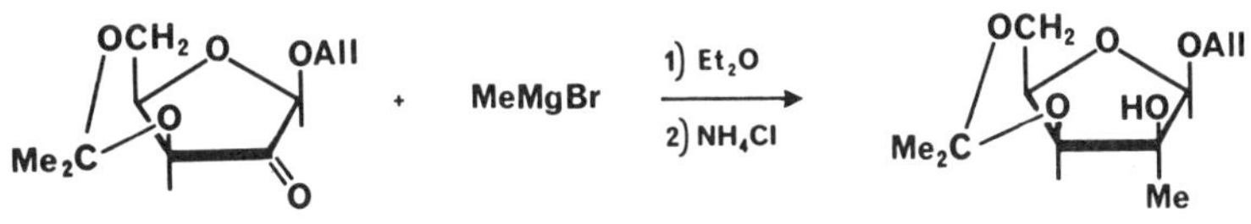

3 4 (2)

polar double bonds represent some of the most valuable synthetic processes.

B. Normal (1,2) and Conjugate (1,4) Addition

When a compound contains two double bonds in conjugation and when this π-system is polarized, the possibility exists for 1,2-(normal) or 1,4-(conjugate or Michael) addition. These two types of processes are illustrated in Figure 1 using as an example the addition of methylmagnesium iodide to an α,β-unsaturated carbonyl compound. A competition sometimes exists between normal and conjugate addition. This competition can result in both types of addition products being isolated from the same

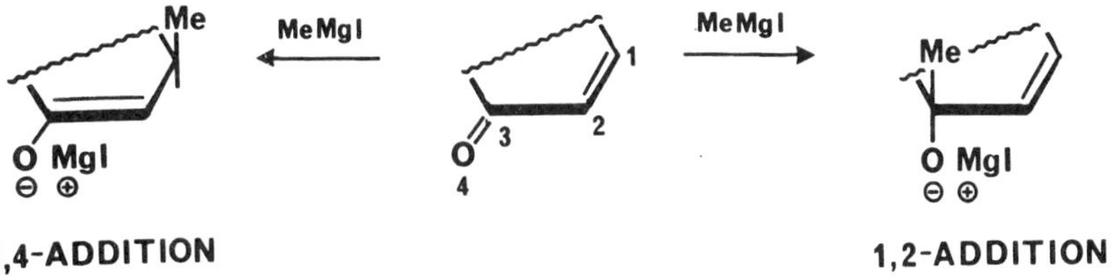

FIG. 1. Normal (1,2) and conjugate (1,4) addition to an α,β-unsaturated carbonyl compound.

reaction mixture. Usually, however, it is possible to select a nucleophilic reagent for which the desired addition will be the major reaction pathway. For example, alkyllithium reagents tend to form 1,2-addition products when reacting with a conjugated π-system (Equation 3) (3) while lithium dialkylcuprates prefer 1,4-addition (Equation 4) (4).

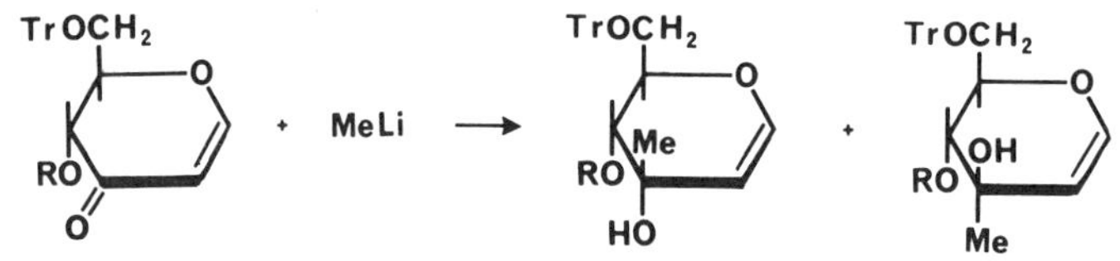

(3)

5 6 7

	6	7
R = H	40 %	60 %
R = MEM	50 %	50 %

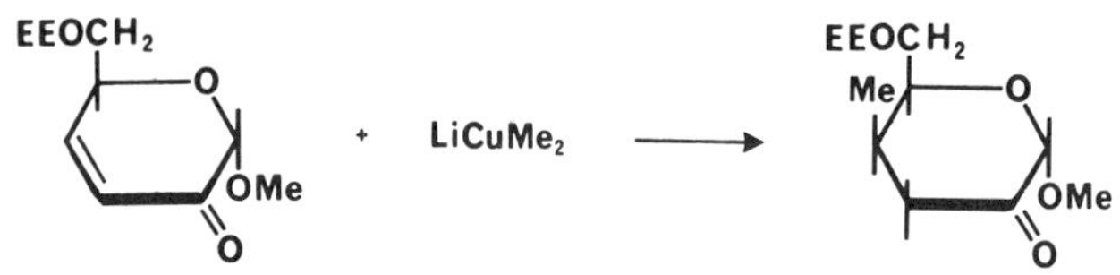

(4)

8 9

EE = $\overset{\displaystyle CH_3}{\overset{|}{C}}HOCH_2CH_3$

C. Carbon Nucleophiles

1. Organolithium Compounds and Grignard Reagents

Reaction of an organometallic reagent with a carbonyl-containing carbohydrate usually is part of a branched-chain sugar synthesis. Since the carbon branch often is a methyl group, methyllithium and methylmagnesium halides are the usual reagents. Alkyllithium compounds are more reactive than Grignard reagents and are less prone to side reactions such as enolate formation (Equation 5); nevertheless, both types of reagents

(5)

generally add to carbonyl compounds to form the desired branched-chain sugars. The reactions of these reagents vary considerably in their stereoselectivity. They can range from being completely stereoselective (Equation 2) to exhibiting no selectivity at all (Equation 3, R = MEM).

In systems where both normal and conjugate addition are possible, methyllithium is the reagent of choice if 1,2-addition is desired (Equation 3). Grignard reagents are much more likely to give 1,4-addition; in fact, the tendency of Grignard reagents toward conjugate addition can be enhanced by the presence of certain catalysts. Combining catalytic amounts of copper(I) salts with Grignard reagents increases the formation of 1,4-addition products. Conjugate addition of Grignard reagents to α,β-unsaturated aldehydes and ketones also increases as the size of the substituents attached to the carbonyl group becomes larger.

The synthesis of carbohydrates from noncarbohydrate precursors has become a topic of considerable recent interest (5). One of the primary carbon-carbon bond-forming reactions in this type of synthesis is the stereoselective addition of organometallic reagents to chiral aldehydes and ketones. An example of this type of addition process is found in the first reaction pictured in Figure 2 where 1,1-diethyl-2-lithio-2-propene [10] stereoselectively adds to 1,2-*O*-isopropylidene-D-glyceraldehyde [11]. Deprotection of the resulting product [12] and cyclization, as shown in Figure 2,

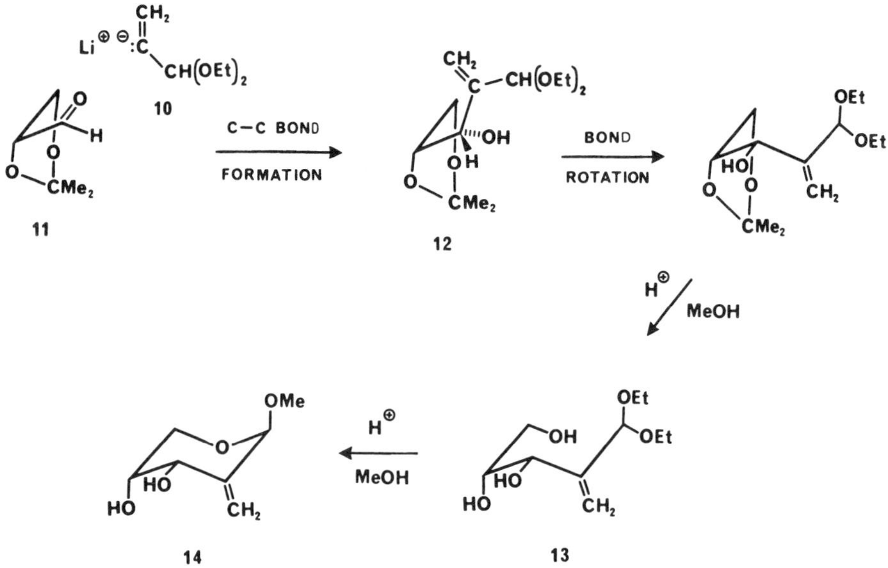

FIG. 2. Addition of an organolithium reagent to 1,2-*O*-isopropylidene-D-glyceraldehyde [11] and subsequent formation of methyl 2-deoxy-2-*C*-methylene-β-D-*erythro*-pentopyranoside [14].

produces methyl 2-deoxy-2-*C*-methylene-β-D-*erythro*-pentopyranoside [14] (6). Quite a variety of Grignard and organolithium reagents (as well as organometallic reagents involving metals such as zinc and tin) have been used in building carbohydrates from noncarbohydrate precursors. Reference 5 contains a recent review of these types of addition reactions.

2. Lithium Dialkylcuprates

When conjugate addition to an unsaturated system is desired, lithium dialkylcuprates are the reagents of choice. A typical reaction is that between methyl 3,4-dideoxy-6-*O*-(1-ethoxyethyl)-α-D-*erythro*-hex-3-enopyranosid-2-ulose [8] and lithium dimethylcuprate, a reaction which gives exclusively 1,4-addition

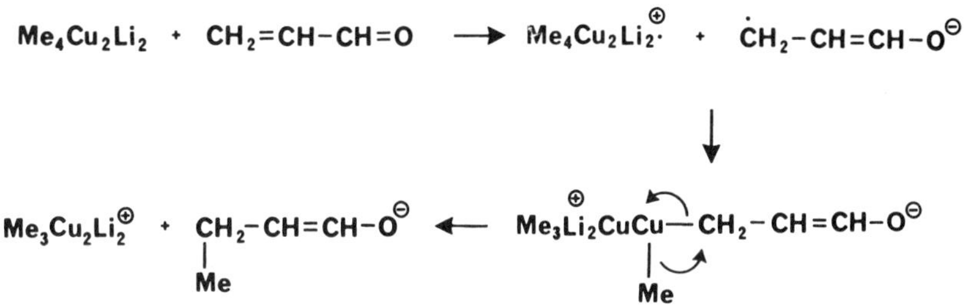

15

FIG. 3. Proposed mechanism for conjugate addition of lithium dimethylcuprate to an α,β-unsaturated carbonyl compound.

(Equation 4) (4). The reason for the preferred 1,4-addition is related to the reaction mechanism, a process which has been extensively studied (7). The proposed mechanism (Figure 3) begins with an electron transfer from the dialkylcuprate, which is thought to exist as a dimer, to the α,β-unsaturated carbonyl compound. Electron transfer is followed by radical coupling to give a transient organocopper compound [15]. Methyl migration from copper to carbon completes the 1,4-addition (7).

3. Anions Derived From Nitro Compounds

a. **Nitromethane.** The anion produced by deprotonation of nitromethane is a nucleophile which adds easily to both aldehydes and ketones. When reaction involves an aldehydo compound (Equation 6) (8), not only is the carbon atom chain in the

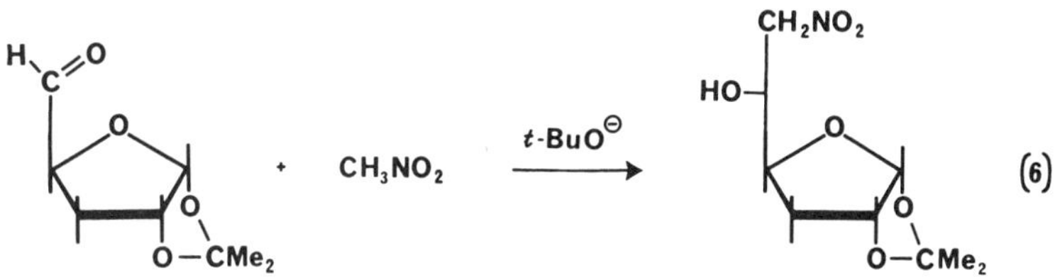

16 17

molecule extended but the newly added carbon atom bears a functional (nitro) group. The presence of the nitro group provides a "handle" for further transformations involving the carbon atom substituted by this group.

If the anion from nitromethane reacts with a ketone, a branched-chain compound is formed (Figure 4) (9). This is not a very common method for introducing a branch chain, however, because in most sugars of interest the entire branch consists of a methyl group. Organometallic reagents are best for introduction of simple alkyl groups. In compounds such as L-streptose [<u>21</u>], where an aldehydo group forms the branching chain, addition of the nitromethane anion can be a means for introducing a carbon branch which can be transformed into the aldehydo group (Figure 4).

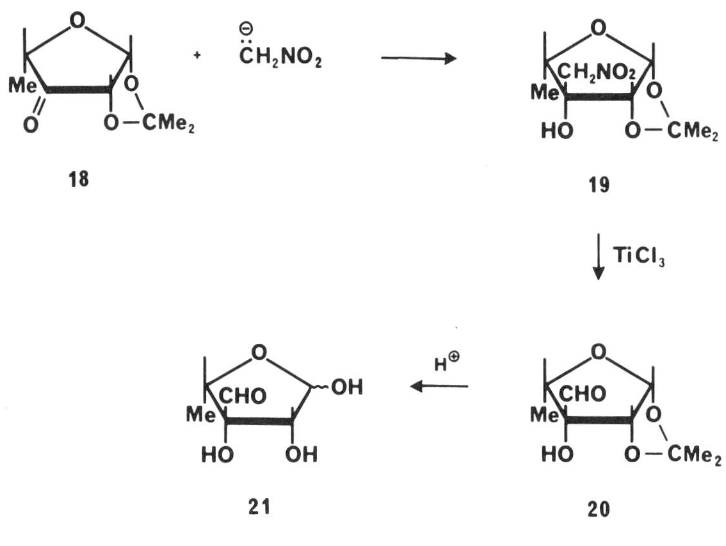

FIG. 4. Addition of the anion derived from nitromethane to 5-deoxy-1,2-*O*-isopropylidene-β-L-*threo*-pentofuranose-3-ulose [<u>18</u>] and subsequent formation of L-streptose [<u>21</u>].

b. Carbohydrates Containing Nitro Groups. If a carbohydrate already contains a nitro group, the nitro-bearing carbon atom can become the nucleophilic center in an addition reaction. For instance, deprotonation of 3-*O*-acetyl-5-deoxy-1,2-*O*-isopropylidene-5-nitro-α-D-ribofuranose [23] with potassium fluoride produces an anion which adds to the aldehyde 22 to couple the two monosaccharide units by a carbon-carbon bond (Figure 5) (10). After coupling, the nitro group can be transformed into a carbonyl group by treatment with potassium permanganate (Figure 5) (10).

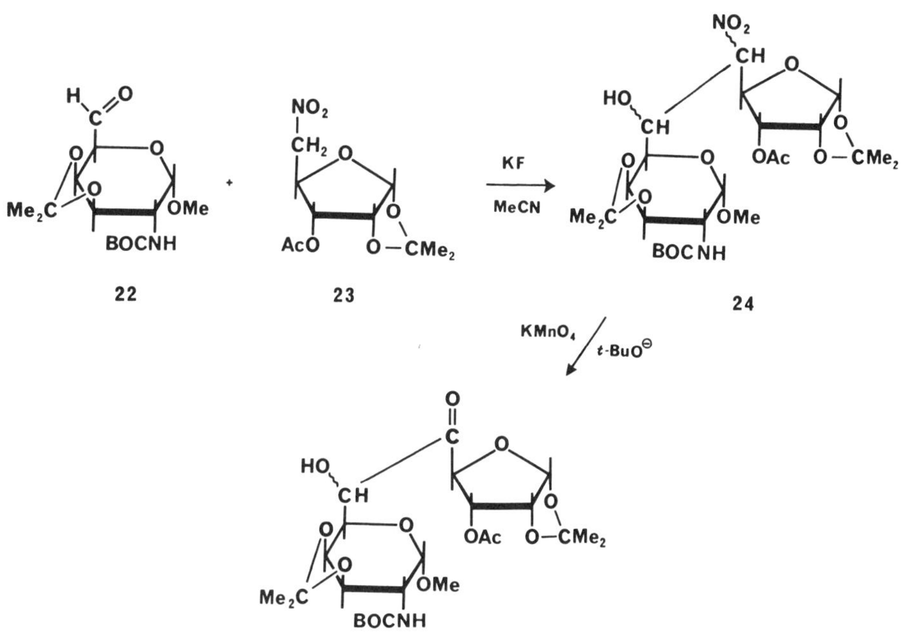

22 23 24

25

FIG. 5. Coupling of two monosaccharide units by nucleophilic addition of a deprotonated nitro compound to a carbonyl group.

4. Wittig Reagents

The Wittig reaction is a powerful method for converting a carbonyl group into a carbon-carbon double bond. The reaction process, outlined in Figure 6, involves the initial formation of the betaine 28 from reaction of a carbonyl compound [26] with a Wittig reagent [27]. The betaine [28] cyclizes to give a second intermediate [29], an oxaphosphetane. The oxaphosphetane [29] is stable at low temperature but decomposes at room temperature to give an alkene 30 and a phosphine oxide [31].

The most common Wittig reagent for the addition of a single carbon atom to a carbonyl-containing carbohydrate is methylenetriphenylphosphorane [27, R = C_6H_5]. Use of a Wittig reagent in synthesizing a branched-chain sugar is of greatest value when the carbon atom at which the branch is to occur does not also contain a hydroxyl group. Such a carbon branch is

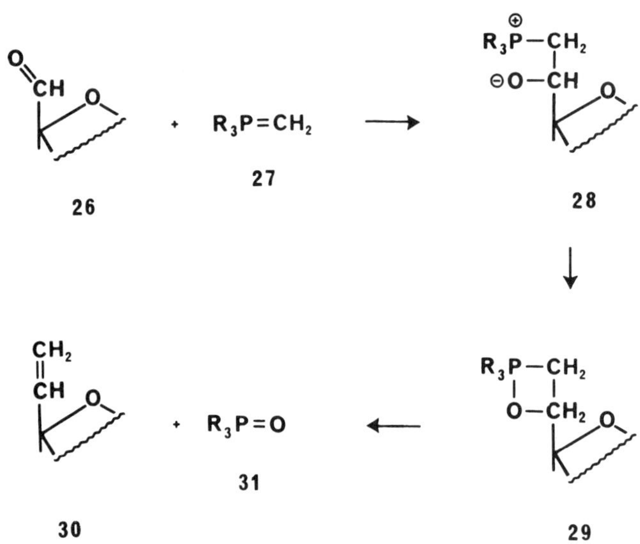

FIG. 6. Proposed mechanism for the Wittig reaction.

readily introduced by a combination of the Wittig reaction followed by catalytic hydrogenation (see Figure 3, Chapter 11). Wittig reagents also are useful for carbon-chain extension; for example, a two-carbon atom unit is added to 1,2:3,4-di-*O*-isopropylidene-α-D-galacto-hexodialdo-1,5-pyranose [1] by reaction with formylmethylenetriphenylphosphorane [32] (Equation 7) (11). Wittig reagents also can be used to extend a

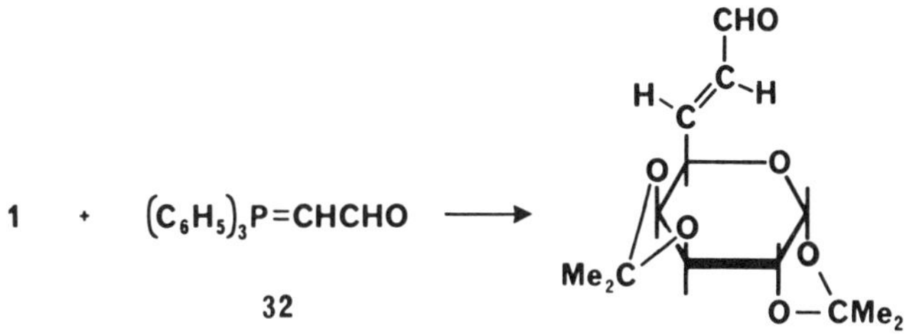

32 33 (7)

carbon atom chain by reaction with the aldehydo group in the open-chain form of a sugar (Equation 8) (12).

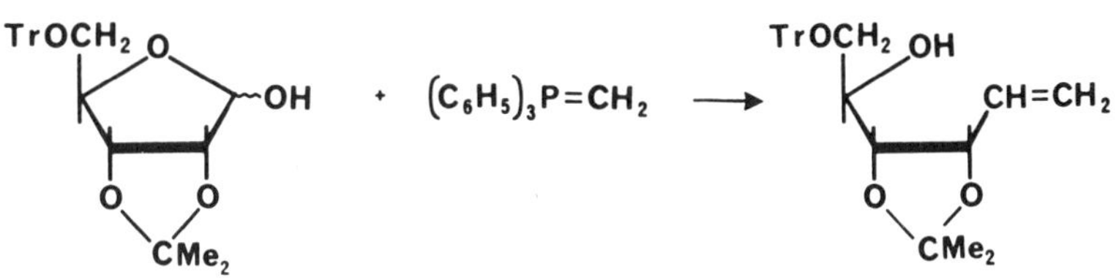

34 35 (8)

5. Cyanide Ion

Cyanide ion is another carbon nucleophile which adds to polar double bonds. Addition occurs to both simple (Equation 9) (13) and conjugated (Equation 10) (14) π-systems. Conjugate addition can involve an α,β-unsaturated carbonyl compound or, as is

36 37 (9)

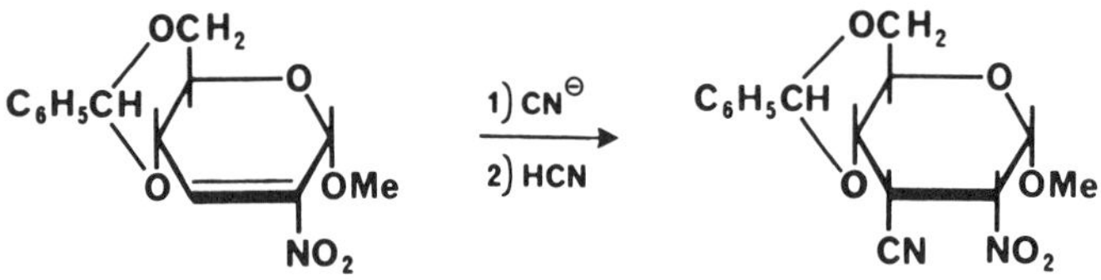

38 39 (10)

illustrated by the reaction shown in Equation 10, conjugate addition can involve an α,β-unsaturated nitro compound. In considering cyanide ion additions, it is useful to recall that such a process was encountered in an earlier chapter. (Reaction of cyanide ion with an aldehydo group in an unprotected sugar to form a cyanohydrin (the Kiliani reaction) was discussed in Chapter 7 (Section III.E.).)

6. The Anion Derived from Dithiane

The lithium salt of 1,3-dithiane is a nucleophile capable of displacement reactions as well as additions to polar double bonds. The compounds formed by addition of this anion can be considered to be "masked" aldehydes. Reaction with boron trifluoride and mercury(II) oxide converts them into the corresponding aldehydo compounds (Figure 7) (15).

D. Hydride Reagents

1. Lithium Aluminum Hydride and Sodium Borohydride

Use of hydride reagents represents the best method for hydrogenation of polar double bonds. The process consists of hydride

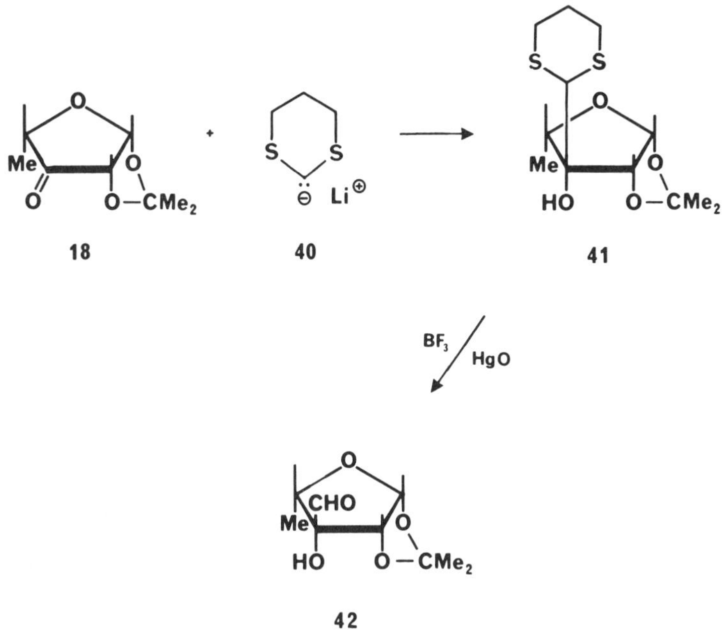

FIG. 7. Addition of the anion derived from dithiane to 5-deoxy-1,2-*O*-isopropylidene-β-L-*threo*-pentofuranose-3-ulose [18].

addition to the positive end of the polar bond and subsequent protonation of the anion thus produced (Figure 8). Protonation does not necessarily take place until an acid is added to the reaction mixture during product isolation. The two common

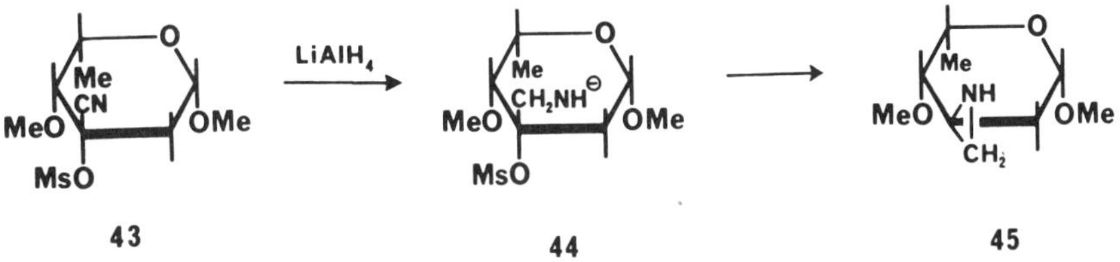

FIG. 8. Proposed mechanism for addition of hydride ion to a carbonyl group.

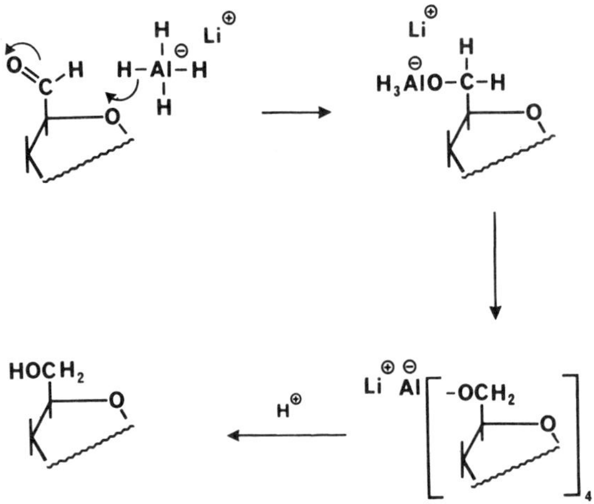

FIG. 9. Reduction of a cyano group in methyl 3-*C*-cyano-2,6-dideoxy-4-*O*-methyl-3-*O*-methylsulfonyl-β-L-*lyxo*-hexopyranoside [43] by lithium aluminium hydride and subsequent mesylate displacement.

hydride reagents are lithium aluminum hydride ($LiAlH_4$) and sodium borohydride ($NaBH_4$). Although both reagents readily reduce aldehydes and ketones, $NaBH_4$ is safer and easier to handle; consequently, it is the reagent of choice in situations where it is capable of causing the desired reaction (e.g., reduction of aldehydes, ketones, and some esters). Lithium aluminum hydride is the reagent chosen when less reactive polar bonds are to be reduced. For example, $LiAlH_4$ is a general reducing agent for cyano groups (Figure 9) (13) as well as esters (Equation 11) (4).

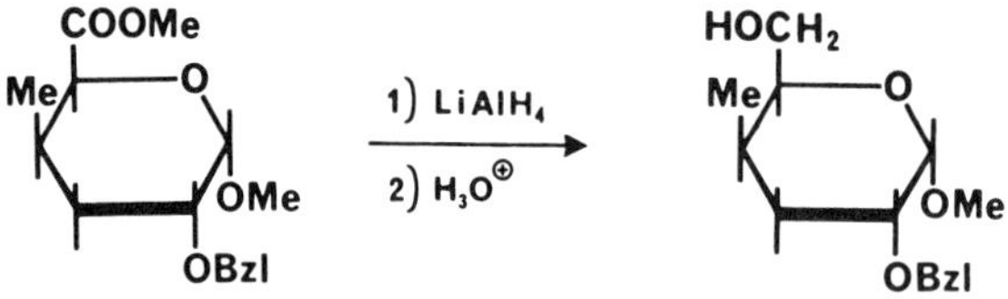

(11)

46 47

Hydride reagents normally do not produce significant amounts of 1,4-addition products when adding to conjugated π-systems; however, sodium borohydride has a greater tendency to undergo conjugate addition than does lithium aluminum hydride. Conjugate addition by $NaBH_4$ can be avoided if reaction is conducted in the presence of lanthanide salts such as cerium(III) chloride (Equation 12) (16).

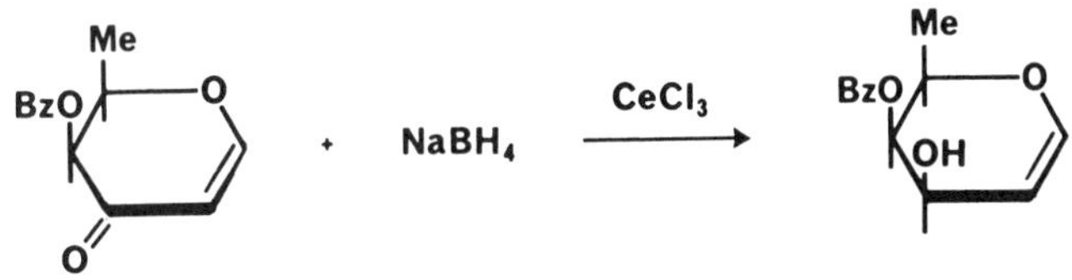

48 49

(12)

2. Hindered Hydride Reagents

Although either $LiAlH_4$ or $NaBH_4$ usually is satisfactory for reduction of a polar double bond, in some cases reagents with special characteristics are desirable. For example, replacement of some of the hydrogen atoms in the borohydride or aluminum hydride ions creates reducing agents which are much more sterically hindered. Reaction of such reagents generally exhibits greater stereoselectivity than reductions with either $LiAlH_4$ or $NaBH_4$. One reagent of this type is lithium tri-*sec*-butylborohydride [50], a compound which normally adds to a carbonyl group in a highly stereoselective fashion; thus, reaction of 50 with methyl 6-deoxy-2,3-*O*-isopropylidene-3-*C*-methyl-α-L-*lyxo*-hexopyranosid-4-ulose [51] produces a single stereoisomer [52] (Equation 13) (17).

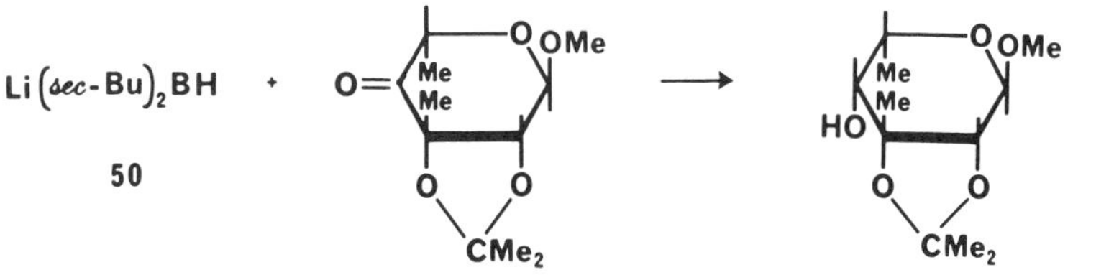

50 51 52 (13)

3. Sodium Cyanoborohydride

Unlike other hydride reagents which decompose rapidly or even violently in acidic solutions, sodium cyanoborohydride [53] can be used under mildly acidic conditions. The ability of this reagent [53] to add to polar double bonds under acidic conditions is particularly useful in the reduction of carbon-nitrogen double bonds because the nitrogen atom can be protonated, a process which dramatically increases the reactivity of the double bond toward hydride addition (Figure 10) (18).

E. Nitrogen-containing Nucleophiles

Since the open-chain form of an unprotected sugar contains a carbonyl group, it is to be expected that some of the reactions of these compounds should involve nucleophilic addition; in fact, this is the case. Two of the reactions of unprotected sugars discussed in Chapter 7, those involving amines (Section III.B.) and those using phenylhydrazine (Section III.F.), fit into this category. These reactions are additions of nitrogen-containing nucleophiles to aldehydo groups in unprotected sugars. In each case the initial addition is followed by elimination of the elements of water to give the observed products, imines and phenylhydrazones, respectively. Hydroxylamine, another nitrogen-containing nucleophile, reacts with carbonyl groups in carbohydrates to form oximes, compounds which can be converted into aminodeoxy sugars (Figure 11) (16).

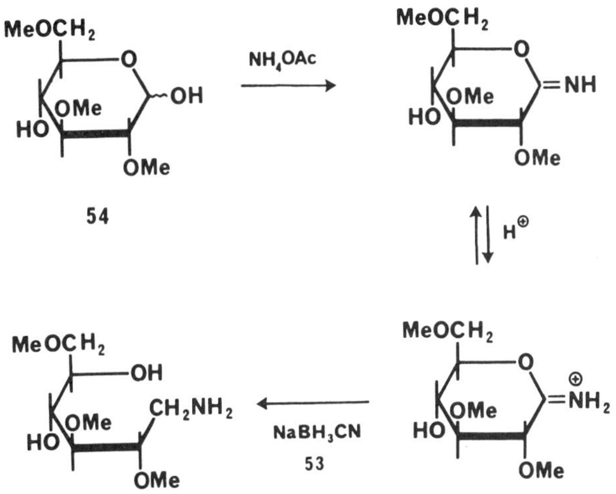

54

56

55

FIG. 10. Proposed mechanism for imine formation from 2,3,6-tri-*O*-methyl-D-glucose [54] and reduction to 1-amino-1-deoxy-2,3,6-tri-*O*-methyl-D-glucitol [56] using sodium cyanoborohydride [53].

F. Oxygen-containing Nucleophiles

The addition of common oxygen-containing nucleophiles, such as water, alcohols, carboxylic acids, and their corresponding anions to polar double bonds is often a reversible reaction. An example of this type of process was encountered earlier in this chapter (Equation 1). Another example is the reaction of carboxylic acid derivatives with alkoxides, the principal method for removal of ester protection (Chapter 8, Section IV.A.).

One oxygen nucleophile which is capable of more than reversible addition is that derived by deprotonation of hydrogen peroxide or alkyl hydroperoxides. These anions can cause epoxidation of α,β-unsaturated carbonyl and nitro compounds. For

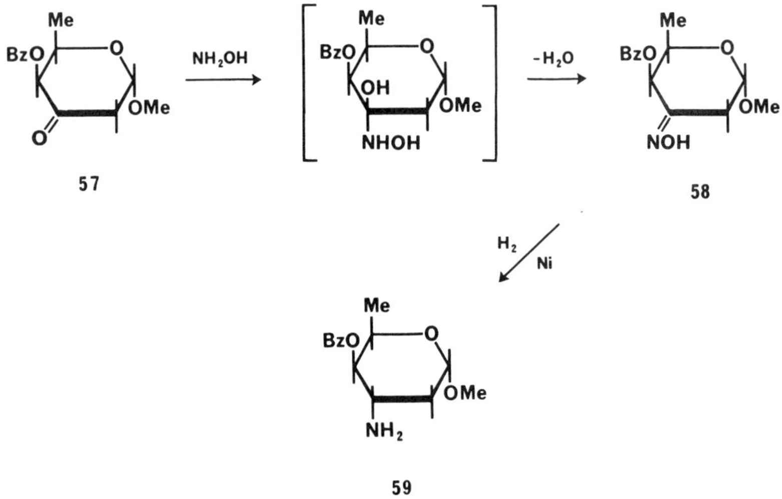

FIG. 11. Reaction of methyl 4-*O*-benzoyl-2,6-dideoxy-α-D-*threo*-hexopyranosid-3-ulose [57] with hydroxylamine to form the oxime 58.

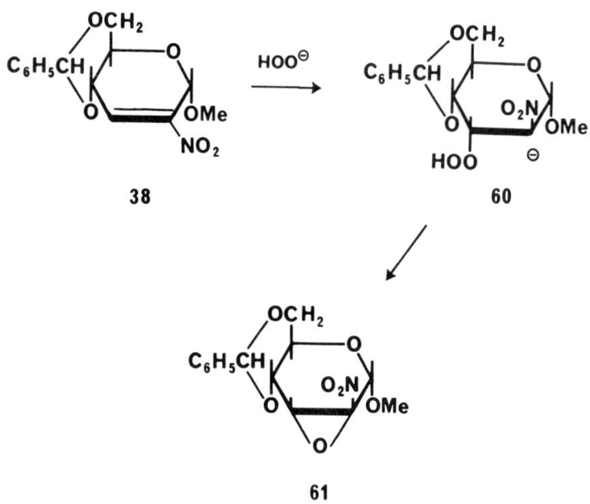

FIG. 12. Reaction of 4,6-*O*-benzylidene-2,3-dideoxy-2-nitro-α-D-*erythro*-hex-2-enopyranoside [38] with the hydroperoxide anion.

example, reaction of methyl 4,6-*O*-benzylidene-2,3-dideoxy-2-nitro-α-D-*erythro*-hex-2-enopyranoside [38] with hydrogen peroxide in the presence of base produces the epoxide 61 (Figure 12) (14). This reaction is believed to proceed by a Michael addition of the hydroperoxy anion to the unsaturated nitro compound [38] and subsequent intramolecular displacement of hydroxide by the anion on the carbon atom bearing the nitro group (Figure 12).

II. ELECTROPHILIC ADDITION REACTIONS

Electrophilic addition reactions are processes initiated by an electrophile interacting with a π-system. Some of these reactions generate cationic intermediates while others actually are concerted processes. Concerted reactions have some of the characteristics of ionic additions and some features associated with cycloaddition reactions. Examples of both concerted and nonconcerted electrophilic addition reactions are found among the halogenation and epoxidation reactions described in this section.

A. Halogenation

Halogen addition typically is initiated by electrophilic attack of a halogen molecule on a π-system to form a pair of halonium ions (Figure 13); thus, reaction of iodine with 3,4-di-*O*-acetyl-1,5-anhydro-2-deoxy-L-*arabino*-hex-1-enitol (L-rhamnal, 62) produces the iodonium ions 63 and 64 (19). Reaction then can follow one of several pathways, depending upon the nature of the nucleophiles present in the reaction mixture. If the unsaturated compound is a glycal and the halonium ions are captured by an alcohol, glycosides are produced (Figure 13) (19). (As will be discussed in Chapter 14, this method for glycoside formation can be used in oligosaccharide synthesis.) If a halonium ion is trapped

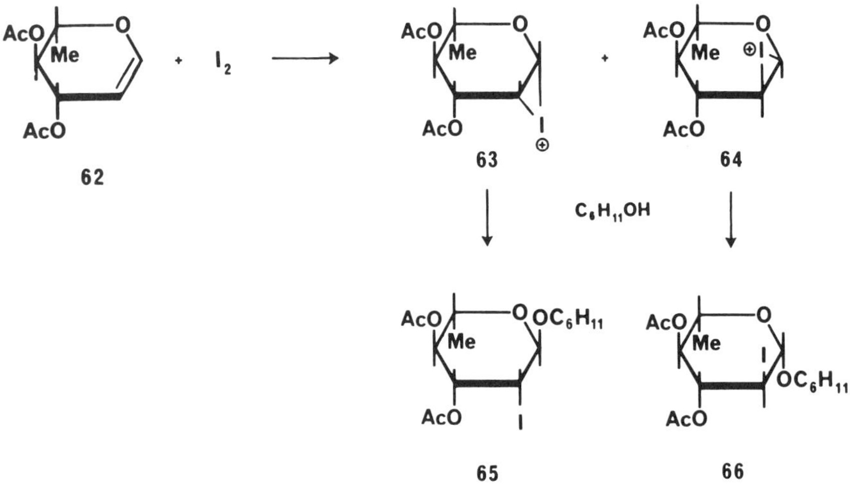

FIG. 13. Proposed mechanism for the reaction of 3,4-di-*O*-acetyl-1,5-anhydro-2-deoxy-L-*arabino*-hex-1-enitol [62] with iodine and cyclohexanol.

intramolecularly, as is the case in the reaction of iodine with 6-*O*-benzyl-1,2-dideoxy-3,4-*O*-isopropylidene-D-allose [67], ring formation results (Figure 14) (20). A third possibility is that a halonium ion would react with a halide ion present in the reaction mixture; however, this type of reaction is not often reported since

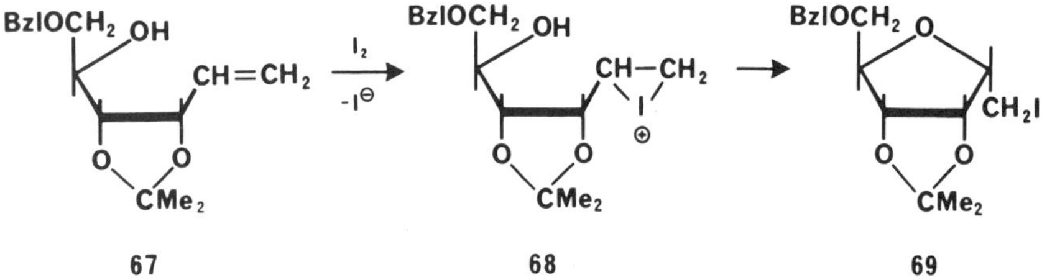

FIG. 14. Cyclization of 6-*O*-benzyl-1,2-dideoxy-3,4-*O*-isopropylidene-D-allose [67] by reaction with iodine.

vicinal dihalides formed from glycals, except for difluorides, usually experience further reaction.

The medical and biological importance of fluorinated carbohydrates has created considerable interest in their synthesis (21). Among the methods for forming deoxyfluoro sugars are electrophilic addition reactions. The fluorine-containing compounds which are of greatest interest are the 2-deoxy-2-fluoro sugars, compounds which can be synthesized by addition of a fluorinating-agent to a glycal and subsequent hydrolysis of the products generated (Figure 15) (22). Reaction of fluorine-containing reagents differs mechanistically from additions involving the larger halogens in that halonium ions are not formed and syn addition is the preferred mode for product formation (Figure 15) (22).

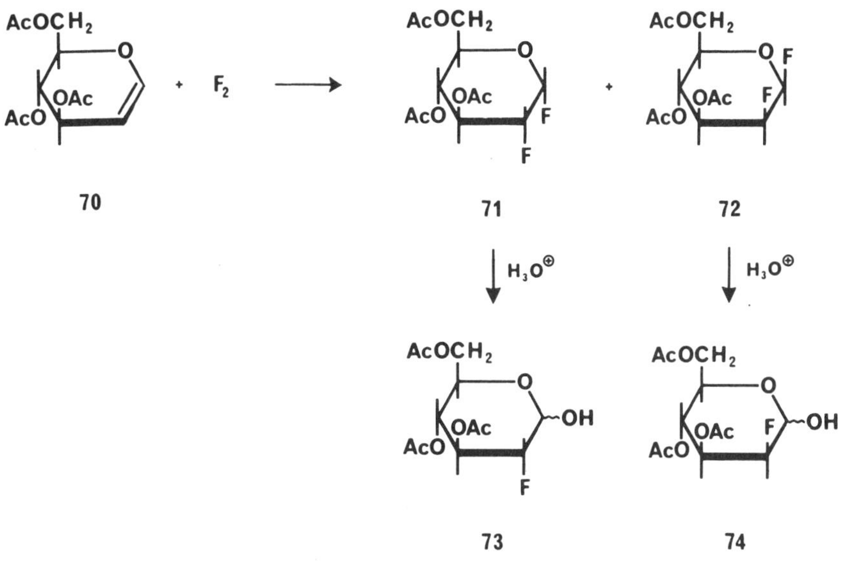

FIG. 15. Addition of fluorine to 3,4,6-tri-*O*-acetyl-1,5-anhydro-2-deoxy-D-*arabino*-hex-1-enitol [70] and hydrolysis of the products.

B. Epoxidation

Epoxide formation can be accomplished by a nucleophilic addition reaction as shown in Figure 12 or by concerted electrophilic addition (Figure 16). In the latter case the epoxidizing agent can be hydrogen peroxide, an alkyl hydroperoxide, or a peroxyacid. Epoxidation reactions have increased considerably in their importance in carbohydrate chemistry in recent years due to their role in the synthesis of carbohydrates from noncarbohydrate starting materials. Metal-catalyzed epoxidation reactions have been found to favor those double bonds with hydroxyl groups which are allylic or homoallylic (Equation 14) (23). Not only are these

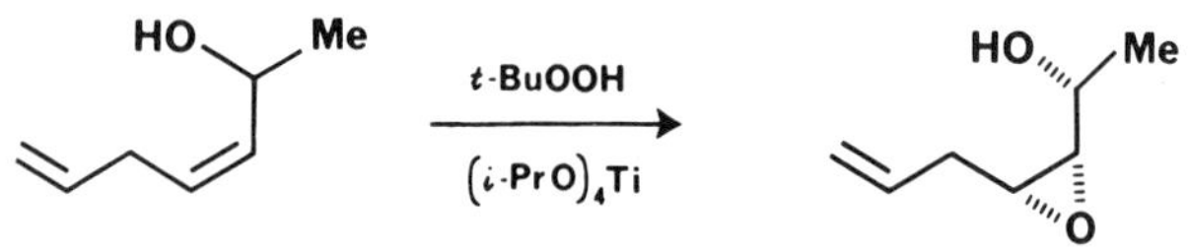

75 76 (14)

epoxidation reactions regioselective but they also are highly stereoselective; thus, reaction of the diene 75 with *t*-butyl hydroperoxide in the presence of titanium tetraisopropoxide produces only the epoxide 76 (Equation 14) (23). In addition, asymmetric epoxidation can be achieved if the reaction with *t*-butyl

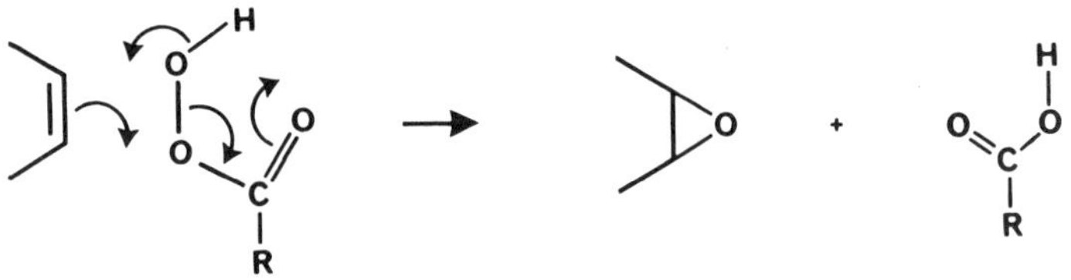

FIG. 16. Proposed mechanism for epoxidation of a double bond using a peracid.

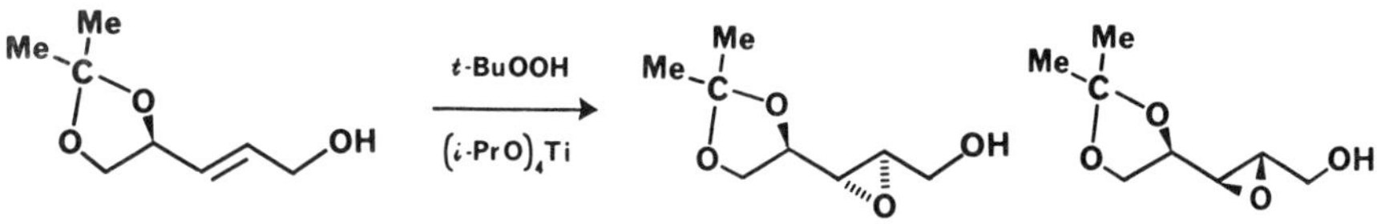

77 78 79

	78	79
(+)-DET	<1	>20
(-)-DET	>20	<1

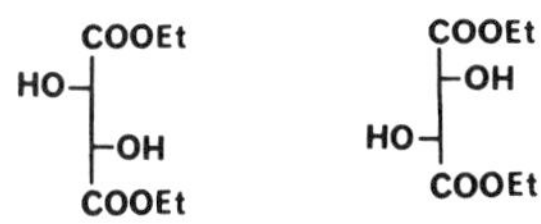

(-)-DET (+)-DET

FIG. 17. Asymmetric epoxidation of a double bond with *t*-butyl hydroperoxide in the presence of titanium tetraisopropoxide and either (+) or (-) diethyl tartrate (DET).

hydroperoxide is conducted in the presence of titanium tetraisopropoxide and an ester of D-(+)- or L-(-)-tartaric acid (Figure 17) (24). This process expands dramatically the access to optically pure materials. Syntheses based upon these reactions have been used to produce a variety of sugars. As one demonstration of the power of this technique, asymmetric epoxidation was used as the "key" reaction in the syntheses of all eight L-aldohexoses from 4-benzyloxy-2-buten-1-ol [80] (25).

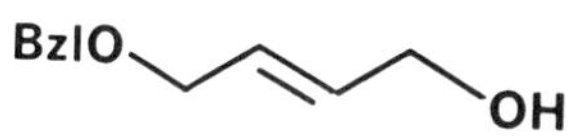

80

C. Mercury-catalyzed Ring Opening (Formation of Substituted Cyclohexanones)

Addition of mercury(II)-containing compounds to π-systems represents another example of an electrophilic addition to a double bond. Although this type of reaction is not widely used, there is one instance in which it offers unique synthetic potential. This transformation, known as the Ferrier reaction, converts certain unsaturated carbohydrates into substituted cyclohexanones (26). The Ferrier reaction is a process which is thought to begin with an electrophilic addition of the mercury(II) ion to an unsaturated carbohydrate [81] which has its double bond between C-5 and C-6 (Figure 18). This addition leads to the following series of events: ring opening to give the carbocation 83, reaction of the carbocation [83] with water to produce the aldehyde 84, and cyclization of 84 to give the substituted cyclohexanone 85 (Figure 18) (26,27). Reaction of methyl 2-azido-3, 4-di-*O*-benzyl-2,6-dideoxy-α-D-*arabino*-hex-5-enopyranoside [86]

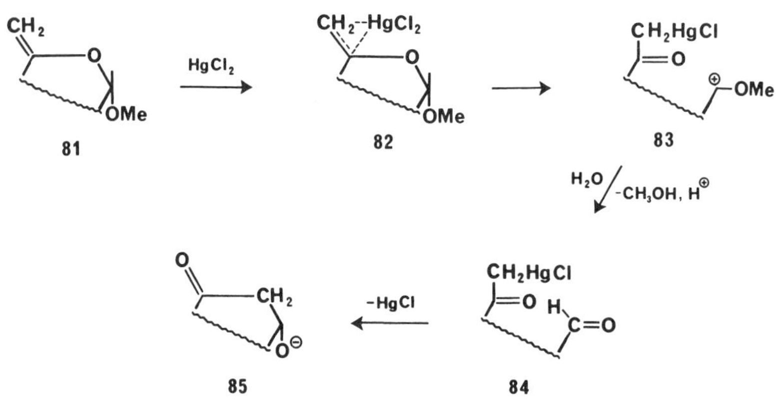

FIG. 18. Proposed mechanism for the Ferrier reaction.

with mercury(II) chloride to give the corresponding cyclohexanone [87] provides an example of the Ferrier reaction (Equation 15) (28).

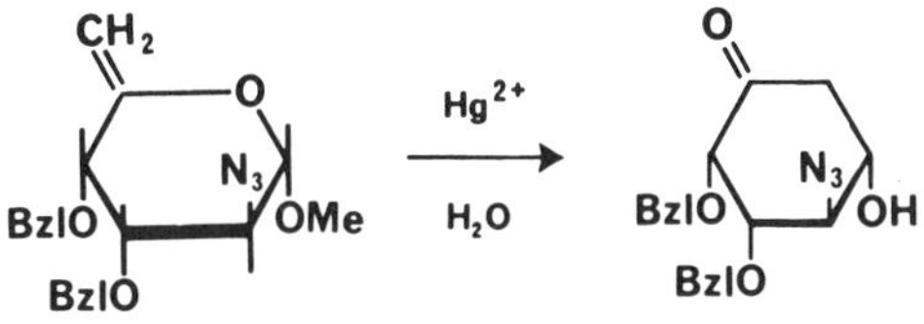

86 87 (15)

III. CYCLOADDITION REACTIONS

A. Diels-Alder Reaction

The Diels-Alder reaction between dienes and aldehydes is a well-recognized process for forming dihydropyran rings (Equation 16).

(16)

Subsequent reactions involving these Diels-Alder adducts can lead to their conversion into carbohydrates (29,30). When highly oxygenated dienes react with aldehydes in the presence of Lewis acid catalysts, compounds are produced which can be readily converted into a variety of monosaccharides and their derivatives. Diels-Alder reactions of this type represent the critical step in another important pathway for synthesis of carbohydrates from noncarbohydrate starting materials (5,30). An example of this type of cycloaddition process involving a highly oxygenated diene [88] is found in the first reaction shown in Figure 19 (16). Treatment of the Diels-Alder adducts with trifluoroacetic acid gives

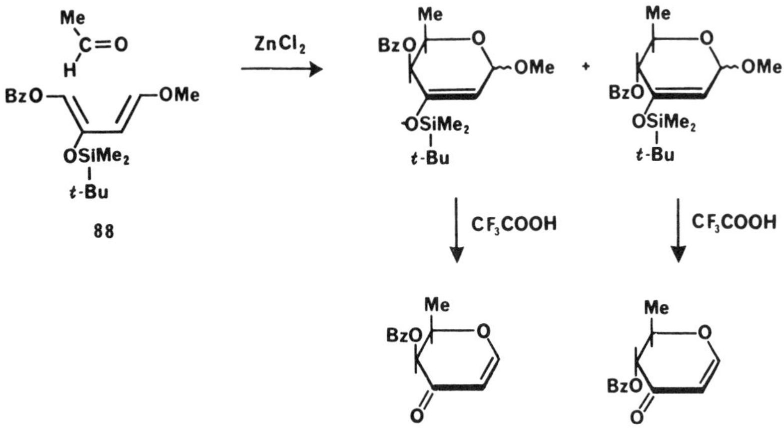

88 48 89

FIG. 19. A synthesis of the anhydro compounds 48 and 89 which begins with a Diels-Alder reaction.

4-*O*-benzoyl-2,6-dideoxy-α-D-*threo*-hex-3-oxo-1-enitol [48] and 4-*O*-benzoyl-2,6-dideoxy-α-D-*erythro*-hex-3-oxo-1-enitol [89] (16). (Racemic mixtures of these two compounds [48 and 89] are formed but only the D-series compounds are shown in Figure 19.) When 48 is treated with sodium borohydride, 1,5-anhydro-4-*O*-benzoyl-2,6-dideoxy-*lyxo*-hex-1-enitol [49] is produced (Equation 12, only D-enantiomers shown) (16).

B. Formation of Diols by Reaction with Potassium Permanganate and Osmium Tetraoxide

Cold, dilute potassium permanganate adds to alkenes to produce vicinal hydroxyl groups. The reaction is a syn addition in which it is believed that a cyclic manganate ester [90] is formed and rapidly hydrolyzed to a diol [91] (Figure 20). The same type of reaction occurs with osmium tetraoxide to produce isolable, cyclic

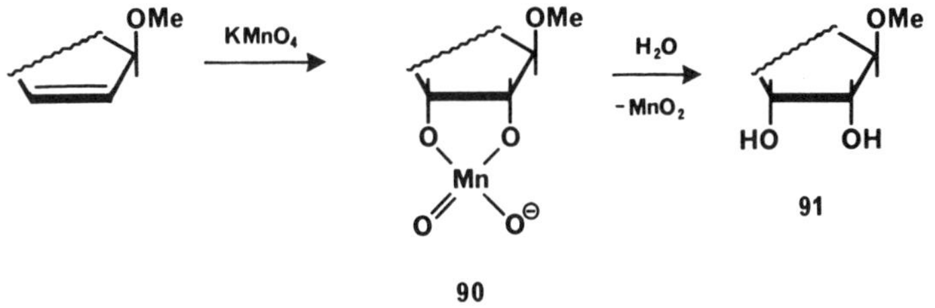

FIG. 20. Proposed mechanism for permanganate oxidation of an alkene to form a diol.

esters [92] which can be converted to the corresponding diols by reaction with hydrogen peroxide. Osmium tetraoxide also can be

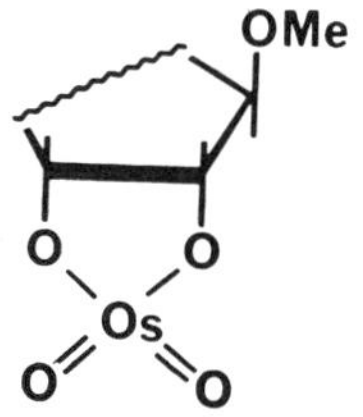

92

used in catalytic amounts if reaction is conducted in the presence of hydrogen peroxide or other oxidizing agents which convert the osmate ester to a *cis* diol and continuously regenerate the osmium tetraoxide. Osmylation of compound 93 produces 8-deoxy-1,2:3,4-di-*O*-isopropylidene-D-*erythro*-α-D-*galacto*-octopyranose [94], thus providing an example of this stereospecific cycloaddition (Equation 17) (11). Reaction of 93 with potassium permanganate parallels its oxidation with osmium tetraoxide in terms of product formation.

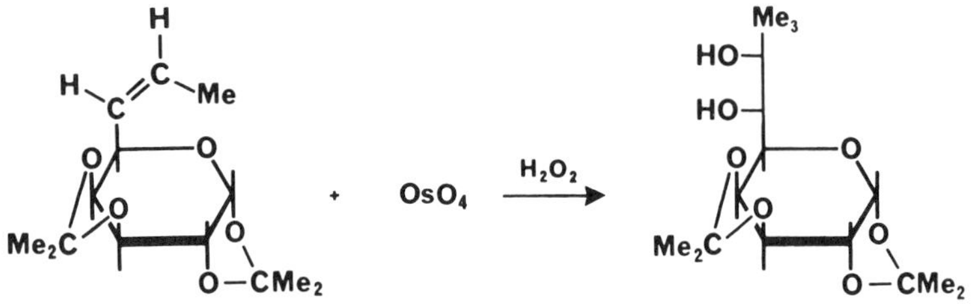

93 94 (17)

REFERENCES

1. D. Horton, M. Nakadate, and J. M. J. Tronchet, *Carbohydr. Res.*, 7, 56 (1968).
2. R. E. Ireland and D. W. Norbeck, *J. Am. Chem. Soc.*, *107*, 3279 (1985).
3. R. D. Dawe and B. Fraser-Reid, *J. Carbohydr. Chem.*, *1*, 21 (1982).
4. Y. Nakahara, A. Fujita, and T. Ogawa, *J. Carbohydr. Chem.*, *3*, 487 (1984).
5. G. J. McGarvey, M. Kimura, T. Oh, and J. M. Williams, *J. Carbohydr. Chem.*, *3*, 125 (1984).
6. J.-C. Depezay and Y. Le Merrer, *Tetrahedron Lett.*, *32*, 2865 (1978).
7. H. O. House, *Acc. Chem. Res.*, *9*, 59 (1976).
8. J. M. J. Tronchet, K. D. Pallie, and F. Barbalat-Rey, *J. Carbohydr. Chem.*, *4*, 29 (1985).
9. R. W. Binkley and J. C. Fan, *J. Carbohydr. Chem.*, *1*, 213 (1982).
10. H. Sasai, K. Matsuno, and T. Suami, *J. Carbohydr. Chem.*, *4*, 99 (1985).
11. J. S. Brimacombe, R. Hanna, and F. Bennett, *Carbohydr. Res.*, *135*, C17 (1985).

12. F. Nicotra, L. Panza, F. Ronchetti, and L. Toma, *Tetrahedron Lett.*, *25*, 5937 (1984).
13. J. Yoshimura, T. Yasumori, T. Kondo, and K.-I. Sato, *Bull. Chem. Soc. Jpn.*, *57*, 2535 (1984).
14. T. Sakakibara, Y. Tachimori, and R. Sudoh, *Carbohydr. Res.*, *131*, 197 (1984).
15. H. Paulsen, V. Sinnwell, and J. Thiem, *Methods Carbohydr. Chem.*, *8*, 185 (1980).
16. S. J. Danishefsky and C. J. Maring, *J. Am. Chem. Soc.*, *107*, 1269 (1985).
17. A. Klemer and H. Stegt, *J. Carbohydr. Chem.*, *4*, 205 (1985).
18. J. A. Bennek, D. Rolf, and G. R. Gray, *J. Carbohydr. Chem.*, *2*, 385 (1983).
19. J. Thiem and P. Ossowski, *J. Carbohydr. Chem.*, *3*, 287 (1984).
20. F. Freeman and K. D. Robarge, *Carbohydr. Res.*, *137*, 89 (1985).
21. P. J. Card, *J. Carbohydr. Chem.*, *4*, 451 (1985).
22. N. Satyamurthy, G. T. Bida, H. C. Padgett, and J. R. Barrio, *J. Carbohydr. Chem.*, *4*, 489 (1985).
23. W. R. Roush and R. J. Brown, *J. Org. Chem.*, *47*, 1371 (1982).
24. T. Katsuki, A. W. M. Lee, P. Ma, V. S. Martin, S. Masamune, K. B. Sharpless, D. Tuddenham, and F. J. Walker, *J. Org. Chem.*, *47*, 1378 (1982).
25. S. Y. Ko, A. W. M. Lee, S. Masamune, L. A. Reed, III, K. B. Sharpless, and F. J. Walker, *Science*, *220*, 949 (1983).
26. R. J. Ferrier, *J. Chem. Soc., Perkin Trans. I*, 1455 (1979).
27. R. Blattner, R. J. Ferrier, and S. R. Haines, *J. Chem. Soc., Perkin Trans. I*, 2413 (1985).
28. F. Chretien and Y. Chapleur, *J. Chem. Soc., Chem Commun.*, 1268 (1984).
29. A. Zamojski, A. Banaszek, and G. Grynkiewicz, *Adv. Carbohydr. Chem. Biochem.*, *40*, 1 (1982).
30. R. R. Schmidt, *Acc. Chem. Res.*, *19*, 250 (1986).

13 Elimination Reactions

I. ACID- VS. BASE-CATALYZED REACTIONS

II. CATEGORIES OF ELIMINATION REACTIONS

- A. E2 Elimination Reactions
  - 1. Reaction Mechanism
  - 2. Leaving Groups
    - a. Sulfonyloxy Groups
    - b. Halogens (Silver Fluoride-catalyzed Dehydrohalogenation)
    - c. Phosphonium Salts
  - 3. Bases
  - 4. The Structure of the Compound Experiencing an Elimination Reaction
- B. Elimination of Adjacent Sulfonyloxy Groups (The Tipson-Cohen Reaction)
- C. Reaction of Halogenated Carbohydrates with Zinc Dust
  - 1. Glycal Formation
  - 2. Reductive Ring Opening

D. Elimination of Neighboring Hydroxyl Groups by Reaction with Triphenylphosphine, Iodine, and Imidazole

E. Free-Radical Elimination

I. ACID- VS. BASE-CATALYZED REACTIONS

The elimination reactions which are most useful in forming unsaturated carbohydrates are those which can be conducted under neutral or basic conditions. Under acidic conditions each hydroxyl group and even many protected hydroxyl groups (e.g., those protected as esters) can be transformed into effective leaving groups by protonation; consequently, although acid-catalyzed elimination reactions occur readily (1), their use in synthesis is rare due to the probability of multiple elimination. Not only does multiple elimination occur under acidic conditions but cationic rearrangements, hydride shifts, and ring contractions can compete with and accompany elimination reactions under these conditions. An example of the extensive dehydration and molecular reorganization which is associated with the heating of sugars in the presence of an acid catalyst is provided by the reaction shown in Equation 1 (2). A second example of an acid-catalyzed reaction

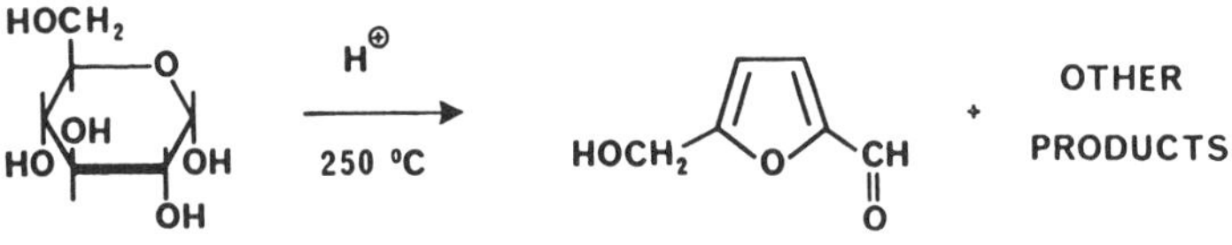

(1)

1 24

which, even though it is conducted under less drastic conditions, still illustrates the combination of molecular changes which can accompany an acid-catalyzed elimination reaction is shown in Equation 2 (3).

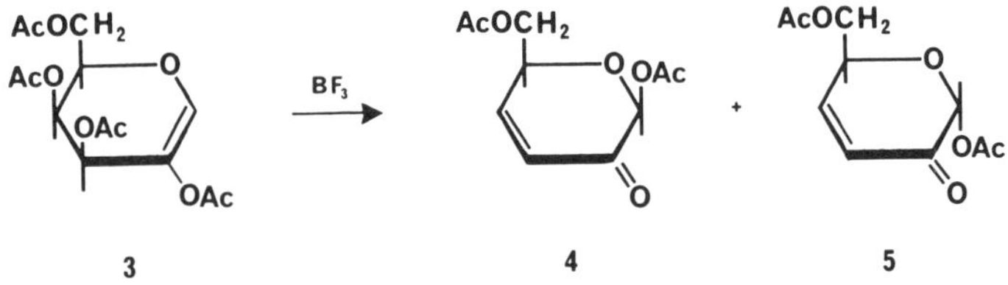

(2)

Although reactions such as rearrangement and ring contraction also can occur under neutral or basic conditions, multiple reactions are much less likely under these conditions because only a single nucleofuge (e.g., halogen or sulfonyloxy group) typically has been introduced into the molecule prior to reaction. Avoiding multiple reactions is nearly always desirable in selecting a process to use as a synthetic tool.

II. CATEGORIES OF ELIMINATION REACTIONS

A. E2 Elimination Reactions

1. Reaction Mechanism

The most important type of elimination process involving carbohydrates is the E2 elimination. An E2 elimination reaction takes place between two neighboring carbon atoms when a base removes a proton attached to one of these atoms (the β carbon atom, Figure 1) while a leaving group is departing from the other (the α carbon atom). The preferred stereochemistry for an E2 reaction has the departing groups coplanar and on opposite sides of the molecule. Stereochemistry is sufficently important that it

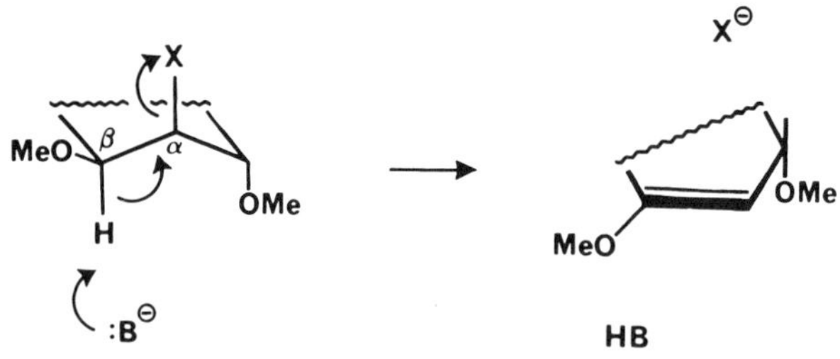

FIG. 1. Proposed mechanism for an E2 elimination.

can determine the direction of the elimination process. For example, when methyl 4,6-*O*-benzylidene-2-*O*-trifluoromethylsulfonyl-α-D-mannopyranoside [6] experiences an elimination reaction, only the hydrogen atom attached to C-3 is removed during double-bond formation (Equation 3) (4). Even though

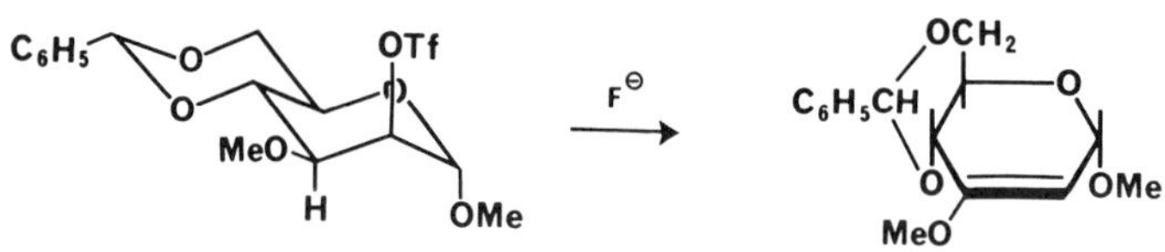

(3)

6 7

there is a hydrogen atom attached to C-1, the proper stereochemistry for reaction involving H-1 is not attained.

2. Leaving Groups

a. Sulfonyloxy Groups. Among the factors which influence the E2 elimination process, none is more important than the nature of the leaving group. The more reactive a leaving group is, the lower the temperature and the weaker the base necessary for reaction to occur. Strong bases at high temperature initiate molecular decomposition and even at room temperature strong bases remove protecting groups such as esters. To avoid these reactions, compounds which contain exceptionally reactive leaving groups are increasingly chosen for elimination processes. For example, elimination of the triflyloxy group from 6 requires only 10 minutes at 50°C (Equation 3) (4). This reaction stands in contrast to that of methyl 2-acetamido-4,6-*O*-benzylidene-2-deoxy-3-*O*-methylsulfonyl-α-D-allopyranoside [8] for which heating at reflux in *N,N*-dimethylformamide for 24 hours is necessary for complete reaction (Equation 4) (5). Although the

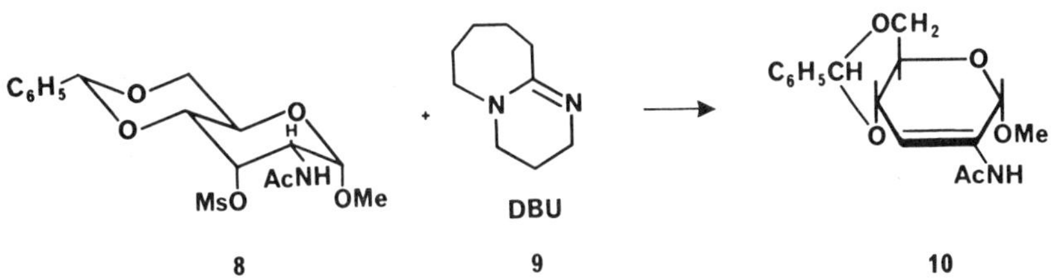

(4)

difference in reactivity between compounds 6 and 8 depends upon a variety of factors, it is reasonable to assume that the primary factor in this dramatic difference is associated with the nature of the leaving groups. When the structure of the molecule and the identity of the bases are changed, the triflyloxy group is still eliminated under mild conditions (room temperature) (Equation 5) (6).

(5)

b. Halogens (Silver Fluoride-catalyzed Dehydrohalogenation). Silver fluoride is able to complex with a halogen atom (usually bromide or iodide) in a deoxyhalogeno sugar to form an excellent leaving group. Dehydrohalogenation usually then takes place under very mild conditions, for example, in pyridine at room temperature (Figure 2) (7).

Silver fluoride-catalyzed elimination has become an integral part of a standard sequence of reactions used to invert configuration at C-5 in monosaccharide derivatives. The first step in this sequence is the ring opening of a 4,6-*O*-benzylidene acetal using *N*-bromo-succinimide to give a 6-bromo-6-deoxy compound (the Hanessian-Hullar reaction, Chapter 9, Section III.B.2). This reaction is followed by a silver fluoride-catalyzed elimination of the elements of hydrogen bromide to introduce a double bond between C-5 and C-6 (Figure 2) (7). Catalytic hydrogenation (Chapter 11, Section II.A.) then produces a 6-deoxy compound which typically has a different configuration at C-5 from that found in the 6-bromo-6-deoxy compound produced by the Hanessian-Hullar reaction (Figure 2) (7).

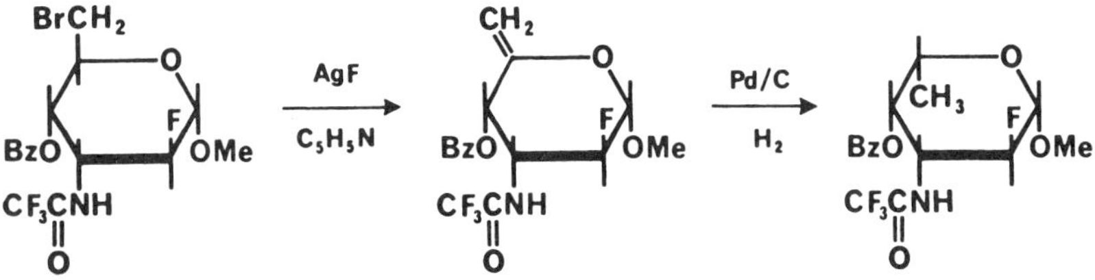

13 14 15

FIG. 2. Reaction sequence for inversion of configuration at C-5 involving a silver fluoride-catalyzed elimination.

c. Phosphonium Salts. Reaction of an unprotected hydroxyl group with triphenylphosphine and diethylazodicarboxylate (DEAD) converts the hydroxyl group into a phosphonium salt, an excellent leaving group. If a good nucleophile is present in the reaction mixture, S_N2 substitution (a Mitsunobu reaction) will take place (Chapter 9, Section III.B.1). If effective nucleophiles are absent, the triphenylphosphonium group can function as the leaving group in an E2 elimination reaction (Equation 6) (8).

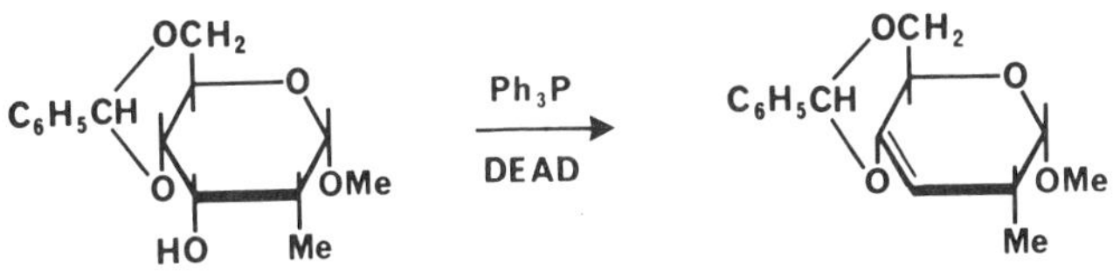

(6)

16 17

3. Bases

One of the advantages in selecting an excellent leaving group for an E2 reaction is that amines, relatively weak bases, are then sufficiently strong to participate in the elimination process. Amines generally do not initiate base-catalyzed, molecular decomposition; further, if a tertiary amine is used, deprotection of base-sensitive protecting groups can be avoided. If the nitrogen atom in a tertiary amine is sufficiently hindered, the possibility of S_N2 substitution with the amine functioning as a nucleophile can be dramatically reduced; in fact, if a very highly hindered tertiary amine, such as 2,6-di-*t*-butyl-4-methylpyridine [18] is used,

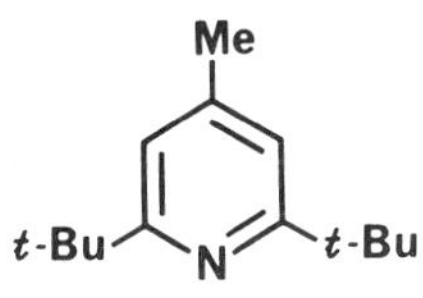

18

all possibility of the amine behaving as a nucleophile can be removed.

The most useful type of amine for elimination reactions is one which is too hindered to participate in S_N2 substitution and also one which maximizes base strength. 1,8-Diazabicyclo[5.4.0] undec-7-ene [9, DBU] is particularly well suited for such reactions (Equations 4 and 5), since it is a nitrogen-containing heterocycle with greater basicity than a typical tertiary amine. The increased basicity is due to the greater resonance stabilization present in the protonated base [20] when compared to that for the unprotonated base [19] (Figure 3). The cation 20 is stabilized by two resonance structures of essentially equal importance, a situation which maximizes resonance stabilization. Resonance stabilization is not as great in 19, however, due to the unequal contribution of the two resonance contributors. One resonance structure has a separation of charge and the other does

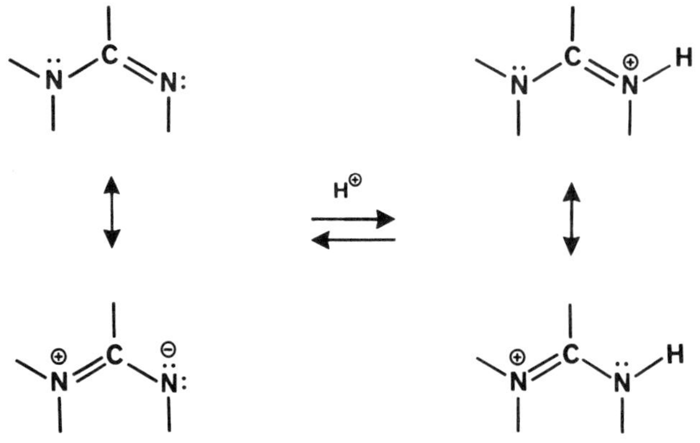

19 20

FIG. 3. Resonance structures for protonated and unprotonated forms of the base 19.

not. Cation stabilization in protonated DBU [20] provides an additional "driving force" for protonation which is not present in a typical tertiary amine.

4. The Structure of the Compound Experiencing an Elimination Reaction

The ease of E2 reaction is dependent upon the structure of the compound from which elimination is taking place. Reaction can be facilitated by steric hindrance if departure of a leaving group and rehybridization of the carbon atom to which the group was attached relieves steric strain. For example, reaction of methyl 2-*O*-benzoyl-4,6-*O*-benzylidene-3-*C*-methyl-α-D-glucopyranoside [21] with methanesulfonyl chloride produces a crowded mesylate [22] which spontaneously eliminates under the reaction conditions (Figure 4) (9).

Elimination reactions take place under much milder conditions and with loss of very poor leaving groups if the elimination process involves the formation of an α,β-unsaturated carbonyl or nitro compound. The stability gained by formation of an α,β-unsaturated π-system is sufficiently great that even nucleofuges as unreactive as acetate easily depart (Equation 7) (10).

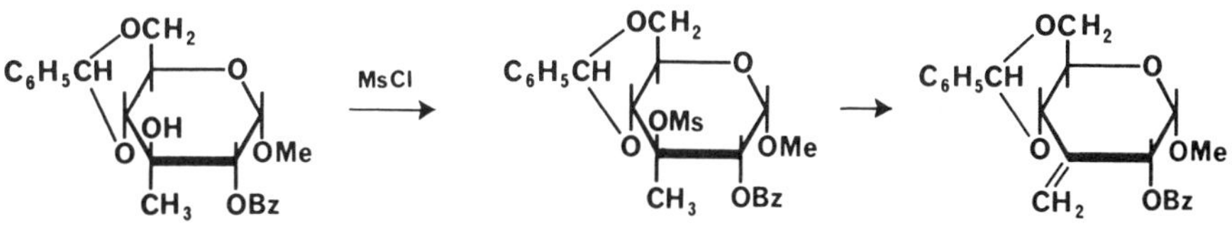

21 22 23

FIG. 4. Reaction sequence for formation of a sterically crowded mesylate [22] and subsequent elimination to give methyl 2-*O*-benzoyl-4,6-*O*-benzylidene-3-*C*-methylene-α-D-*ribo*-hexopyranoside [23].

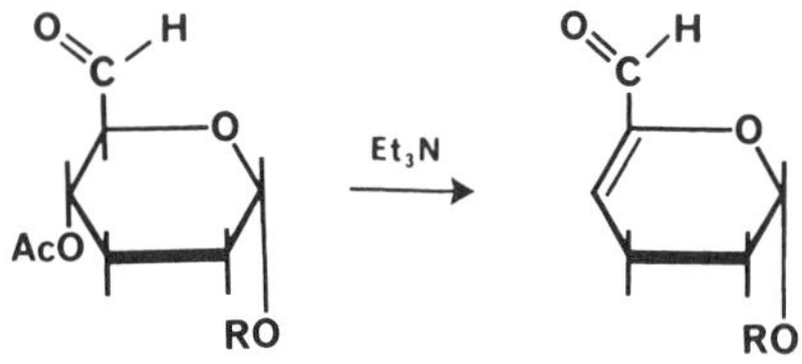

24 25 (7)

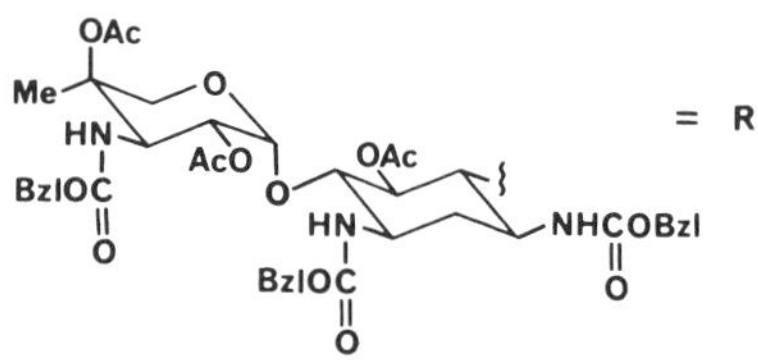

B. Elimination of Adjacent Sulfonyloxy Groups (The Tipson-Cohen Reaction)

The Tipson-Cohen reaction is an elimination process which involves both nucleophilic substitution and electron transfer from a metal (zinc). In order to be able to experience the Tipson-Cohen reaction, a compound must have sulfonyloxy groups on adjacent carbon atoms. When this type of disulfonate is heated in a solvent such a DMF with a mixture of sodium iodide and zinc dust, elimination will take place if one of the sulfonyloxy groups can be displaced by iodide; that is, reaction begins by sulfonyloxy displacement to give an intermediate deoxyiodo compound [27] (Figure 5). This intermediate then reacts with zinc, probably by way of an organozinc compound [28], to form an unsaturated carbohydrate. An example of the Tipson-Cohen reaction is shown in Equation 8 (11).

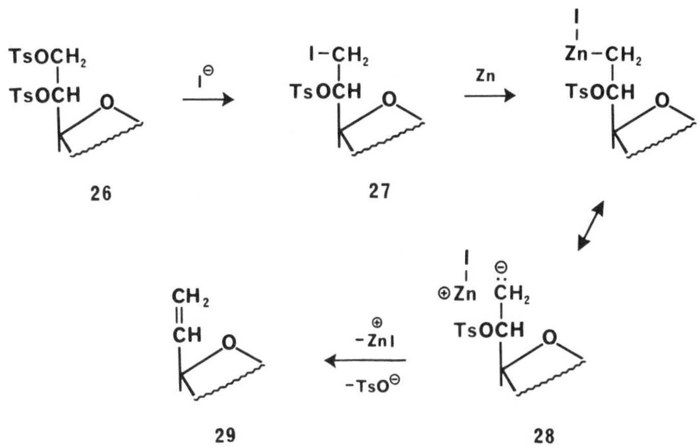

FIG. 5. Proposed mechanism for the Tipson-Cohen reaction.

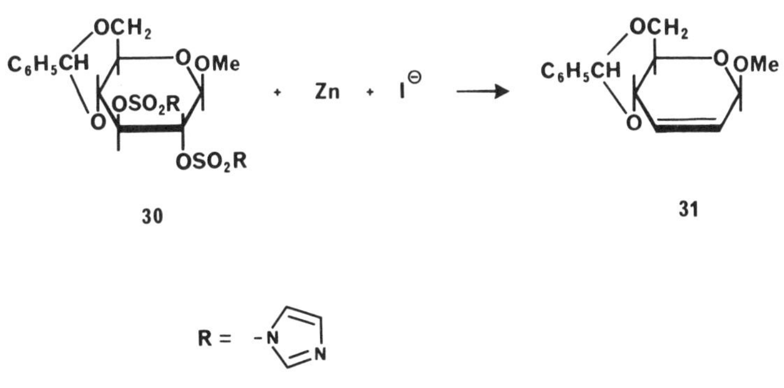

(8)

C. Reaction of Halogenated Carbohydrates with Zinc Dust

1. Glycal Formation

An elimination reaction that bears mechanistic similarity to the Tipson-Cohen reaction is used to form glycals from glycosyl halides (Equation 9) (12). This process involves transfer of

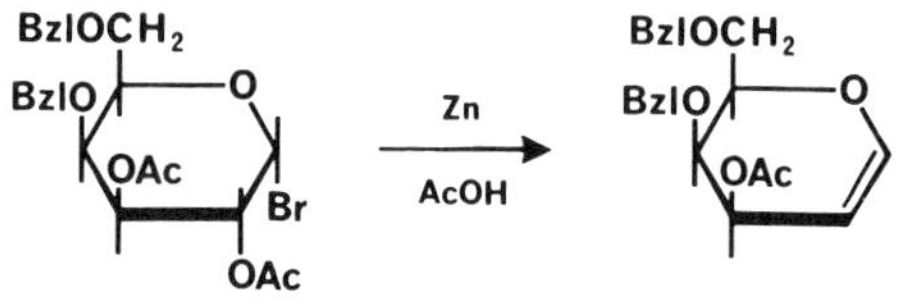

32 33 (9)

electrons from a metal to an acetylated glycosyl halide. Electron transfer leads to elimination of halide and carboxylate anions resulting in formation of a glycal. Reaction of 2,3-di-*O*-acetyl-4,6-di-*O*-benzyl-α-D-galactopyranosyl bromide [32] with zinc dust in acetic acid provides an example of this type of glycal synthesis (Equation 9) (12).

2. Reductive Ring Opening

Ring opening takes place when a 6-deoxy-6-iodo compound such as methyl 6-deoxy-6-iodo-3,4-di-*O*-methyl-2-*O*-(2,3,4-tri-*O*-methyl-β-D-xylopyranosyl)-α-D-galactopyranoside [34] is heated in a hydroxylic solvent in the presence of zinc dust (Figure 6) (13). This reaction, as with those just described (Equations 8 and 9), is an oxidation-reduction process. Oxidation of metallic zinc to zinc(II) provides a source of electrons for formation of iodide and alkoxide ions during the ring-opening process (Figure 6). Two reaction mechanisms are indicated in Figure 6. The nonconcerted process suggests that ring opening precedes loss of methoxide. The concerted mechanism requires these events to be simultaneous. Recent evidence favors the concerted pathway (14).

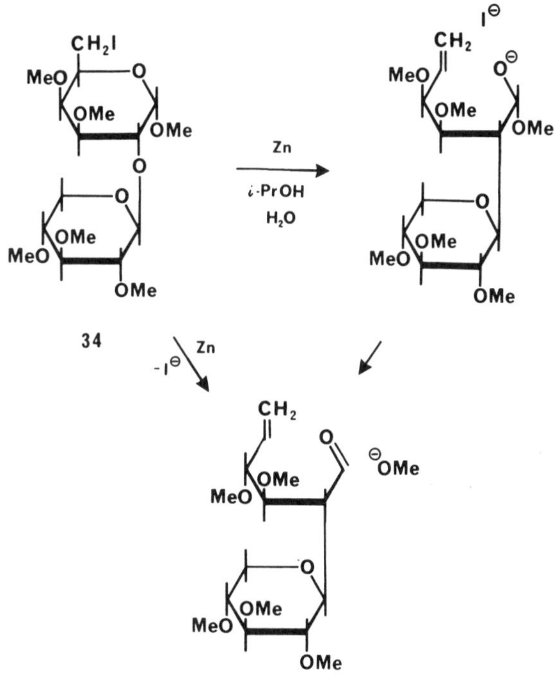

FIG. 6. Sequence of reaction steps involved in reductive elimination from compound 34.

D. Elimination of Neighboring Hydroxyl Groups by Reaction with Triphenylphosphine, Iodine, and Imidazole

If a molecule contains two adjacent hydroxyl groups, it can be converted into a bisphosphonium salt [37] by reaction with triphenylphosphine and imidazole (15). If iodide ion also is present in the reaction mixture, displacement of one of the phosphonium groups to give an intermediate deoxyiodo compound [38] can occur. This intermediate [38] is not isolated but rather experiences a reaction which is analogous to an E2 elimination

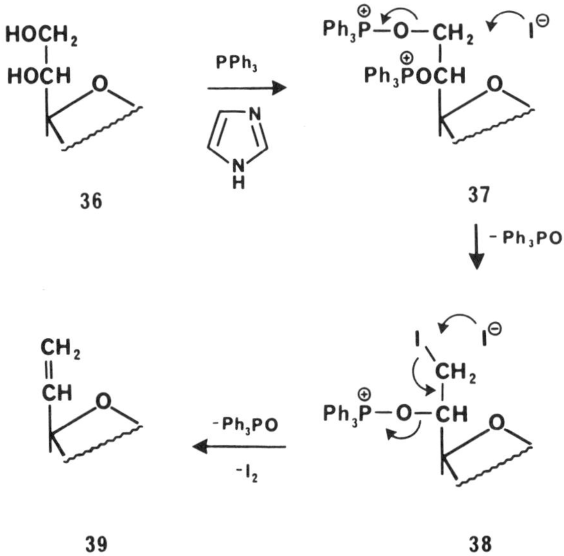

FIG. 7. Proposed mechanism for formation of a double bond by reaction of a diol with triphenylphosphine, imidazole, and iodine.

(Figure 7) (14). An example of this type of reaction is the generation of methyl 2,6-di-*O*-benzoyl-3,4-dideoxy-α-D-*erythro*-hex-3-enopyranoside [<u>41</u>] from methyl 2,6-di-*O*-benzoyl-α-D-glucopyranoside [<u>40</u>] (Equation 10) (16).

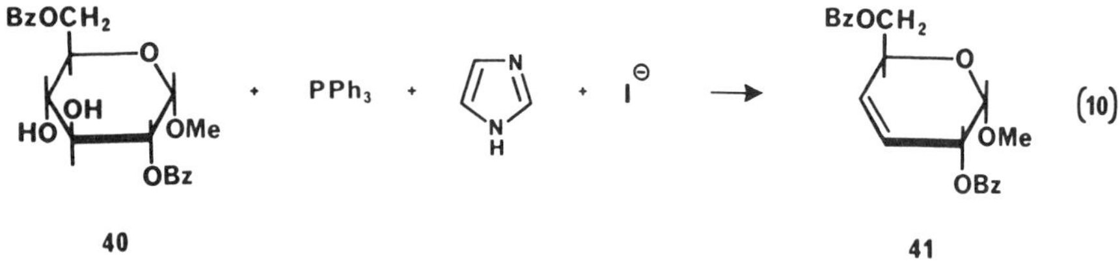

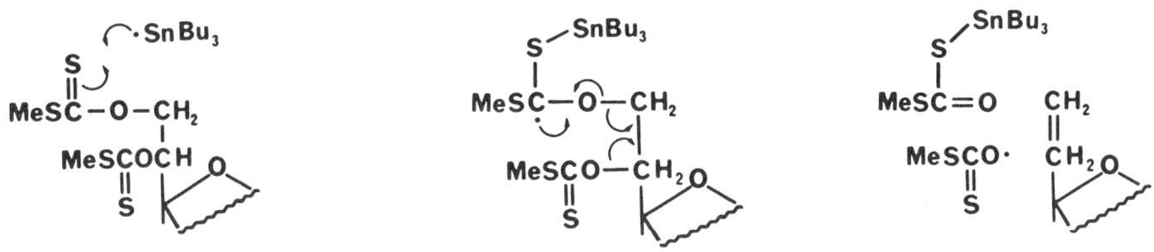

FIG. 8. Proposed mechanism for reaction of tri-*n*-butyltin hydride with a bisdithiocarbonate. (Arrows indicate movement of single electrons.)

E. Free-Radical Elimination

The free-radical reaction between a bisdithiocarbonate (a di-*O*-(methylthio)carbonyl compound) and tri-*n*-butyltin hydride is a fundamentally different type of elimination process from those already considered. In this reaction the tri-*n*-butyltin radical adds to one of the carbon-sulfur double bonds to initiate a series of reactions, shown in Figure 8, which lead ultimately to the formation of a carbon-carbon double bond (17). An example of this type of reaction is shown in Equation 11 (18).

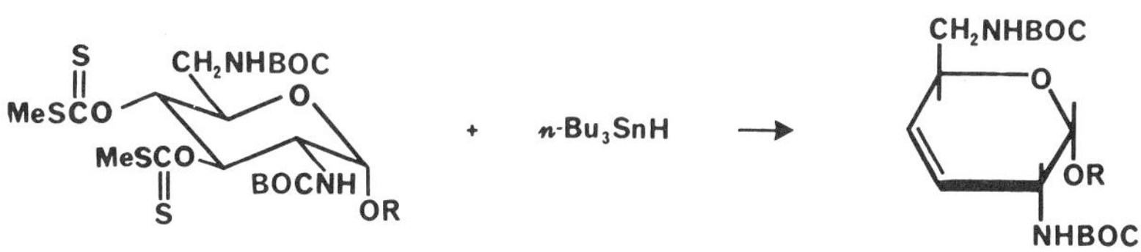

40 41 (11)

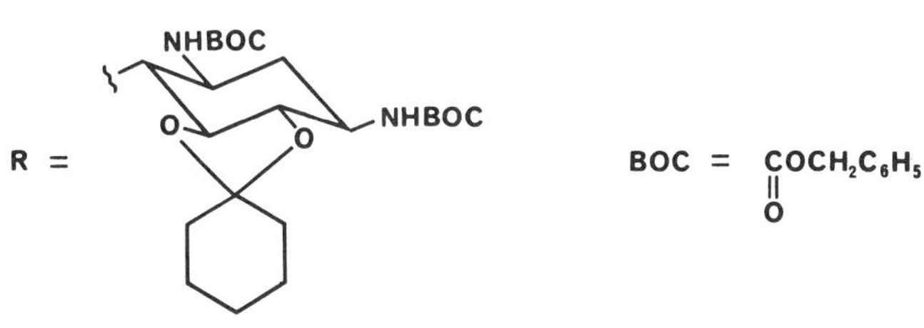

REFERENCES

1. M. S. Feather and J. F. Harris, *Adv. Carbohydr. Chem. Biochem.*, *28*, 161 (1973).
2. M. S. Feather and J. F. Harris, *Carbohydr. Res.*, *15*, 304 (1970).
3. P. Koll, K. Klenke, and D. Eisermann, *J. Carbohydr. Chem.*, *3*, 403 (1984).
4. T. Haradahira, M. Maeda, Y. Kai, H. Omae, and M. Kojima, *Chem. Pharm. Bull.*, *33*, 165 (1985).
5. A. Calvo-Mateo, M. J. Camarasa, and F. G. De las Heras, *J. Carbohydr. Chem.*, *3*, 461 (1984).
6. T. W. Flechtner, *Carbohydr. Res.*, *77*, 262 (1979).
7. D. Picq and D. Anker, *J. Carbohydr. Chem.*, *4*, 113 (1985).
8. Y. Chapleur, F. Germain, A. Aubry, and D. Bayeul, *J. Carbohydr. Chem.*, *3*, 443 (1984).
9. J. Yoshimura, N. Kawauchi, T. Yasumori, K.-I. Sato, H. Hashimoto, *Carbohydr. Res.*, *133*, 255 (1984).
10. J. J. K. Wright and P. Lee, *J. Carbohydr. Chem.*, *1*, 157 (1982).
11. K. Bock and M. Meldal, *Acta Chem. Scand.*, *B 38*, 225 (1984).
12. N. V. Bovin, S. E. Zurabyan, and A. Y. Khorlin, *J. Carbohydr. Chem.*, *2*, 249 (1983).
13. G. O. Aspinall, D. Chatterjee, and L. Khondo, *Can. J. Chem.*, *62*, 2728 (1984).
14. B. Bernet and A. Vasella, *Helv. Chim. Acta*, *67*, 1328 (1984).
15. P. J. Garegg and B. Samuelsson, *Synthesis*, 469 (1979).
16. P. J. Garegg, R. Johansson, and B. Samuelsson, *J. Carbohydr. Chem.*, *3*, 189 (1984).
17. A. G. M. Barrett, D. H. R. Barton, and R. Bielski, *J. Chem. Soc. Perkin Trans. I*, 2378 (1979).
18. D. H. R. Barton, D.-K. Zheng, and S. D. Gero, *J. Carbohydr. Chem.*, *1*, 105 (1982).

14 Oligosaccharide Syntheses

I. THE "STATE OF THE ART" IN OLIGOSACCHARIDE SYNTHESES

The primary challenge in the syntheses of oligosaccharides is achieving stereoselective glycoside formation. A host of reactions and reagents have been used for this process and high stereoselectivity has been achieved in many cases; nevertheless, no single type of reaction or set of reaction conditions is successful in every instance. The situation with respect to oligosaccharide syntheses is concisely summarized in the following way: "Although we now have learned to synthesize oligosaccharides, it should be emphasized that each oligosaccharide synthesis remains an independent problem, whose resolution requires considerable systematic research and a good deal of knowledge. There are no universal conditions for oligosaccharide syntheses" (1).

II. GENERAL FACTORS RELATING TO REACTANT AND CATALYST REACTIVITY IN OLIGOSACCHARIDE SYNTHESES

A. Nucleophilicity of Hydroxyl Groups in Partially Protected Carbohydrates

Although the ability of a particular hydroxyl group to participate as a nucleophile in oligosaccharide synthesis cannot always be predicted, some useful guidelines concerning hydroxyl group reactivity do exist. Probably the most important factor governing relative reactivity of these groups is steric hindrance. Since axial groups generally have a more crowded environment than equatorial ones, glycoside formation usually occurs more rapidly when an equatorial hydroxyl group is involved. An example of this difference in reactivity is provided by the reaction of 3,4,6-tri-*O*-acetyl-2-deoxy-2-phthalimido-β-D-glucopyranosyl bromide [1] with benzyl 3,6-di-*O*-(*p*-methoxybenzyl)-α-D-mannopyranoside [2] (Equation 1) (2). Even though there is an unprotected axial hydroxyl group at C-2 in compound 2, reaction occurs regioselectively with the group at C-4.

(1)

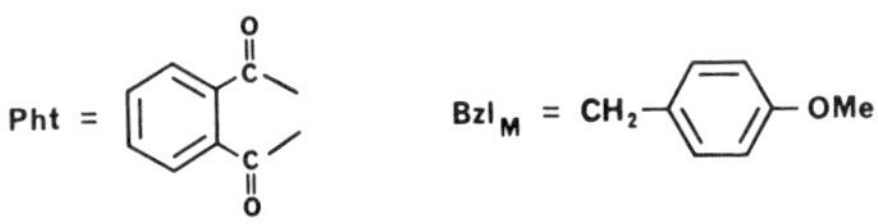

When all hydroxyl groups attached to C-2, C-3, and C-4 in an aldohexopyranoside have an equatorial orientation, the general order of reactivity in forming glycosidic linkages is: 6-OH >> 3-OH > 2-OH > 4-OH (3). An example of this difference in reactivity is provided by the reaction of the disaccharide 1,2,3,6-tetra-*O*-acetyl-4-*O*-(2,3-di-*O*-acetyl-α-D-glucopyranosyl)-β-D-glucopyranose [5] with 2,3,4,6-tetra-*O*-benzyl-α-D-glucopyranosyl bromide [4] to give the trisaccharide 6 (Equation 2)

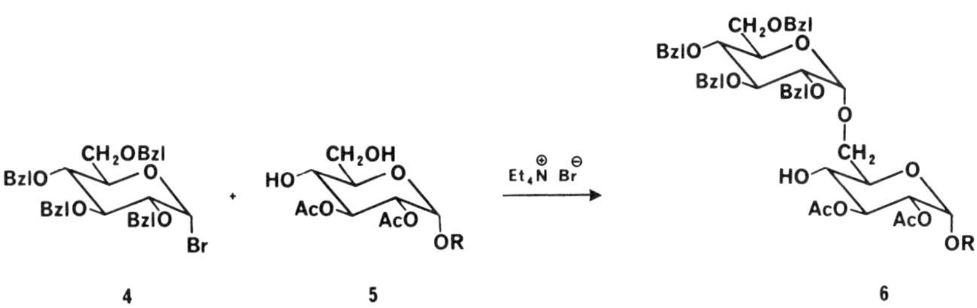

(2)

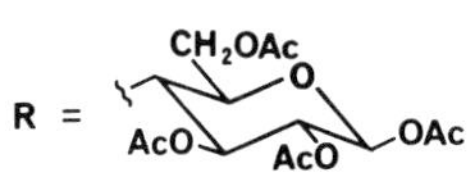

(4). The new glycosidic linkage formed in this process is produced by reaction of the C-6 hydroxyl group in 5 even though one at C-4 is unprotected.

B. Activating Effect of Silver- and Mercury-containing Catalysts: The Koenigs-Knorr Reaction and the Helferich Procedure

Oligosaccharide formation using glycosyl halides is often catalyzed by silver or mercury salts. When reaction takes place between a glycosyl halide and a partially protected carbohydrate in the presence of a silver salt, it is called an Koenigs-Knorr reaction. Even though this reaction was one of the first discovered for glycoside synthesis, it remains, along with its several modifications (described at appropriate places in this chapter), the most widely used reaction for oligosaccharide synthesis. When the reaction between a glycosyl halide and a partially protected sugar is catalyzed by a mercury-containing compound (usually mercury(II) cyanide), the reaction is described as being conducted under Helferich conditions or according to the Helferich procedure (5). Helferich conditions usually include nitromethane as a solvent or co-solvent. Since silver salts are more active catalysts than those containing mercury and since the identity of the anion plays a role in catalytic activity, mercury- and silver-containing compounds provide a broad spectrum of catalytic ability. The order of reactivity among common catalysts is $AgOTf/Ag_2CO_3 > AgClO_4/Ag_2CO_3 > HgBr_2 > Hg(CN)_2/HgBr_2 > Hg(CN)_2$ (3).

C. The Reactivity of Leaving Groups Attached to the Anomeric Carbon Atom

1. The Influence of Protecting Groups

a. Protecting Group Identity. The reactivity of a leaving group attached to an anomeric carbon atom (e.g., a halogen atom in a glycosyl halide) is a function of several factors, including the type of protecting groups present in the molecule. Acetoxy, benzoyloxy, and other electron-withdrawing groups tend to decrease reactivity, while reactivity is increased by benzyl and similar electron-donating groups. These activating and deactivating effects exist because most reactions leading to glycoside formation require at least some accumulation of positive charge at the anomeric carbon atom during reaction. If electron-withdrawing groups are present, the formation of a partial or full positive charge at the anomeric center is made more difficult by inductive electron withdrawal from the ring. In contrast, groups which inductively donate electrons to the ring assist the departure of a leaving group by helping to stabilize a developing positive charge.

b. Protecting Group Location. The location of a protecting group as well as its identity can influence the ease of cation formation at C-1. Neighboring group participation by substituents attached to C-2 (discussed in Section III.A.) represents the most important of the influences exerted by a protecting group; however, substituents at relatively remote positions also can have a significant role. The effect of remote substituents in a ring system is attributable, at least in part, to through-bond interactions (6,7). For example, electron-withdrawing C-4 substituents inhibit ionization at C-1 in a glycosyl halide by removal of electron density from the ring oxygen atom through delocalization (Figure 1). This delocalization reduces the ability of the oxygen atom to assist in stabilizing a developing positive charge at C-1; that is, the relative importance of the resonance contributor 7 is increased by an electron-withdrawing R group (Figure 1). Increased importance of 7 makes more difficult the departure of a leaving group from C-1 to produce a positive charge on this carbon atom.

2. The Nature of the Leaving Group Attached to the Anomeric Carbon Atom

a. Halogens. The identity of the leaving group attached to the anomeric carbon atom is a significant factor in glycoside synthesis. This is particularly true for halogens, the most common type of

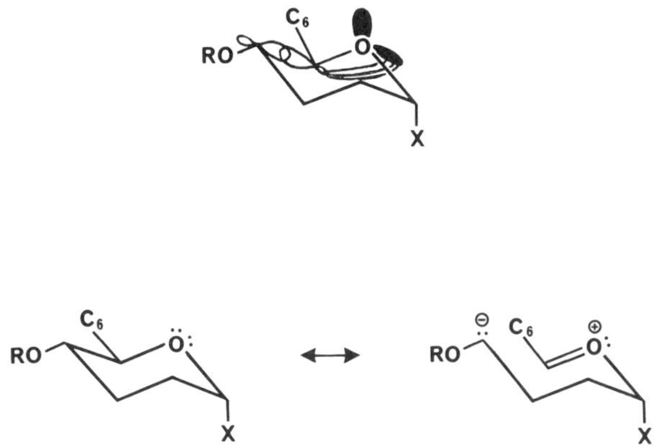

8 7

FIG. 1. Orbital interaction indicating electron withdrawal from the ring oxygen atom by delocalization (top). Resonance structures indicating electron withdrawal from the ring oxygen atom (bottom).

leaving group in these reactions. Glycosyl bromides are more reactive that chlorides, although chlorides are usually sufficiently reactive if a catalyst containing a silver or mercury salt is present. Glycosyl iodides tend to be too unstable to be synthetically useful. A fluorine atom attached to the anomeric carbon atom is relatively unreactive and requires a combination of catalysts (e.g., silver perchlorate and tin(II) chloride) in order to participate in glycoside formation (8). The stability of glycosyl fluorides can be an advantage because of the difficulties sometimes encountered in handling the often readily hydrolyzed (and otherwise unstable) chlorides and bromides.

Even with the considerable success enjoyed by glycosyl halides in oligosaccharide synthesis, there are disadvantages to their use. Among these disadvantages are the relative instability of these halides and the sometimes drastic conditions required for their preparation. The silver salts used as catalysts are costly and the mercury salts are highly toxic. This combination of disadvantages has led to a continuing search for replacements for halogens as leaving groups in oligosaccharide synthesis. The common alternatives to glycosyl halides are described below. Although each has some attractive characteristics, none appears likely to replace glycosyl halides.

b. Sulfonyloxy Groups. Since sulfonyloxy groups are effective leaving groups in both S_N1 and S_N2 reactions, their displacement from an anomeric center can lead to glycoside formation. Glycosyl 2,2,2-trifluoroethanesulfonates (glycosyl tresylates), such as 2-*O*-benzoyl-3,6-di-*O*-benzyl-4-*O*-(chloroacetyl)-1-*O*-(2,2,2-trifluoroethylsulfonyl)-α-D-galactopyranose [9], are more reactive than tosylates or mesylates but less reactive than triflates. This intermediate level of reactivity appears well suited for oligosaccharide synthesis (Equation 3) (9). Displacement of a

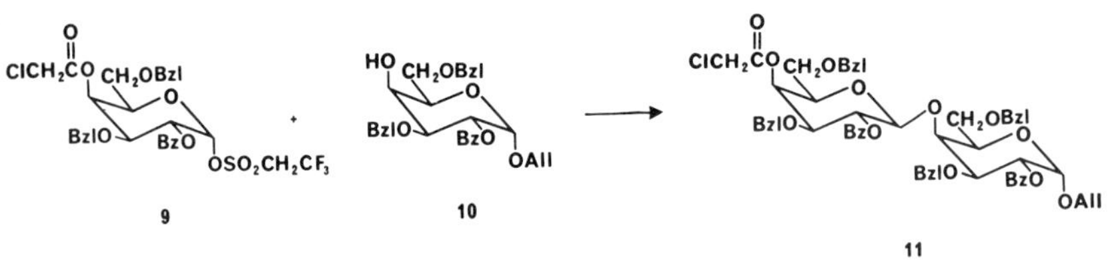

(3)

sulfonyloxy group from an anomeric carbon atom is not widely used in glycoside synthesis, however, because glycosyl sulfonates are more difficult to prepare than other sulfonates (9) and they lack the range of reactivity of glycosyl halides because they cannot be activated by silver and mercury-containing salts.

c. Imidates and Trichloroimidates. Amides react with glycosyl halides in the presence of silver salts to form imidates (Equation 4) (10), compounds which can be used in the synthesis of

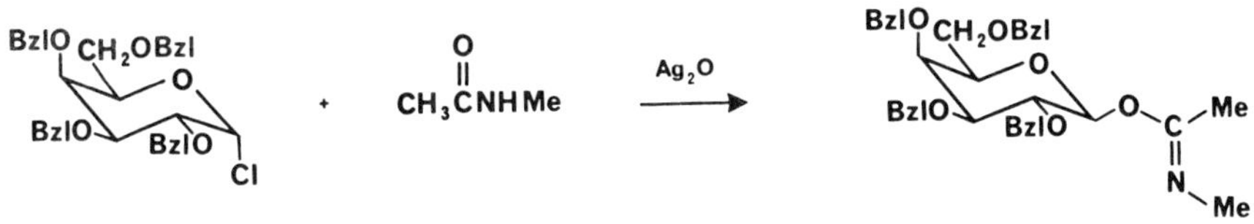

12 13 (4)

oligosaccharides. This type of synthesis requires protonation of the imidate to convert it into an effective leaving group. The protonated imidate then reacts with a partially protected carbohydrate to form a new glycosidic linkage. Thus, when *O*-(2,3,4,6-tetra-*O*-benzyl-β-D-galactopyranosyl)-*N*-methylacetimidate [13] is protonated by *p*-toluenesulfonic acid, it reacts with the disaccharide 14 (regioselectively with the more reactive hydroxyl group at C-3) to form the trisaccharide 15 (Equation 5) (10). The

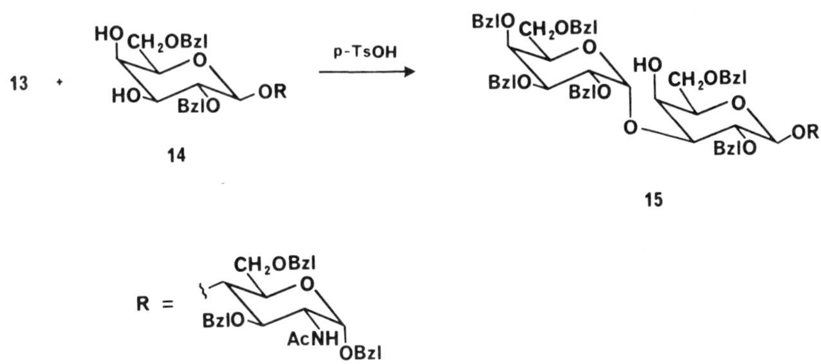

14 15 (5)

reactivity of typical imidates is relatively low; in fact, *O*-benzoyl or *O*-acetyl protecting groups cannot be used because they reduce imidate reactivity too greatly.

Trichloroimidates, which can be prepared as shown in Equation 6, have the advantage of being more reactive than imidates themselves (11). Use of trichloroimidates in glycoside synthesis is

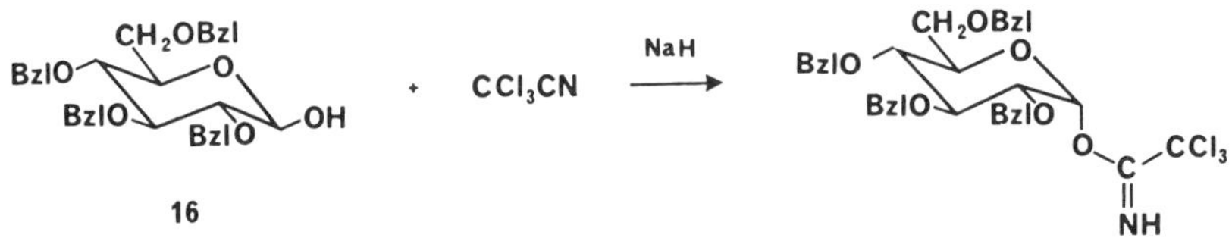

(6)

typically done in the presence of catalytic amounts of a strong acid, but reactions are otherwise conducted under mild conditions. Yields are generally high and often favor formation of a single anomer (12).

d. Activated Thioalkyl and Thioaryl Groups. When a thioglycoside reacts with *N*-bromosuccinimide (NBS) (13) or a thiophilic metal ion such as mercury (14), the sulfur atom and its attached substituent become an excellent leaving group. This reactive intermediate can be used to form an oligosaccharide (Equation 7) (14). One advantage to this approach to glycoside

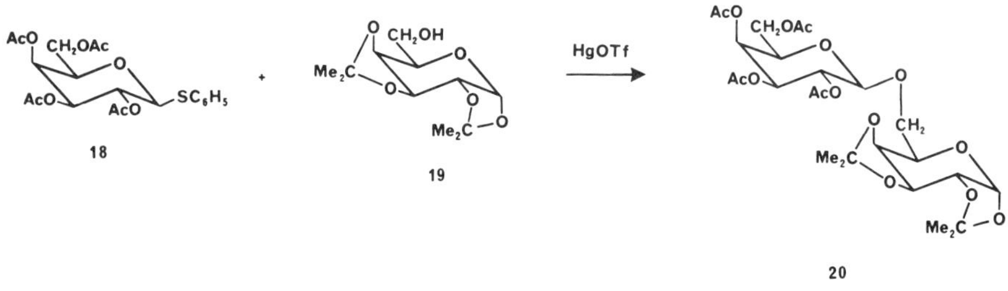

(7)

synthesis when compared to the Koenigs-Knorr reaction is that the thioglycosides are much more stable than are typical glycosyl halides. Also, if activation is done with NBS, heavy metal salts can be avoided.

3. The Effect of the Solvent

The nature of the solvent can be important in determining the mechanistic pathway adopted for glycoside formation. A relatively nonpolar solvent such as dichloromethane will suppress ionization and thus favor S_N2 reaction, while a polar solvent such an *N,N*-dimethylformamide will encourage ionization by solvating charged species.

III. BASIC APPROACHES TO THE STEREOSELECTIVE SYNTHESES OF OLIGOSACCHARIDES

Several basic approaches have been found to be successful in the stereoselective syntheses of oligosaccharides. Each of these is described below in a general form and illustrated with specific examples. Some recent review articles on oligosaccharide syntheses are found in references 1, 12, and 15.

A. Neighboring Group Participation

1. Reaction Mechanism

The greatest stereoselectivity is generally attained when a neighboring group participates in glycoside formation. For example, in the reaction of a glycosyl halide [21], a 2-*O*-acetyl substituent can determine stereoselectivity by interacting with the initially formed cation [22] to produce an acetoxonium ion [23] (Figure 2). The approach of a "free" hydroxyl group in a partially protected carbohydrate to the acetoxonium ion [23] to produce a glycosidic linkage [24] can be successful only from the side of the molecule opposite to that containing the participating group. If the approach of a partially protected carbohydrate to the other side of the molecule results in reaction, an ortho-ester [25]

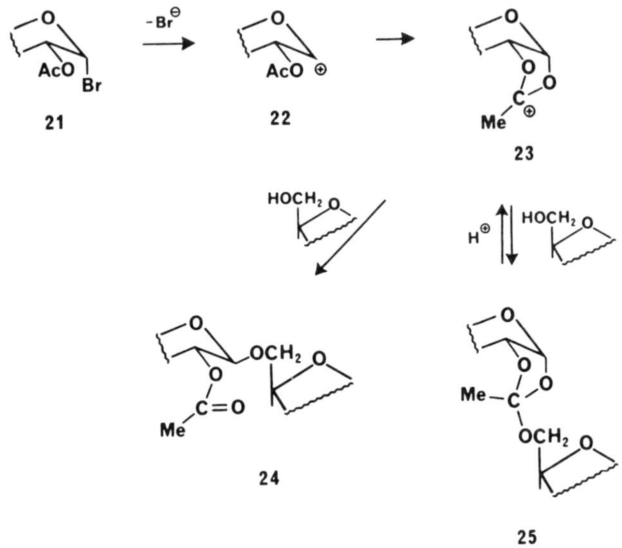

FIG. 2. Proposed mechanism for acetoxy group participation in glycoside formation.

will be formed (Figure 2). Ortho-ester formation may be a common event during oligosaccharide synthesis (16) but since the formation of these compounds usually is reversible under the reaction conditions, generation of an ortho-ester normally represents only a temporary diversion along the pathway to glycoside synthesis.

2. Neighboring Groups

a. Esters. Neighboring group participation by a 2-*O*-benzoyl group is the primary factor responsible for formation of a β-glycosidic linkage when 2,3,4,6-tetra-*O*-benzoyl-α-D-galactopyranosyl bromide [<u>26</u>] and 1-*O*-acetyl-2,4,6-tri-*O*-benzoyl-β-D-galactopyranose [<u>27</u>] are coupled in a Koenigs-Knorr reaction using silver triflate and *s*-collidine (Equation 8) (17). Repetition of

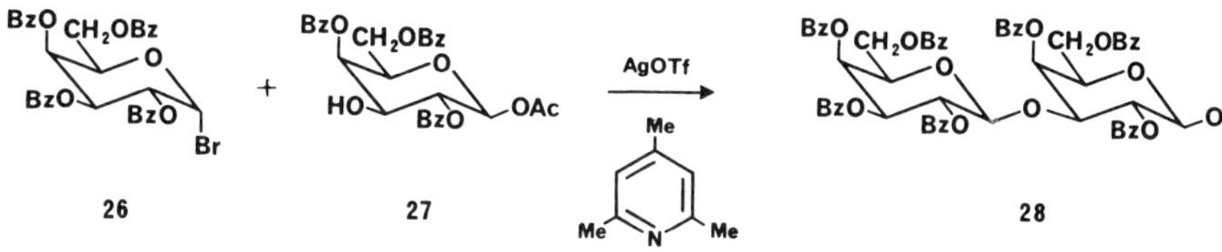

(8)

this coupling process, when accompanied by appropriate protection and deprotection reactions, produces a fully β-linked heptasaccharide (17). Neighboring group participation by an acetoxy group is responsible for the stereoselective formation of the β-glycoside 31 as a result of reaction of methyl 2,3-di-*O*-benzyl-4-deoxy-4-fluoro-β-D-galactopyranoside [29] with 2,3,4,6-tetra-*O*-acetyl-α-D-galactopyranosyl bromide [30] under Helferich conditions (Equation 9) (18). As a final example, one encountered

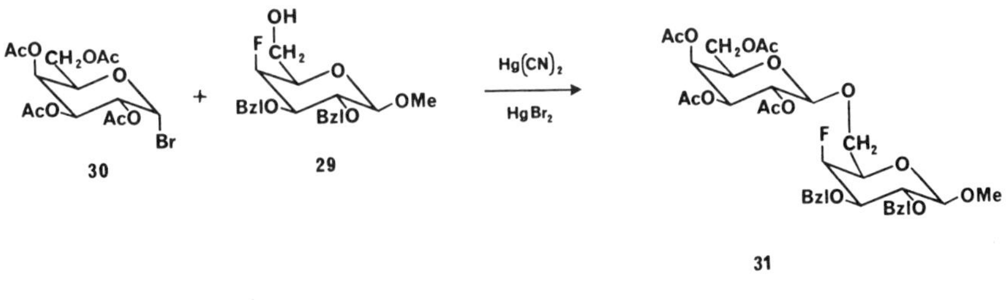

(9)

earlier in this chapter, the stereoselectivity in the reaction of the tresylate 9 (Equation 3) is also a result of neighboring group participation.

If the C-2 neighboring group in a glycosyl halide has an orientation relative to the anomeric center different from that in compounds 9, 26, and 30, participation can result in formation of an oligosaccharide in which the new glycosidic linkage (or linkages) has the α-configuration (Equation 10) (12).

Although the typical participating group is attached to the 2-position in a pyranose ring, participation from groups attached to other ring carbon atoms is possible. For example, when ethyl

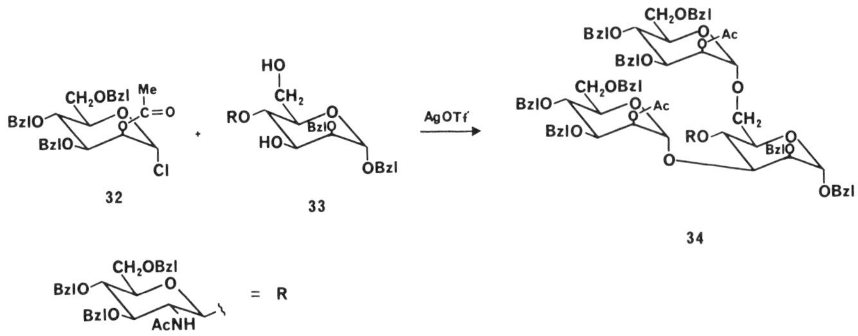

(10)

2,6-dideoxy-3,4-di-*O*-(*p*-methoxybenzoyl)-1-thio-α-D-*ribo*-hexopyranoside [35] reacts with the disaccharide 37, a trisaccharide [38] is formed which has a new β-linkage (Figure 3) (19). The participation of the *O*-(*p*-methoxybenzoyl) group at C-3 in 35 is presumably responsible for the stereoselectivity in this reaction.

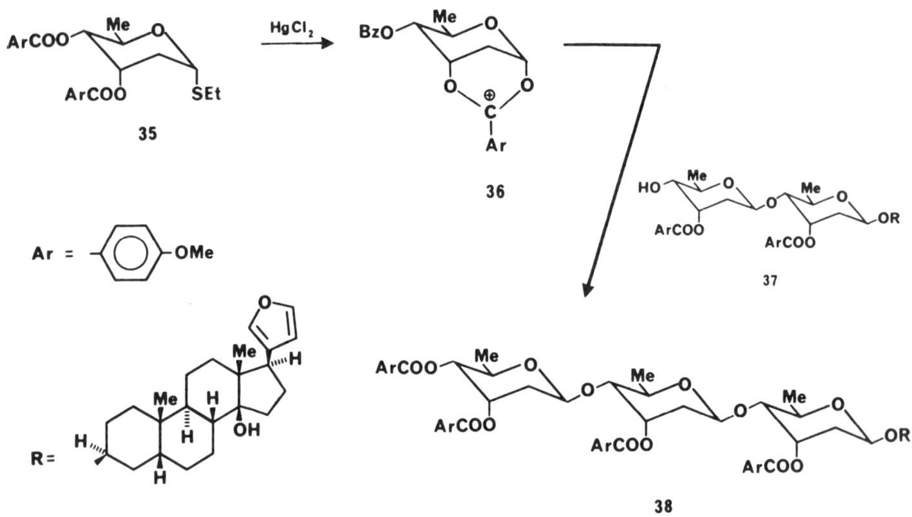

FIG. 3. Proposed participation of the 4-*O*-*p*-methoxybenzoyl group in ethyl 2,6-dideoxy-3,4-di-*O*-(*p*-methoxybenzoyl)-1-thio-α-D-*ribo*-hexopyranoside [35] in glycoside formation.

In general, however, participation by groups not at the 2-position requires a molecule to assume an energetically unfavorable conformation; consequently, participation by such groups is generally more difficult to achieve.

b. Amides and Imides. Although esters usually are the neighboring groups which participate in oligosaccharide formation, amido, imido, and other groups also can be involved. For example, neighboring group participation by the phthalimido group is responsible for the exclusive formation of a β-glycoside in the reaction shown in Equation 1.

c. Halogens. Even halogen participation can exert a directive influence on the stereoselectivity of a reaction; thus, when benzyl 4-*O*-benzoyl-2-bromo-2,6-dideoxy-β-D-glucopyranoside [41] and 4-*O*-benzoyl-2-bromo-2,6-dideoxy-3-*O*-formyl-α-D-glucopyranosyl bromide [39] react in the presence of silver triflate, the ratio (6.5/1) of β- to α-glycosides [42 and 43, respectively) produced indicates that participation of the bromine atom attached to C-2 in the glycosyl halide influenced glycoside formation (Figure 4) (20).

B. Rapid Anomerization of Glycosyl Halides (Halide-catalyzed Glycosidation)

Another approach favoring formation of a particular stereoisomer during an oligosaccharide synthesis takes advantage of the generally greater reactivity of β-glycosyl halides [44] when compared to their α-anomers [45] (Figure 5). α-Halides are less reactive because they are stabilized by the anomeric effect (Chapter 6, Section I.A.2.). The anomerization approach (sometimes called halide-catalyzed glycosidation) (21) depends upon rapid equilibration between glycosyl halides, a process which can be catalyzed by halide ion. Existence of an equilibrium between α- and β-glycosyl

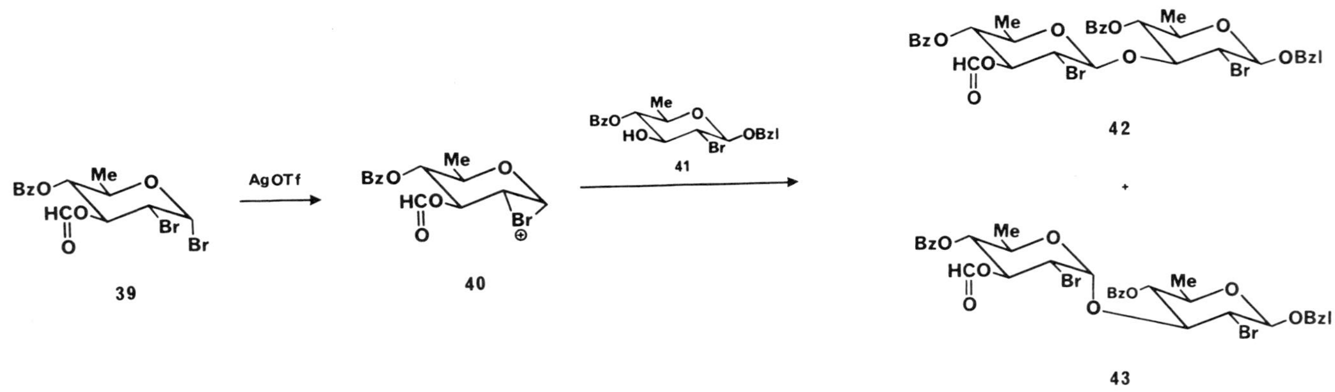

FIG. 4. Proposed participation of the 2-bromo group in 4-*O*-benzoyl-2-bromo-2,6-dideoxy-3-*O*-formyl-α-D-glucopyranosyl bromide [39] in glycoside formation.

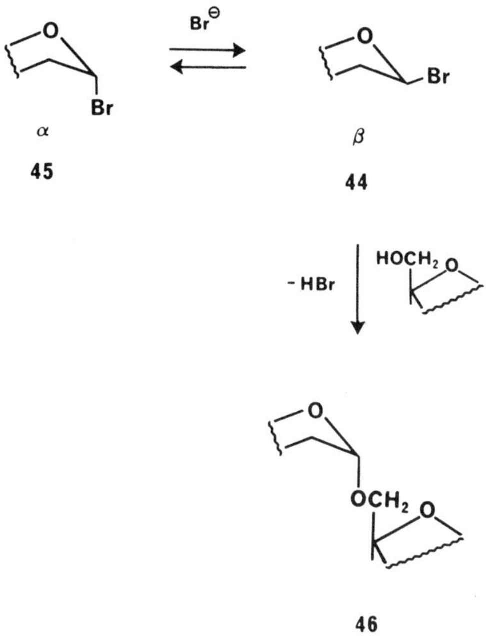

FIG. 5. Proposed mechanism for glycoside synthesis involving rapid anomerization of glycosyl halides (halide-catalyzed glycosidation).

halides allows oligosaccharide formation to take place primarily through reaction of the β-halide, since this more reactive halide will be continuously replenished by the equilibration process (Figure 5). Because the glycosyl halide is reacting without the catalytic effect of a silver or mercury salt, the β-halide must be a reasonably reactive one to produce an acceptable reaction rate. The preferred pathway for approach of the partially protected sugar will be to the side of the pyranoid ring opposite the departing halide ion; thus, displacement of a halogen atom from a β-halide [44] will lead preferentially to formation of an α-glycoside [46].

Interconversion between α- and β-glycosyl bromides during glycoside formation can be accomplished by conducting the reaction in the presence of a tetraalkylammonium salt. An example of the application of this approach to glycoside formation is found in the reaction shown in Equation 11, in which

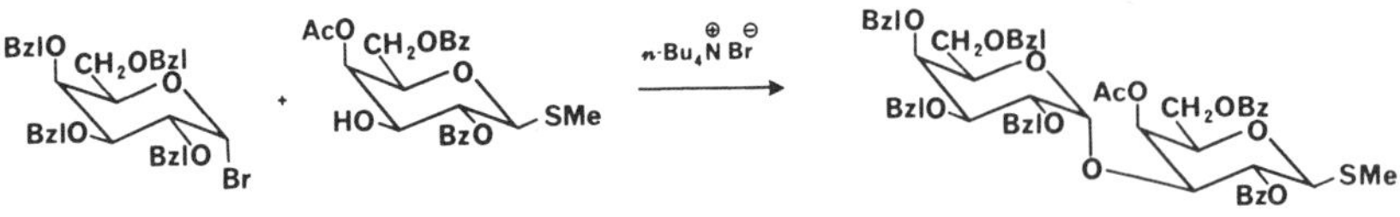

47 48 49 (11)

2,3,4,6-tetra-*O*-benzyl-α-D-galactopyranosyl bromide [47] and methyl 4-*O*-acetyl-2,6-di-*O*-benzoyl-1-thio-β-D-galactopyranoside [48] react in the presence of tetrabutylammonium bromide to give the α-disaccharide 49 (22). Another example of halide-catalyzed glycosidation was seen earlier in this chapter in the reaction shown in Equation 2.

C. Reaction Without Glycosyl Halide Anomerization

A third method for stereoselective glycoside formation depends upon absolute minimization of glycosyl halide anomerization. This process requires either than S_N2 substitution occurs at the anomeric center or that if an S_N1 reaction is taking place, the nucleophile begins bonding with the anomeric carbon atom soon enough after ionization that preferential attack is on the side of the molecule opposite the departing leaving group (Figure 6). A substitution reaction in either situation has its primary value in the synthesis of β-glycosides since α-glycosyl halides are the usual reactants.

One set of conditions which promotes reaction without glycosyl halide anomerization is created by an active but insoluble catalyst

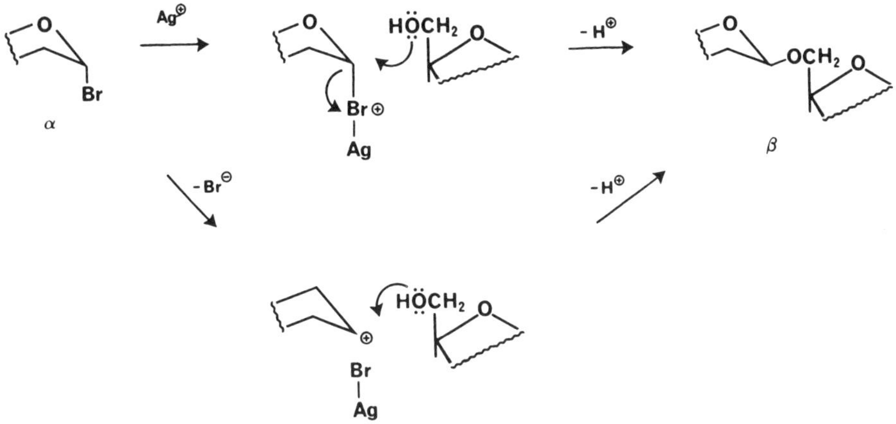

FIG. 6. Proposed mechanism for glycoside formation without glycosyl halide anomerization.

present in a reaction mixture. When reaction of the glycosyl halide takes place, it occurs on or near the surface of the insoluble material. Approach of a partially protected carbohydrate to the anomeric center in a reacting glycosyl halide will be primarily from the side of the molecule opposite to the departing halide ion which is interacting with the insoluble catalyst (Figure 7). An example of this type of reaction is found in the glycoside formation between 6-*O*-acetyl-2,3,4-tri-*O*-benzyl-α-D-mannopyranosyl

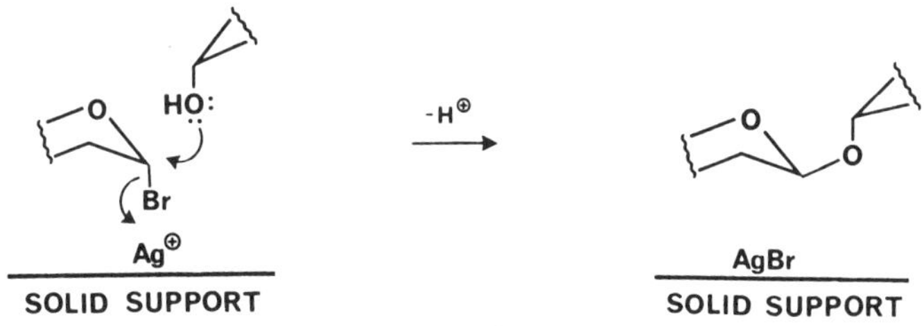

FIG. 7. Proposed mechanism for glycoside formation using an insoluble silver catalyst.

bromide [50] and the 1,6-anhydro azide 51 in the presence of the insoluble catalyst silver silicate (silver ion deposited on silica gel) to produce the β-linked disaccharide 52 (Equation 12) (23).

50 51 52 (12)

D. Cyclic Intermediates Involving the Anomeric Carbon Atom

1. Ortho-esters

Although ortho-esters [25] can be intermediates in glycoside formation in which ester groups are participating (Figure 2), they usually are not isolated; rather reaction is allowed to continue until glycoside formation is complete. It is possible, however, to form a simple ortho-ester (not the reaction intermediate 25 shown in Figure 2 but rather an ortho-ester derived from a simple alcohol) and use it in place of a glycosyl halide in oligosaccharide synthesis (24,25,26). One problem with the intermediacy of a simple ortho-ester is that the alcohol released by cleavage of such a compound can become involved in glycoside formation; however, this problem can be minimized by use of a *t*-butyl ortho-ester which liberates the sterically hindered and thus weakly nucleophilic *t*-butyl alcohol. When, for example, methyl 3-*O*-benzyl-4-*O*-(chloroacetyl)-β-L-idopyranuronate 1,2-(*t*-butyl orthoacetate) [53] reacts with benzyl 6-*O*-acetyl-3-*O*-benzyl-2-(benzyloxycarbonyl)amino-2-deoxy-α-D-glucopyranoside [54] in the presence of 2,6-dimethylpyridinium perchlorate [55], the disaccharide 56 is formed (Equation 13) (27).

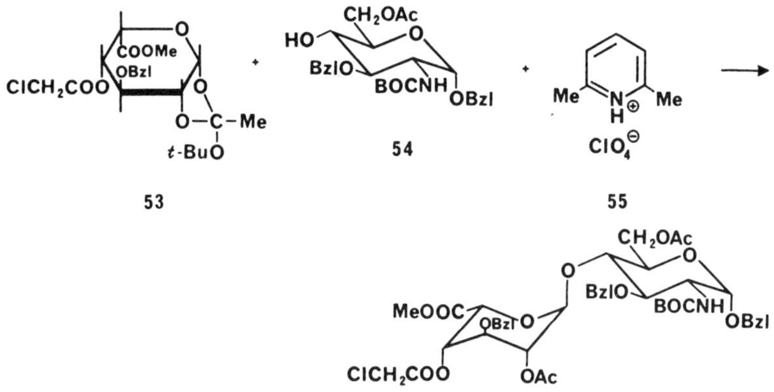

53 54 55

56

(13)

2. Oxazolines

A type of compound which is, in some ways, analogous to an ortho-ester is an oxazoline. Like ortho-esters, oxazolines form glycosides when they react with appropriate partially protected carbohydrates in the presence of an acidic catalyst; thus, when the oxazoline 57 reacts with benzyl 2-acetamido-2-deoxy-3-*O*-(2,3,4-tri-*O*-acetyl-6-deoxy-β-D-galactopyranosyl)-α-D-galactopyranoside [58] in the presence of *p*-toluenesulfonic acid, the trisaccharide 59 is formed (Equation 14) (28).

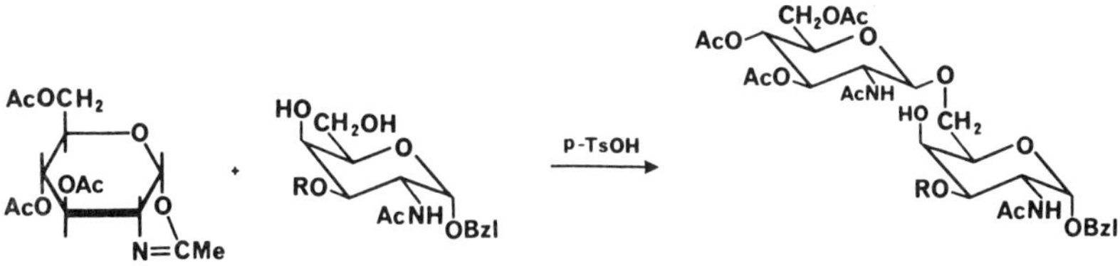

(14)

57 58 59

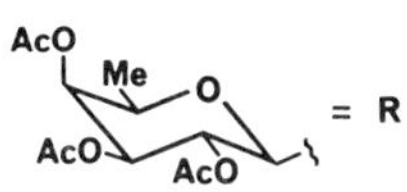

E. Electrophilic Addition to a Glycal

Electrophilic addition to a glycal represents another method for oligosaccharide synthesis. One way to accomplish glycoside formation by electrophilic addition is to form an intermediate halonium ion which then reacts with a partially protected sugar derivative. (Use of this type of addition reaction to form simple glycosides was discussed in Chapter 12, Section II.A.). Addition of iodine to 4-*O*-acetyl-1,5-anhydro-2,6-dideoxy-3-*C*-methyl-L-*arabino*-hex-1-enitol [60] to form the intermediate iodonium ion 61 and subsequent reaction with benzyl 4-*O*-benzoyl-3-*O*-(2,6-dideoxy-β-D-*arabino*-hexopyranosyl)-2,6-dideoxy-β-D-*arabino*-hexopyranoside [62] gives the corresponding trisaccharide [63] (Figure 8) (29).

F. Alkylation of Alkoxy Anions

A somewhat different approach to oligosaccharide synthesis involves the alkylation of compounds such as 2,3,4,6-tetra-*O*-benzyl-1-*O*-sodio-β-D-glucopyranose [64] with active alkylating agents. For example, treatment of 64 with benzyl 2,3,4-tri-*O*-benzyl-6-*O*-(trifluoromethylsulfonyl)-α-D-glucopyranoside [65] results in S_N2 substitution to produce the β-glycoside 66 with high stereoselectivity (Equation 15) (30). In glycoside syntheses

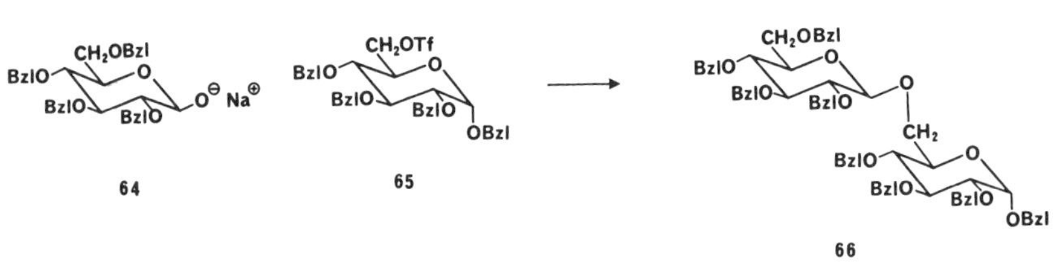

64 65 66 (15)

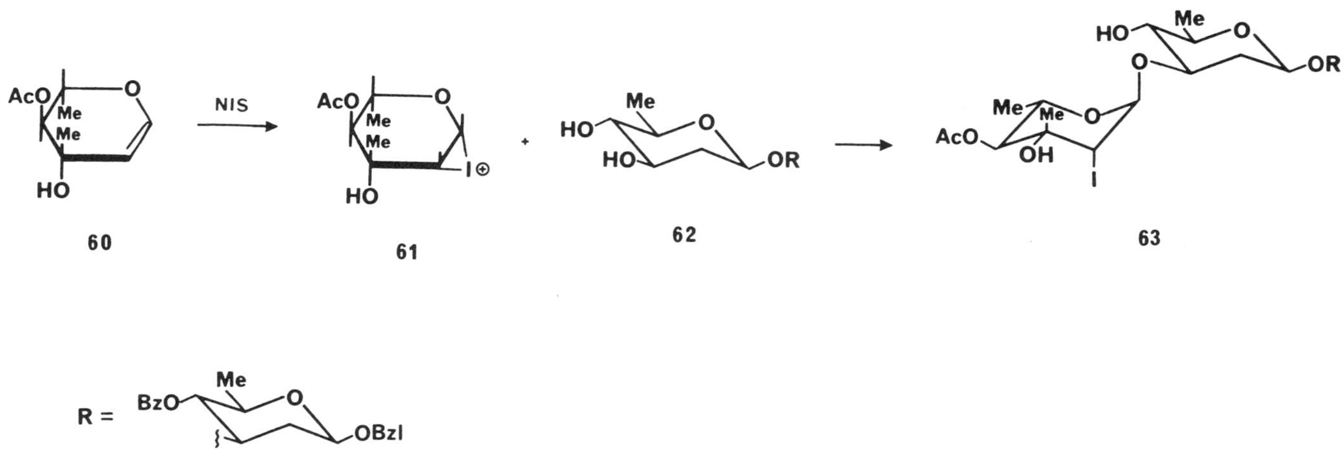

FIG. 8. Proposed mechanism for glycoside formation by electrophilic addition to 4-*O*-acetyl-1,5-anhydro-2,6-dideoxy-3-*C*-methyl-L-*arabino*-hex-1-enitol [60].

with compounds such as 64, in which the alkoxy substituent is at C-1 in a pyranoid ring, the stereochemistry at the anomeric carbon atom in the newly formed glycosidic linkage appears to be determined primarily by the relative nucleophilicities of the α and β alkoxy anion anomers and their rate of interconversion. Also a critical factor in determining the stereochemistry in the oligosaccharide being synthesized is the manner in which the cation (e.g., sodium, potassium, or lithium) coordinates with its associated anion (12). If, for example, complexation of a sodium ion is particularly strong with a sugar when the alkoxy group at the anomeric position has a β configuration, a β-anomeric linkage will be formed preferentially. A significant limitation to this approach to glycoside synthesis is that reaction of secondary triflyloxy groups to form disaccharides is not particularly effective (12).

REFERENCES

1. H. Paulsen, *Angew. Chem. Int. Ed. Engl. 22*, 156 (1982).
2. J. Arnarp, M. Haraldsson, and J. Lonngren, *J. Chem. Soc. Perkin Trans. I.*, 535 (1985).
3. Reference 1, p. 158.
4. K.-I. Takeo and K. Shinmitsu, *Carbohydr. Res.*, *133*, 135 (1984).
5. B. Helferich and J. Zirner, *Chem. Ber.*, *95*, 2604 (1962).
6. C. A. A. van Boeckel, T. Beetz, and S. F. van Aelst, *Tetrahedron*, *40*, 4097 (1984).
7. C. A. A. van Boeckel and T. Beetz, *Recl. Trav. Chim. Pays-Bas*, *104*, 171 and 174 (1985).
8. R. E. Dolle and K. C. Nicolaou, *J. Am. Chem. Soc.*, *107*, 1695 (1985).

9. H. A. El-Shenawy and C. Schuerch, *Carbohydr. Res.*, *131*, 227 (1984).
10. P. Sinay, *Pure Appl. Chem.*, *50*, 1437 (1978).
11. R. R. Schmidt and J. Michel, *J. Carbohydr. Chem.*, *4*, 141 (1985).
12. R. R. Schmidt, *Angew. Chem. Int. Ed. Engl.*, *25*, 212 (1986).
13. K. C. Nicolaou, S. P. Seitz, and D. P. Papahatjis, *J. Am. Chem. Soc.*, *105*, 2430 (1983).
14. P. J. Garegg, C. Henrichson, and T. Norberg, *Carbohydr. Res.*, *116*, 162 (1983).
15. H. Paulsen, *Chem. Soc. Rev.*, *13*, 15 (1984).
16. P. J. Garegg, P. Konradsson, I. Kvarnstrom, T. Norberg, S. C. T. Svensson, and B. Wigilius, *Acta Chem. Scand. B*, *39*, 569 (1985).
17. P. Kovac, R. B. Taylor, and C. P. J. Glaudemans, *J. Org. Chem.*, *50*, 5323 (1985).
18. P. Kovac and C. P. J. Glaudemans, *J. Carbohydr. Chem.*, *3*, 349 (1984).
19. K. Wiesner, T. Y. R. Tsai, and H. Jin, *Helv. Chem. Acta*, *68*, 300 (1985).
20. J. Thiem and M. Gerken, *J. Carbohydr. Chem.*, *1*, 229 (1982).
21. R. U. Lemieux, K. B. Hendricks, R. V. Stick, and K. James, *J. Am. Chem. Soc.*, *97*, 4056 (1975).
22. P. J. Garegg and S. Oscarson, *Carbohydr. Res.*, *136*, 207 (1985).
23. H. Paulsen, R. Lebuhn, and O. Lockhoff, *Carbohydr. Res.*, *103*, C7 (1982).
24. N. K. Kochetkov, A. F. Bochkov, T. A. Sokolovskaya, and V. J. Snyatkova, *Carbohydr. Res.*, *16*, 17 (1971).
25. N. K. Kochetkov, A. J. Khorlin, and A. F. Bochkov, *Tetrahedron*, *23*, 693 (1967).

26. N. K. Kochetkov, A. J. Khorlin, and A. F. Bochkov, *Tetrahedron Lett.*, 289 (1964).
27. J.-C. Jacquinet, M. Petitou, P. Duchaussoy, I. Lederman, J. Choay, G. Torri, and P. Sinay, *Carbohydr. Res.*, *130*, 221 (1984).
28. C. F. Piskorz, S. A. Abbas, and K. L. Matta, *Carbohydr. Res.*, *131*, 257 (1984).
29. J. Thiem and M. Gerken, *J. Org. Chem.*, *50*, 954 (1985).
30. R. R. Schmidt, M. Reichrath, and U. Moering, *J. Carbohydr. Chem.*, *3*, 67 (1984).

Index